MATERIALS SCIENCE RESEARCH
Volume 10

SINTERING AND CATALYSIS

MATERIALS SCIENCE RESEARCH

A Continuation Order Plan is available for this series. A continuation order will bring delivery of each new volume immediately upon publication. Volumes are billed only upon actual shipment. For further information please contact the publisher.

MATERIALS SCIENCE RESEARCH • Volume 10

SINTERING
AND CATALYSIS

Edited by
G. C. Kuczynski
University of Notre Dame

PLENUM PRESS · NEW YORK AND LONDON

Library of Congress Cataloging in Publication Data

International Conference on Sintering and Related Phenomena, 4th, University of
   Notre Dame, 1975.
   Sintering and catalysis.

   (Materials science research; v. 10)
   Includes bibliographical references.
   1. Sintering—Congresses. 2. Catalysis—Congresses. I. Kuczynski, George Czeslaw,
1914-   II. Title. III. Series.
TN695.I56 1975   671.3'7   75-35639
ISBN 978-1-4684-0936-9   ISBN 978-1-4684-0934-5 (eBook)
DOI 10.1007/978-1-4684-0934-5

Proceedings of the Fourth International Conference on Sintering and Related Phenomena, held at the University of Notre Dame, Notre Dame, Indiana, May 26-28,
1975.

© 1975 Plenum Press, New York
Softcover reprint of the hardcover 1st edition 1975

A Division of Plenum Publishing Corporation
227 West 17th Street, New York, N.Y. 10011

United Kingdom edition published by Plenum Press, London
A Division of Plenum Publishing Company, Ltd.
Davis House (4th Floor), 8 Scrubs Lane, Harlesden, London, NW10 6SE, England

Preface

The proceedings of the 4th International Conference on Sintering and Related Phenomena, contained in this volume, have been broadened in scope to include the phenomena of sintering and coalescence of catalytic materials dispersed upon refractory oxides. For it has long been recognized within the circles of chemists and chemical engineers working in the field of catalysis that one of the chief causes of the decline in heterogeneous catalytic activity and/or selectivity is, indeed sintering, or perhaps using a better term, coalescence of the supported catalytic metal and compounds thereof. Essentially catalytic deactivation by sintering is now well recognized as Ostwald ripening; which of course is a phenomenon familiar to scientists grappling with the problem of sintering of powder compacts. The 4th Conference at Notre Dame marks the first occasion at which scientists and engineers of each discipline were assembled in the same room to exchange views on these phenomena of mutual concern.

In the wake of the Conference at Notre Dame, all parties acknowledged the synergistic benefit which issued from this exchange, both at the formal and informal level. All were persuaded that signal benefits will be realized by a continuation of this collaboration in the form of future sintering conferences in which both powder metallurgists and catalytic scientists and engineers would participate.

Although the presence of catalytic workers at this conference gave it a definite new flavor, the fundamental problems of the mechanisms of particle coalescence as developed during the past twenty-six years by the common effort of metallurgists, ceramists, physicists and chemists have not been neglected. This volume contains a series of excellent papers reporting on most recent research on the various aspects of this difficult problem.

The Notre Dame Conference is one of the series of University Conferences on Ceramic Science organized yearly by a more or less informal confederation of four institutions; North Carolina University at Raleigh, N. C.; The University of California at Berkeley, California; Alfred University at Alfred, New York, and the University of Notre Dame, Notre Dame, Indiana.

We wish at this point to express our gratitude for the generous support of E.R.D.A. without which this conference could not have become a reality. This organization in the very early months of its existence, happily saw fit to generously support our enterprise. A donation from the American Oil Company is also gratefully acknowledged.

Our gratitude is due to all participants, especially to those who prepared the papers, thus contributing to the success of our conference.

It is my pleasant duty to thank my colleagues, Professor J. J. Carberry of the Department of Chemical Engineering and A. Miller of the Department of Metallurgical Engineering of Materials Science for serving on the organizing committee and aiding me in various aspects of my endeavor.

Many thanks are also due to the graduate students of the Department of Metallurgical Engineering and Materials Science for the numerous chores they undertook before, during and after the Conference. I also thank Mrs. J. Peiffer and Mrs. H. Deranek for their help in editing the Conference papers.

Last but not least, I am grateful to the Staff of the Center of Continuing Education of the University of Notre Dame for smoothly running our Conference.

NOTRE DAME   G. C. Kuczynski
July 1975

Contents

MECHANISMS OF SINTERING

PRESSURE SINTERING

TRANSMISSION ELECTRON MICROSCOPY — SOME TECHNIQUES USEFUL IN SINTERING STUDIES

A.F. Moodie and C.E. Warble

Division of Chemical Physics, CSIRO

P.O. Box 160, Clayton, Victoria, Australia 3168

INTRODUCTION

Since they are, in general, very stable, ceramic materials have been intensively investigated by means of electron microscopy and electron diffraction. Some properties of these materials, however, depend on their detailed morphology, their surface structure, and their interactions at high temperatures on a scale of about 10 Å. Detail of this kind is important in sintering reactions, and to observe it in the electron microscope requires some modifications to standard technique. It is the purpose of this communication to describe these modifications, to illustrate application with several examples, and to outline an approximation which can be useful in interpretation.

INTERPRETATIVE METHODS AND EXPERIMENTAL REQUIREMENTS

Problems in the present experiments fall into two classes: those of maintaining the specimen clean on a molecular scale, and those of combining the operation of a goniometer furnace with high resolution.

The requirements for cleanliness are set by the same theoretical considerations that permit the direct observation of surface structure. These can be understood in terms of the following approximation, which can also be used in interpreting most of the results in the present experiments.

Under the conditions of these particular experiments the specimen can be thought of as acting on the electron beam as a phase object, the phase change being proportional to the potential

1

of the specimen projected in the direction of the incident electron
beam. (Cowley & Moodie, 1957). Numerical values are such that a
few Angstroms in thickness are sufficient to produce an appreciable
phase change. This phase change must now be made visible by means
of an electron optical system. The approximations used in
Zernike's light optical phase-contrast microscope are in general
not valid for electrons, and further, the construction of an
electron quarter wave plate would be extremely difficult. However,
it can be shown that a controlled underfocussing of the objective
lens transforms the phase changes due to potential into contrast
changes proportional to projected charge density (PCD) (Cowley &
Moodie, 1960; Lynch, Moodie & O'Keefe, 1975). A sensitive technique for the detection of, say, single unit cell steps on a
crystal surface, is therefore available.

A complementary disadvantage arises, since even molecular contamination will, by the same process, perturb or obliterate detail.
The specimen environment must therefore be at a much lower partial
pressure of contaminant than in standard electron microscopic
practice. Contaminant here includes water vapour, since many
ceramics are etched in the presence of the electron beam at low
partial pressures, and the severity of this etching increases with
temperature.

The problems in high-temperature work can be summarised in the
following requirements:

1.   Dimensions must be compatible with the volume available at the
high-resolution position in the bore of a short focal length lens.
In practice this volume is about 0.2 ml.

2.   Goniometric control must be available.

3.   Electron optical aberrations must not be increased by the
method of heating.

4.   Drift rates must be low, no more than 6 $\overset{\circ}{A}$ min^{-1}.

5.   Protective shielding must be provided since the bore of a short
focal length pole piece is easily damaged.

EXPERIMENTAL METHODS

A Hitachi HU/125-S electron microscope was used in the present
experiments. In order to improve the vacuum the following modifications were carried out:

1.   An automatic control system was installed so that the microscope can be pumped continuously for a week. At the end of the
week all traps are baked and the cycle is started again.

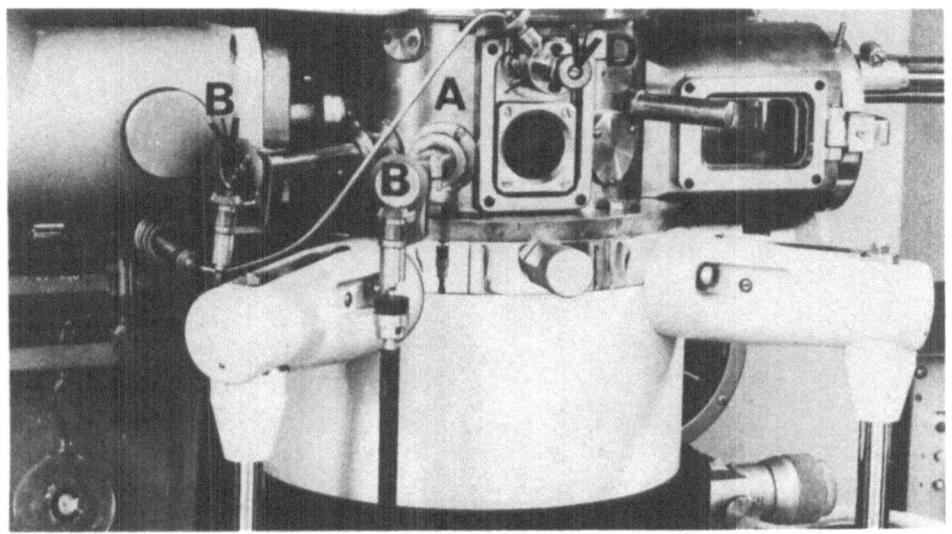

Fig.1. The special-purpose stage (A), installed in a Hitachi HU/125-S electron microscope. Tilt controls (B), gas storage (C) and gas inlet valve (D) are indicated.

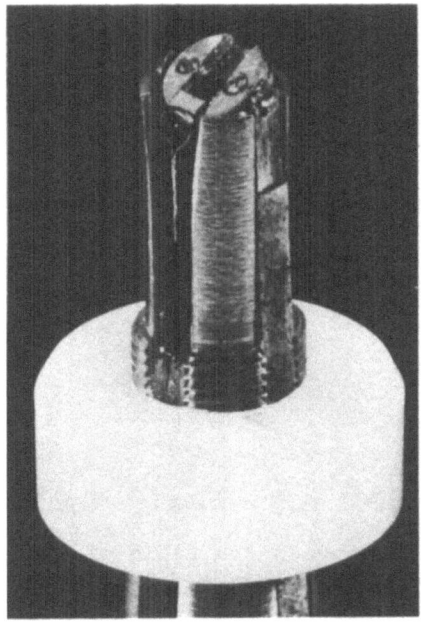

Fig.2. Furnace goniometer with protective cap removed. The diameter of the specimen mounting cup is 3.2 mm.

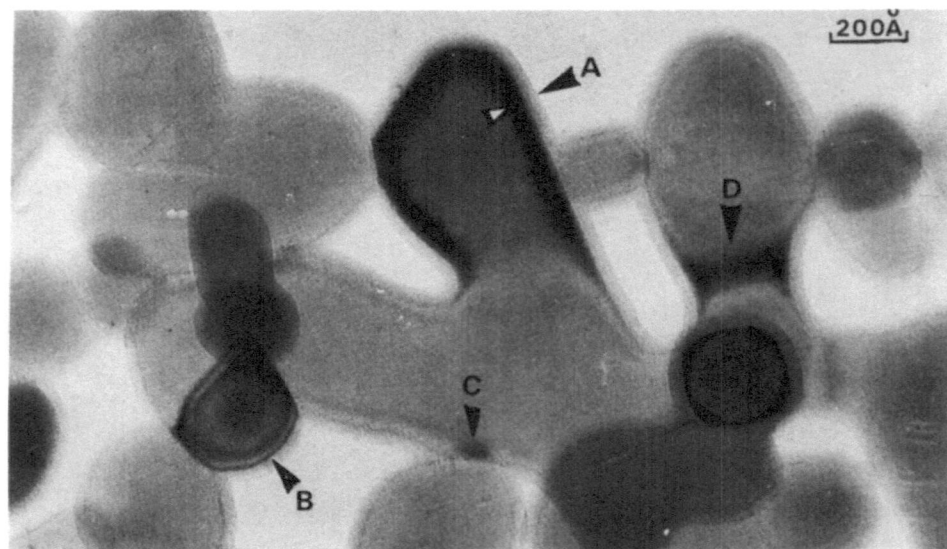

Fig. 3. MgO sintered at 900°C, mounted on an amorphous carbon film
and imaged in a conventional stage. Contamination is pronounced
(A), but contrast effects due to thickness fringes (B), dislocations (C) and strain (D) are evident. All surface structure is
obscured.

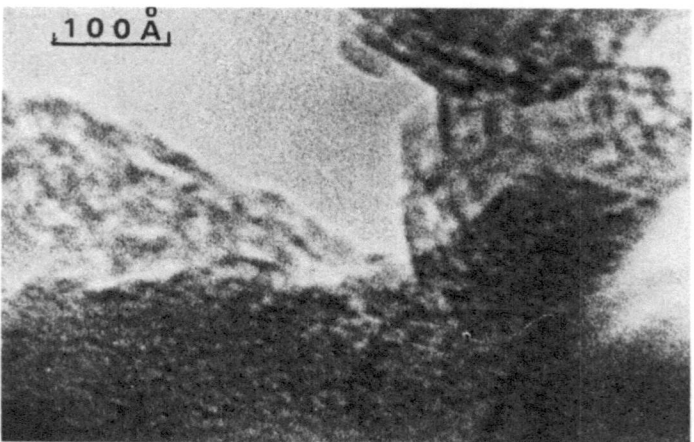

Fig.4. Contamination invading a field of sintered material and
obliterating surface morphology.

2. Sintered alumina traps and liquid nitrogen traps were fitted to the backing pumps.

3. The air bleed to the electron gun was removed.

4. Dry Viton A seals were used and where necessary gasket seatings were repolished.

5. An additional liquid nitrogen trap was installed in the column in order to intercept contaminants from the photographic plates.

6. An ionisation gauge was installed in the pumping line to the gun and a Minimass mass spectrometer was fitted to the pumping line to the specimen chamber.

With these modifications the gun can be maintained at a pressure of approximately 2×10^{-7} Torr. At this pressure ion damage is greatly reduced, focussed contamination cannot be detected, the stability of operation at 125 kV is substantially improved and filament life is about six months. The contamination rate is, however, still too high to allow reliable and reproducible observation of surface detail. This is overcome by enclosing the specimen in a cryopumped cone. The specimen stage designed to achieve this is also fitted with heating, cooling, goniometric and gas-handling facilities (Mills & Moodie, 1968, 1974). The installation of this stage in the microscope is shown in Fig. 1, and details of specimen mounting in Fig. 2.

With this stage the main sources of contamination are the heating grid and the specimen. The obscuring effect of contamination can be seen in Figs. 3 and 4. Palladium grids are normally used, and these are etched, baked and cut into heating strips. Removal of contamination from the specimen itself presents the greatest problem. No general satisfactory solution to the problem has been found but, where the chemistry permits, ultrasonic disperson in acetone followed by fogging and prolonged baking is a standard first step. Amorphous carbon films are not normally used in this work since they produce contrast effects similar to ground glass screens in optical phase-contrast microscopy, and thus obscure the relevant detail (Fig. 5.).

OPERATION OF THE MICROSCOPE

The specimen, mounted in the cone assembly, is fitted in a grease-free air lock and prepumped against liquid nitrogen. The cone is then moved into the column and, when the solid state chemistry permits, baked at 200°C - 300°C before engaging the cryopump.

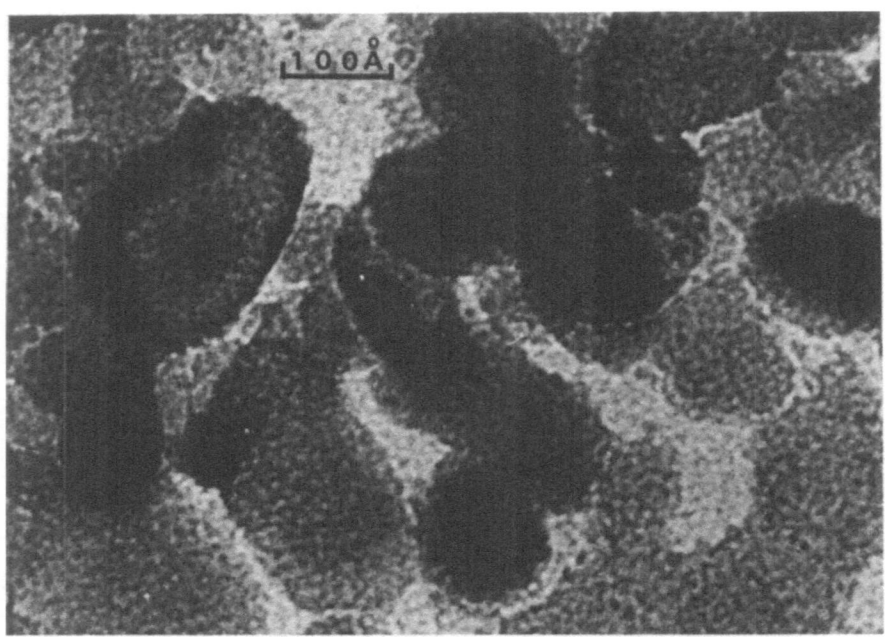

Fig.5. MgO sintered at 700 C, imaged under clean conditions, but
mounted on an amorphous carbon film. The 'ground glass screen' effect
of the film obliterates nearly all surface morphology.

After engaging the cryopump the temperature is set to the required
value and standard electron microscopic procedures started. In
order to approach the conditions in which the PCD approximation
has some validity, the specimen is oriented as far as possible
from satisfying any Bragg condition, i.e., so that no thickness
fringes are visible. With a sufficiently large objective aperture
the overall contrast then falls nearly to zero. An under-focus
setting from several hundred to several thousand Angstroms brings
up PCD contrast of workable levels, the precise setting chosen
depending on the resolution aimed at. Examples of the different
results obtained by the standard and the modified techniques are
shown in Figs. 3 and 6.

APPLICATIONS

Surface Structure

     The validity of the PCD approximation depends on the defect
of focus, the thickness, the angle of incidence and the resolution.
This has been discussed quantitatively, and the calculations compared with observation for periodic objects, that is, lattice

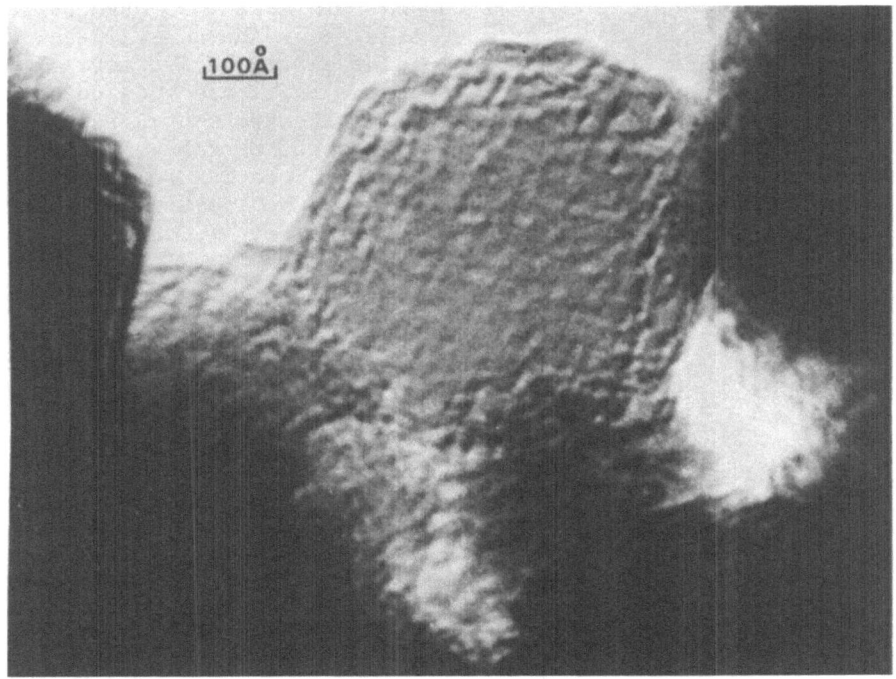

Fig.6. The same material as in Fig. 3 imaged in the special-purpose
stage under clean conditions and at an orientation appropriate to
the PCD approximation. Surface structure is apparent, as is the
complex structure of the junction areas.

images, by Lynch, Moodie & O'Keefe (1975),   Substantially the
same results can be derived for non-periodic objects and a number
of experimental criteria are available as tests of validity. If
the contrast of certain detail reverses through the Gaussian focus
the aperture is effectively infinite at the resolution of that
detail. If, as will normally be the case at high resolution, the
contrast does not reverse, or only reverses approximately, either
the aperture is too small for that detail, or the spherical aber-
ration is too high, or both. When the contrast does not reverse,
qualitative interpretation may still prove possible on considering
the aperture function to act directly on the charge density. This,
in itself is, however, a fairly severe approximation and more
elaborate procedures are generally necessary.

     In practice a reasonable qualitative interpretation can often
be obtained from a through-focus series. An example is provided
by Fig. 7. This shows part of a field of MgO crystals which

thickness-fringe observations show to be internally perfect. Under
standard conditions the smaller cubes would appear to have perfect
flat bounding faces, and when the larger cubes have, as in this
field, gross steps, these steps would appear to be perfect. In
fact, in the absence of contamination the faces are seen to be
covered with a complicated pattern of growth steps of a few unit
cells in height, and parallel to the axes of the cubes. While such
a surface is complicated, it is relatively smooth in comparison
with the vastly more complicated surface of MgO particles sintered
at. for instance, 900°C (Fig. 8). Again in Fig. 7, on the scale of

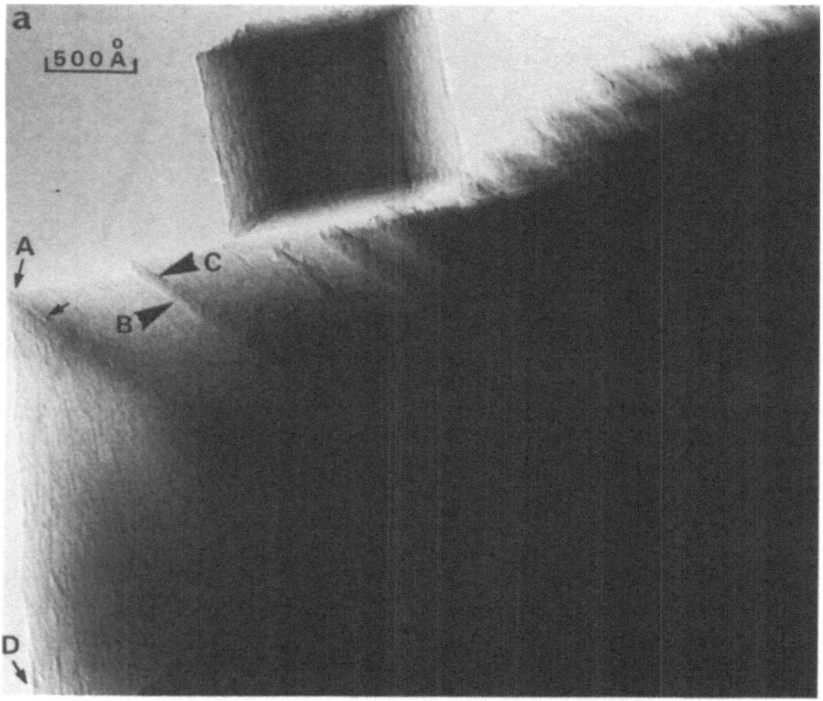

Fig. 7a. Growth steps on internally perfect crystals of MgO. Three
distinct levels in detail can be seen: large steps of 40-100 A in
height, perturbations of about 20 A on the faces of these steps,
and finally a pattern of steps of 1-3 unit cells in height over the
entire surface of the cubes. The larger cube is sufficiently large
to run into the Gaussian focus, where PCD contrast disappears. On
the scale of a few unit cells the edges are imperfect (A). Edges
formed by the intersection of planes forming concave figures relative to the incident beam are white (B), while those forming convex
figures are black (C). Unexplained features such as some prominent
black dots (D) indicate concentrations of charge smaller than one
unit cell.

Fig. 7b. Same field of view as in Fig. 7a, but imaged at greater underfocus. The surface contrast effects are exaggerated, the resolution is reduced, and the PCD approximation is no longer valid at resolutions beyond about 10 A.

a few unit cells the edges are imperfect (for instance A). As predicted theoretically, edges formed by the intersection of planes forming a concave figure relative to the incident beam are white (B), while the edges on a convex figure are black (C). Contrast alternates as black and white on a flight of steps, and the correctness of the interpretation can be checked by the profile. It must be kept in mind that the charge density is projected, so that in general a superposition of detail on entrance and exit faces is obtained. In order to separate this detail a number of images may be required. Sufficient detail can be seen to enable comparison to be made both with the classical predictions for step growth (Burton, Cabrerra & Frank, 1951) and with more recent computer-simulated predictions (for instance, Jackson, 1974). No detailed comparison with the latter prediction has been made, but preliminary work has been attempted (Moodie & Warble, 1967).

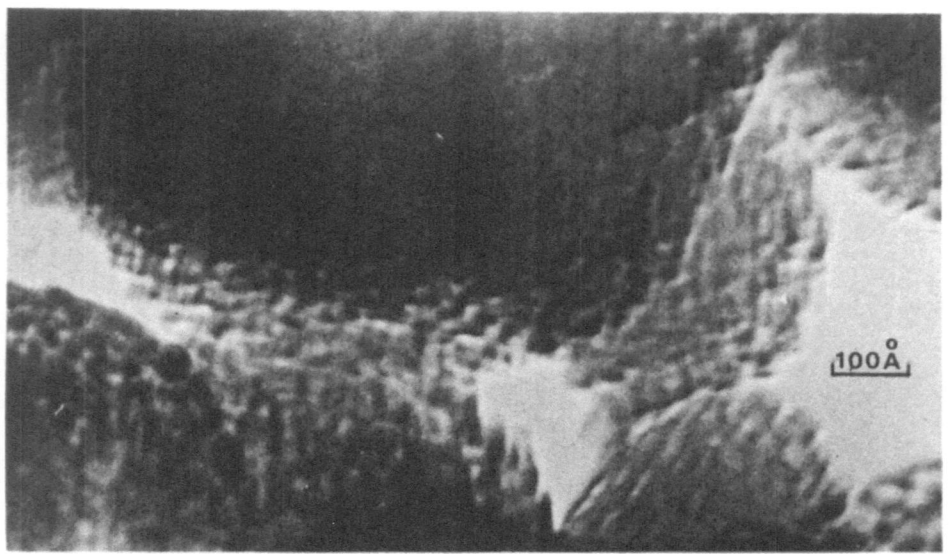

Fig. 8. Surface structure in neck areas between MgO crystallites
sintered at 900°C.

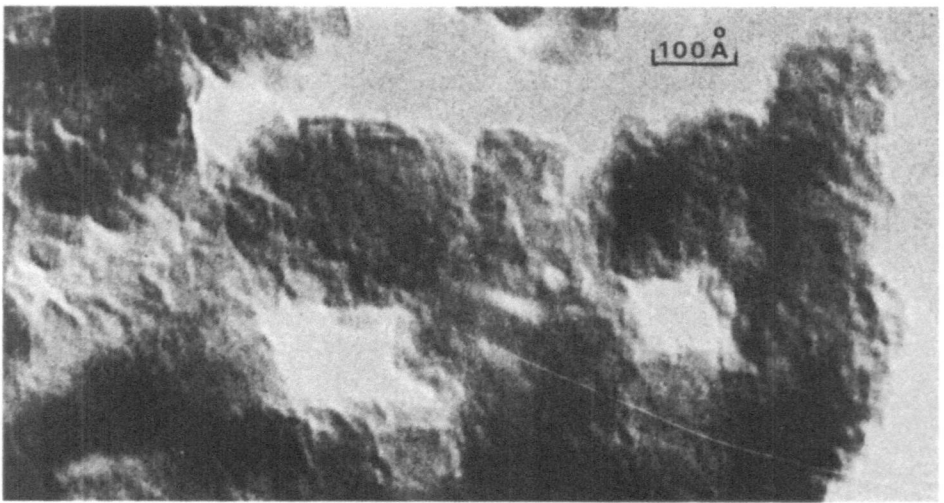

Fig. 9. Step structure on commercial Linde B alumina.

The field of Fig. 7a. is shown in Fig. 7b. at an underfocus
setting which exceeds the limits of validity of the PCD approximation at this resolution in most areas. Images of this type can
be useful in directing attention to prominent features in the pattern which can then be examined in detail in images nearer to the
Gaussian focus. The contrast distribution in Fig. 7b may be considered to be, approximately, the PCD exaggerated and spread by
Fresnel diffraction, with consequent loss of resolution.

There are several unexplained features in these patterns. In
particular, some of the prominent black dots (D) in Fig. 7b. remain
unresolved in Fig. 7a. and must derive from small concentrations of
charge, certainly smaller than one unit cell. We have, so far,
been unable to determine what these are.

High-Temperature Observations

In the current experiments, high temperature refers to the
range 900°C - 1400°C. The upper limit is set currently by the
difficulty in fabricating a suitable heating element.

In initial experiments, direct observation of MgO smoke particles at high temperature was not possible, so the sample was
observed at liquid nitrogen temperature, the beam switched off,
the specimen brought to temperature in situ, cooled, and again
observed. In this way it was hoped to follow details in the sintering process, thus supplementing observations on samples prepared
outside the microscope (Figs. 4,8 & 9) (Moodie & Warble, 1971;
Stringer, Warble & Williams, 1969). In fact, most fields of view
contained few ceramic crystals, and these exhibited profiles which
differed from any in our previous observations. A possible explanation lay in the etching process mentioned previously, but the
step structure made this implausible. An improvement in technique
allowed observation at high resolution during heating. It could
then be seen that the metal grid reacted with the ceramic through
an intermediate liquid-like phase (Figs. 10 and 11). This has been
made the basis for a practical metal-ceramic bonding process (De
Bruin, Moodie & Warble, 1972). With resolutions of about 12 - 15 Å
the dynamics of this process can be recorded cinematographically
(Lugton and Warble, 1970). Many of the relevant contrast effects
in these observations do not fall within the range of validity of
the PCD approxiamtion, but are readily interpretable in terms of
other standard approximations. For instance, the persistence of
straight thickness fringes up to the region of reaction (Fig. 11),
even after quenching, shows that negligible strain is transmitted
to the ceramic, an observation that is reflected in macro bond-
strength measurements (Bailey & Black, 1974).

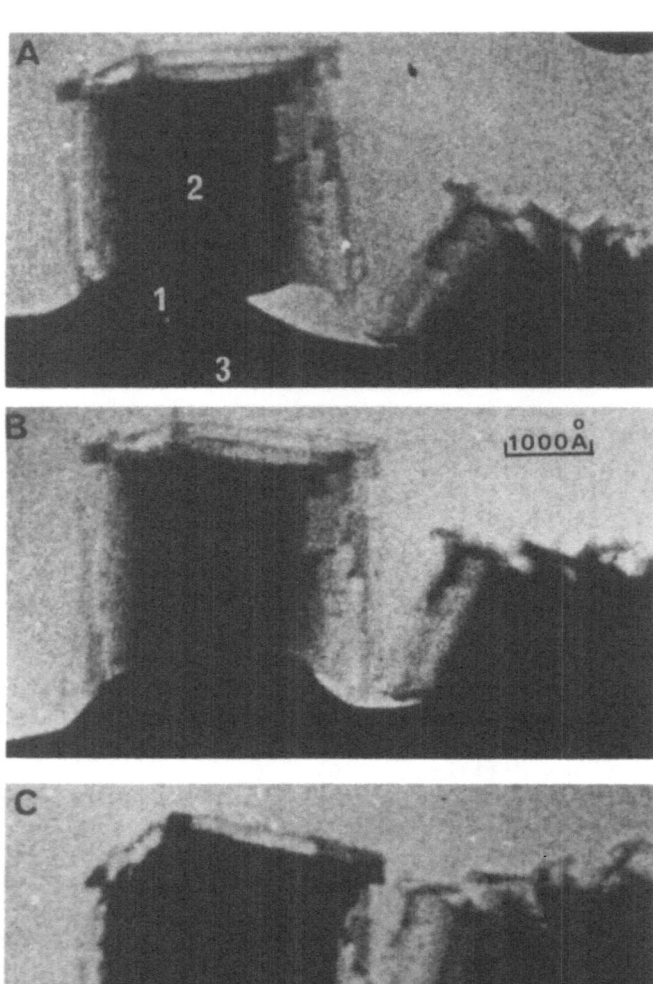

Fig. 10. Three stages in the reaction at approximately 1100°C between Pd and MgO. The images are taken from a cinematographic record,
so that the resolution is reduced to approximately 15 A. The liquid-
like phase (1) in this sequence retracts from the crystal (2) toward
the grid (3), leaving an etched surface. In C the crystal has
twisted so that a Bragg condition is nearly satisfied and contrast
due to thickness fringes is emerging.

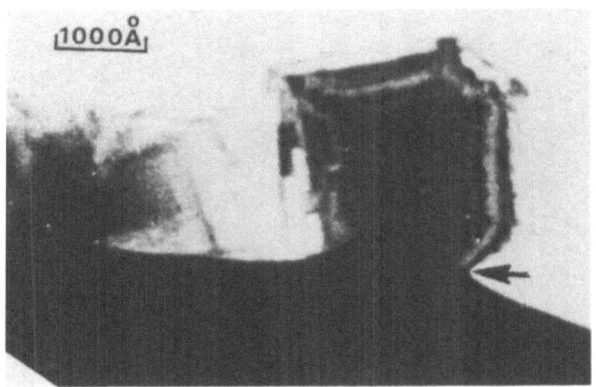

Fig. 11. A cinematographic image of the Pd-MgO reaction at 1100°C.
The MgO crystal almost exactly satisfies a Bragg condition and the
thickness fringes can be seen to run up to the reaction zone, showing that the crystal suffers negligible strain.

The reactivity of many metals and ceramics at high temperatures
is in itself interesting, but it constitutes a severe problem for
the observation in situ of, say, the sintering of magnesia or
alumina in furnaces of the present design. No completely satisfactory solution has been found, but, at some inconvenience, observations can be made using carbon fibres. Commercial material is
ground, dispersed in acetone, collected on the heating grid and
cleaned by baking. The specimen is then collected on the grid in
the standard way and observations made on those particles adhering
to the carbon fibres. This is only practicable again when the partial pressure of water vapour has been reduced to a low value, since
otherwise carbon etches relatively rapidly in the beam.

Adsorption

The experiments which, so far, have constituted the most severe
test of the various techniques have been the attempts to observe
the adsorption of small molecules on magnesia. Phase changes
through small molecules of low atomic number are not very great and
at the same time lateral size, although within the resolving capa-
bilities of the better contemporary microscopes, is still quite
small. The greatest difficulties, however, arise in the maintenance
of an acceptable level of cleanliness, and the selection of substrates of sufficient perfection not to obscure contrast due to the
adsorbed species. It has proved possible to achieve this (Moodie
& Warble, 1974), but the technique is not yet established as
routine. Crystals are observed at a temperature of 200°C - 300°C
and a suitable field selected. Gas is then admitted into the cone,
the temperature reduced and observations recorded. The temperature

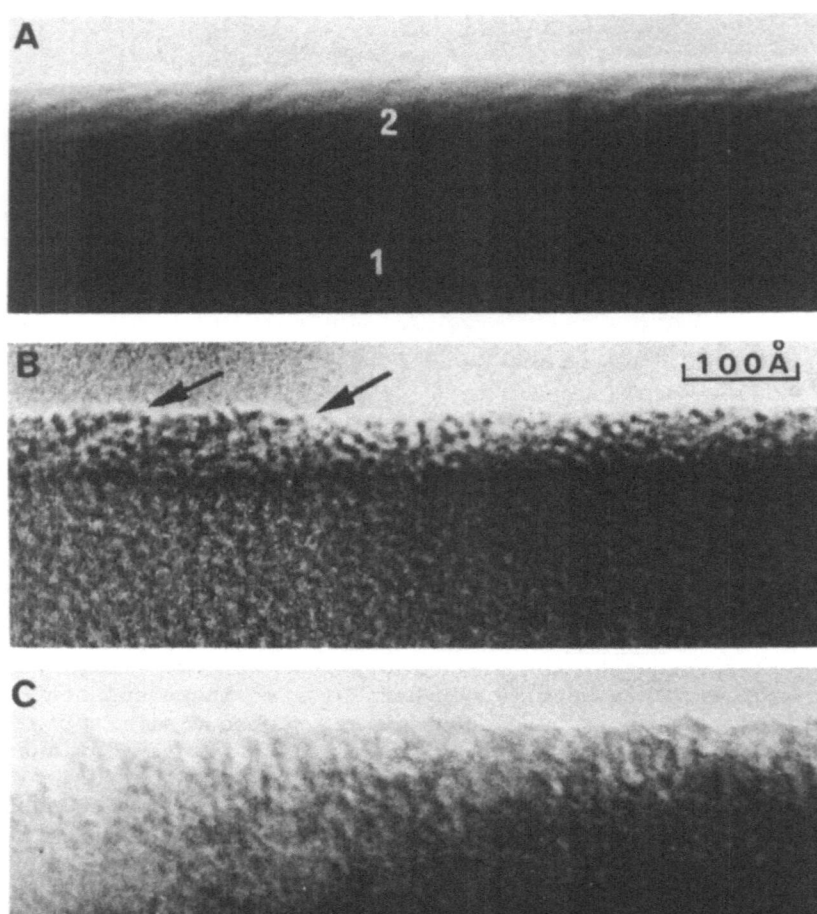

Fig. 12.
(A) Part of a plate of MgO showing clean, step-free face (1), and
     clean stepped face (2).

(B) Gas covered plate of MgO. Steps can be observed along narrow
     plate edges (arrows).

(C) Contaminated plate of MgO.

is then raised to verify that desorption takes place. It has been
found that, even with random coverage, contrast due to the adsorption of gas differs appreciably from that due to the various types
of contamination (Fig. 12). While at the moment molecules can be
most easily detected on nearly flat perfect surfaces they can, in
fact, be observed in sufficiently prominent positions on the more
complicated sinter surfaces.

Individual molecules can frequently be observed at low coverages, but these are not resolved in a Rayleigh sense; rather, their
presence is indicated by a disc, the shape of which is largely determined by the aberrations of the electron microscope. Nevertheless some species can be observed to adsorb preferentially along
edges at low coverage, and to polymerize in the beam at high
coverages, although the latter contrast cannot yet be interpreted.

If the adsorbed species is of appropriate size, higher coverages of gas will often reflect higher densities of steps. We have
found that with MgO all faces are made up of (100) faceted steps
(Moodie & Warble, 1971). Adsorption measurements incorporating
the use of argon or other inert gas, at least on MgO, can then
reflect a 'face per unit area' effect. At present a limit to the
technique has been set by the background pressure of non-condensed
gas in the instrument.

CONCLUSIONS

Current equipment and interpretative techniques are adequate
for the direct observation of the surface structure and high-temperature characteristics of a number of ceramic materials. Although
not yet established as routine, it is possible to observe directly
adsorbed molecules on crystal surfaces. The principal experimental
requirement is the maintenance of cleanliness on a molecular level.

REFERENCES

Bailey, F.P. and Black, K.J.T., CSIRO Division of Chemical Physics, Melbourne, Private Communication, 1974.

Burton, N.K., Cabrerra, N. and Frank, F.C., Phil. Trans. A, $\underline{243}$, 299-358, 1951.

Cowley, J.M. and Moodie, A.F., Acta. Cryst. $\underline{10}$, 609-619, 1957.

Cowley, J.M. and Moodie, A.F., Proc. Phys. Soc. $\underline{76}$, 378-384, 1960.

De Bruin, H.J., Moodie, A.F. and Warble, C.E., J. Mat. Sci. $\underline{7}$, 909-918, 1972.

Jackson, K.A., J. Cryst. Growth, $\underline{24/25}$, 130-136, 1974.

Lugton, F.D. and Warble, C.E., Rev. Sci. Inst. $\underline{41}$ (12), 1793-1797, 1970.

Lynch, D., Moodie, A.F. and O'Keefe, M., n-Beam Lattice Images V, Acta. Cryst. A., in press, 1975.

Mills, J.C. and Moodie, A.F., Rev. Sci. Inst., $\underline{39}$, (7), 962-969, 1968.

Mills, J.C. and Moodie, A.F., Eighth Inter. Conf. on Electron Microscopy, Canberra, Vol. 1, 182-183, 1974.

Moodie, A.F. and Warble, C.E., Phil. Mag. $\underline{16}$, (143), 891-904, 1967.

Moodie, A.F. and Warble, C.E., J. Cryst. Growth, $\underline{10}$, 26-38, 1971.

Moodie, A.F. and Warble, C.E., Eighth Inter. Conf. on Electron Microscopy, Canberra, Vol. 1, 230-231, 1974.

Stringer, R.K., Warble, C.E. and Williams, L.S. in Materials Science Research, Vol. 4, Eds. T.J. Grey and V.D. Fréchette (Plenum Press, N.Y. 1969). pp.53-95.

STRUCTURE AND THERMODYNAMIC PROPERTIES OF MICROCLUSTERS

J. J. Burton

Exxon Research Corporation

Linden, New Jersey 07036

Extensive experimental and theoretical evidence suggest that the packing of atoms in microclusters does not correspond to any lattice structure. Rather the atoms are arranged in an icosahedral packing. The evidence for this packing and the reasons for its occurance are reviewed. In addition, some unexpected properties of microclusters are discussed.

Small clusters, often containing fewer than 1000 atoms, are widely used as catalysts. Their activity and sintering (agglomeration) have been widely studied. Most of this work has been highly phenomenological and has not utilized information on the properties of the microclusters. In this paper, we review data on the structure of microclusters and conclude that the normal growth habit of microclusters is based on some seed crystal possessing a five fold symmetry axis. We present the theoretical bases for the stability of microclusters having a five fold symmetry axis. Finally, we present the results of some theoretical studies of the thermodynamic properties of microclusters.

EXPERIMENTAL RESULTS

Small clusters of metal atoms can be readily produced either by vapor phase deposition on a suitable substrate such as NaCl or mica or by gas phase condensation. Figure 1 shows an electron micrograph of Au particles produced by Ogawa and Ino[1] on a NaCl substrate. Five well defined particle shapes are discernable -

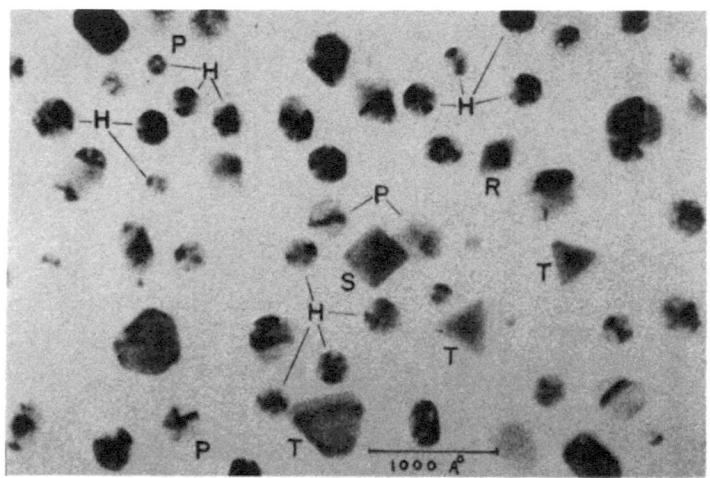

Figure 1.   Electron micrograph of Au particles produced by vapor
deposition on a NaCl substrate.

hexagonal (H), pentagonal (P), rhombahedral (R), square (S), and
triangular (T). Figure 2 shows some of the hexagonal particles
more clearly, and Figure 3 shows a beautiful pentagonal Ag particle
obtained by Kimoto and Nishida.[2] Detailed analysis of the electror

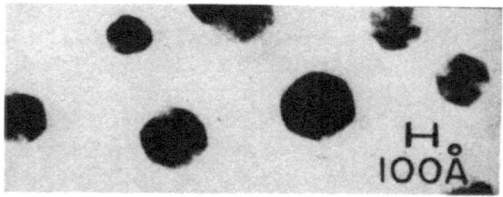

Figure 2.   Electron micrograph of hexagonal Au particles.

Figure 3.   Electron micrograph of pentagonal Ag particle.

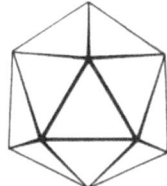

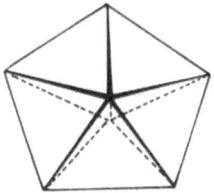

Figure 4. Drawing of icosahedron   Figure 5. A pentagonal
showing its hexagonal shape.   bipyramid.

diffraction patterns from these clusters[1] indicates that hexagonal pictures are actually of icosahedral particles, Figure 4,
which have a five fold symmetry axis. The pentagonal pictures
are obtained from pentagonal bipyramid particles, Figure 5. The
rhombahedral pictures are a special view of a pentagonal particle, and the square and triangular particles are normal face
centered cubic particles with (100) and (111) orientations respectively. No very small (≾80 A diameter) square or traingular particles have ever, to our knowledge, been observed and very large
pentagonal and icosahedral particles are also not observed. It
would thus appear that structures having five-fold symmetry axis
(pentagonal and icosahedral particles) are the preferred structures
of very small microclusters, though not, of course, of macroscopic
crystals. Five-fold symmetric particles have been reported in a
very wide range of materials including Ag,[1,2] Ar,[3] Au, [1]
C,[4] Co,[2] Cu,[5] F,[6] Ge,[7] Ni[1,6] Pd,[1] Pt,[6] Rh[8] and
Zn.[9]

STABILITY OF MICROCLUSTERS WITH FIVE FOLD ROTATIONAL SYMMETRY AXIS

Experimental evidence indicates that particles with five fold
symmetry axis are formed in both vapor phase condensation and deposition experiments, rather than particles with normal lattice
structures. These non-lattice structures must either be more stable
than the corresponding lattice structures or, at least, be the result of growth of an abnormal stable seed.

Extensive theoretical calculations[10-18] have demonstrated
that the 13 atom icosahedral cluster, Figure 6, is more stable than
the corresponding 13 atom face centered cubic (fcc) cluster, Figure

Figure 6. 13 atoms in an icosahedron packing.

Figure 7. 13 atoms in a
fcc packing.

7, or hexangonal close-packed clusters. The source of this stability is readily apparent from simply counting the number of nearest
neighbor bond contacts in Figures 6 and 7. The icosahedral packing has 42 "bonds" while the fcc packing has 36 "bonds". Quantum
mechanical calculations[18] also support the likelihood that the
icosahedral 13 atom packing is more stable.

These arguments indicate that the 13 icosahedron should be
lower in energy (i.e. more stable) than the corresponding fcc
cluster. Hoare and Pal[14] have also worked out the growth sequence
for the atom-by-atom growth of a microcluster and have shown that
icosahedral structure is naturally formed during crystal growth.
The growth sequence is as follows: Two atoms come together to form
a dimer, Fig. 8a. When a third atom is added to form a trimer, the
equilateral triangle, Fig. 8b is the stable structure. Addition of
a fourth atom produces a tetrahedron, Fig. 8c. At this point the
growth deviates from normal fcc packing which would require an extra
atom as shown in Fig. 8d. This configuration is unstable and re-
laxes into the trigonal bipyramid, Fig. 8e. Addition of another
atom to the 5-atom trigonal bipyramid produces the 6-atom tripyramid, Fig. 8f. Addition of another atom to the 6-atom tripyramid
produces a pentagonal bipyramid, Fig. 8g. The 7-atom pentagonal
bipyramid is the smallest cluster which shows the 5 fold rotational
symmetry axis which is characteristic of the icosahedral packing
sequence. Addition of five 5 atoms on the five upper faces of the
pentagonal bipyramid pictured in Fig. 8g plus one further atom on
the five fold symmetry axis produces the 13 atom icosahedron, Fig.
6. Therefore, the 13 atom icosahedron is the result of the natural
atom-by-atom growth of the microcluster. Burton[19] has proven that
even if the 13 fcc cluster were formed by some unknown growth sequence it would be unstable and would spontaneously shear into the
icosahedral packing. Hoare and Pal[14] have examined the further
atom-by-atom growth from the 13 atom icosahedron and have shown

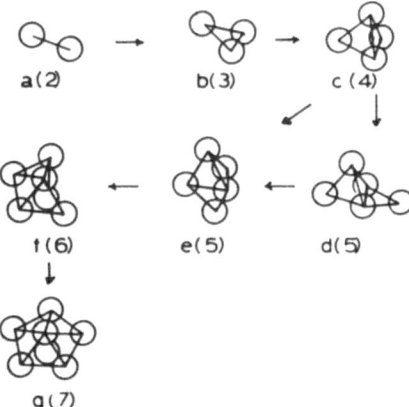

a(2)   b(3)   c(4)

f(6)   e(5)   d(5)

g(7)

Figure 8. Growth sequence to a 7-atom pentagonal bipyramid. The numbers of atoms in the clusters are indicated in the figure. The 5 atom fcc structure, d, is unstable, and deforms to e, which is the smallest stable non-fcc cluster.

that a 55 atom icosahedron, Figure 9, will form. Note that the 55 atom icosahedron has triangular close packed (111) faces of 6 atoms per face. Therefore, Hoare and Pal[14] suggest that the larger pentagonal particles sometimes seen in the electron microscope, Figure 3, may result from growth from these faces.

While the smallest microclusters are therefore likely to have five fold symmetry, large crystals always exhibit normal lattice symmetries. Therefore, it is reasonable to ask whether the several hundred angstrom particles which have been observed, Figure 1, are merely the result of growth of the icosahedral seed but are metastable, or whether they are themselves minimum energy structures. Ino[1,20,21] has carried out a careful analysis of icosahedral particles, including elastic strain energy effects, and has calculated the minimum diameters at which various metals should become

Figure 9. 55 atom icosahedron.

TABLE I

Calculated Minimum Diameters for Conversion
of Icoashedral Particles to Normal fcc Particles

Material	Diameter (A)
Au	107
Ag	76
Cu	68
Ni	43
Pd	50
Pt	56

more stable in the fcc configuration than in the icosahedral configuration. Some of his results are shown in Table 1. Ino has
shown that these minimum diameters for fcc packing would be reduced
by about 5A if the particle is adsorbed on a substrate.

THERMODYNAMIC PROPERTIES

It is not possible, at this time, to directly measure the
thermodynamic properties of small clusters of atoms. Rather recourse
must be had to some theoretical model to predict cluster properties
from bulk properties. The so called liquid drop model[22] (or
capillarity approximation) has been widely used. According to the
drop model, the surface energy of a cluster of radius r is $4\pi r^2 \sigma$
where σ is the surface energy of bulk material. (If we speak of
the cluster having i atoms rather than radius r, than the surface
energy of the cluster according to the drop model is proportional
to $i^{2/3}$)

Briant[23,24] has performed extensive molecular dynamics calculations on clusters, using a Lennard-Jones potential appropriate
to argon, in order to test the validity of the drop model. Two of
Briant's results are of particular importance:

1. The melting temperature of microcluster is well below the
melting temperature of a large sample. In a molecular dynamics
simulation of a small cluster two criteria must be met to recognize
a melting transition. Firstly, and most importantly, a phase transition must be observed. This is clearly seen in a plot of the
calculated average kinetic energy temperature as a function of total
cluster energy, Fig. 10, where a sudden drop in cluster temperature
with increasing energy is seen. Secondly, it is necessary to show
that the low temperature-low energy phase is an ordered solid while

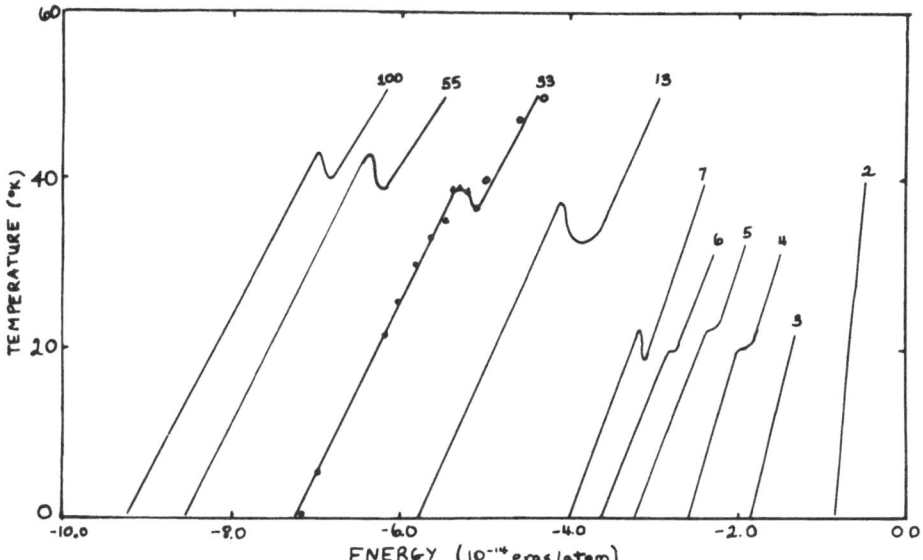

Fig. 10. Molecular dynamics data for the calculated temperature as a function of the total energy per atom for argon microclusters. (24) The numbers of atoms in the clusters are indicated.

the high temperature-high energy phase is a disordered liquid. Fig. 11a shows a computer drawing of a 13 atom cluster at 20°K, while the 13 atom cluster at 60°K is shown in Figure 11b. Clearly, at 20°K the cluster is our familiar ordered 13 atom icosahedron while at 60°K the cluster is highly disordered. Atomic self diffusion coefficients calculated for the high temperature-high energy phase(23,24) are of the order of 10^{-5} cm^2/sec, showing that this

(a)   (b)

Fig. 11. Computer drawings(25) of a 13 atom argon microcluster at 20°K (a) and 60°K (b).

phase is a liquid. From Fig. 10, one sees that the melting phase
transition in argon microclusters containing fewer than 100 atoms
occurs at temperatures below ~42°K, that is at below half of the
bulk melting temperature, 84°K.

    2. The surface energies of microclusters are incorrectly des-
cribed by the drop model.[24] The surface energy of a microcluster
of i atoms at temperature T, E^s (i, T) can be calculated from molecular dynamics data from

$$E^s (i, T) = E_c (i, T) - i E_b (T) + 6 kT$$

where E_c(i, T) is the energy of the cluster and E_b is the energy
per atom in the bulk phase;[24] the factor 6 kT adjusts the energy
of the bulk for the missing 6 degrees of freedom in the cluster.
In Fig. 12 are plotted Briant's results for the surface energies,

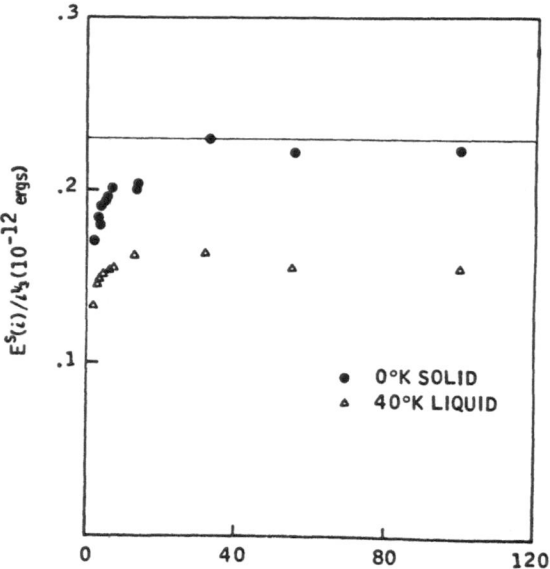

Figure 12. Surface energies, E^s(i), of argon microcluster of i
atoms.[24] The predictions of the drop model (line line) in the
solid phase are shown for comparison with the exact results (points).

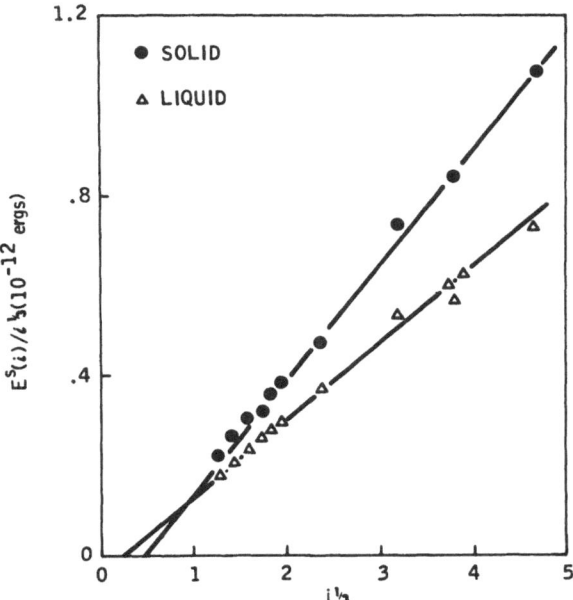

Figure 13. Surface energies, $E^S(i)$ of argon microclusters of i atoms.[24] According to Tolman's theory[26] $E^S(i)/i^{1/3}$ should be a straight line when plotted against $i^{1/3}$.

$E^S(i)$ of argon microclusters containing i atoms in both the solid and liquid phase. According to the drop model, $E^S(i)/i^{2/3}$ should be constant. The values of $E^S(i)/i^{2/3}$ predicted for solid argon microclusters by the drop model are shown for comparison with the exact results. Clearly, the drop model does not adequately describe the data.

Tolman[26], has suggested that a more appropriate expression for the surface energy of a microcluster than that of the drop model is $4\pi r^2 \sigma (1 + 2\delta/r)$ where δ is the distance between the physical and thermodynamic dividing surface. (If we speak of a cluster having i atoms rather than radius r, than the surface energy according to Tolman's model should have the form $Ai^{2/3} + Bi^{1/3}$). According to Tolman, a plot of $E^S(i)/i^{1/3}$ against $i^{1/3}$ should give a straight line. Such a plot is shown in Figure 13.

Apparently, the Tolman equation does adequately describe the surface energy of, at least, argon microclusters. Whether it is valid for more complex systems, such as metals, is not known.

CONCLUSIONS

Detailed atomistic studies of microclusters have yielded three fundamental results which may have considerable importance in sintering and catalysis.

1. The structure of a microcluster is not based on a lattice packing.[17]

2. The melting temperature of a microcluster is significantly below that of bulk material.[23,24]

3. The surface energy of a microcluster is not correctly described by the classical drop model.[24]

The practical implications of these results are not clear. They indicate that our conception of a small cluster, upon which we have relied for many years, is fundamentally incorrect.

REFERENCES

1. S. Ogawa and S. Ino, J. Crystal Growth 13/14, 48 (1972).

2. K. Kimoto and I. Nishida, J. Phys. Soc. Japan 22, 940 (1967).

3. J. Farges, B. Raoult, G. Torchet, J. Chem. Phys. 59, 3454 (1973).

4. R. H. Wentorf, The Art of Science of Growing Crystals, (J. Gilman, ed.) p. 176 (Wiley, New York 1963).

5. N. Wada, Japan J. Appl. Phys. 7, 1287 (1968).

6. A. J. Melmed and D. O. Hayward, J. Chem. Phys. 31, 545 (1969).

7. S. Mader, J. Vac. Sci. Tech. 8, 247 (1971).

8. E. B. Prestridge and D. J. C. Yates, Nature 234, 345 (1971).

9. J. D. Eversole and H. P. Broida, J. Appl. Phys. 45, 596 (1974).

10. J. G. Allpress, and J. V. Sanders, Australian J. Phys. 23, 23 (1970).

11. J. J. Burton, Nature 229, 335 (1971).

12. M. Hoare and P. Pal, Adv. Phys. 20, 161 (1971).

13. M. Hoare and P. Pal, Nature Physical Science 230, 5 (1971).

14. M. Hoare and P. Pal, J. Crystal Growth 17, 77 (1972).

15. M. Hoare and P. Pal, Nature Physical Science 236, 35 (1972).

16. J. K. Lee, J. A. Barker, and F. F. Abraham, J. Chem. Phys. 58, 3166 (1973).

17. J. J. Burton, Catalysis Reviews 9, 209 (1974).

18. J. G. Fripiat, K. T. Chow, M. Boudart, J. B. Diamond, and K. H. Johnson, to be published.

19. J. J. Burton, J. Chem. Phys. 52, 345 (1970).

20. S. Ino, J. Phys. Soc. Japan 26, 1559 (1969).

21. S. Ino, J. Phys. Soc. Japan 27, 941 (1969).

22. L. M. Skinner and J. R. Sambles, Aerosol Science 3, 199 (1972).

23. C. L. Briant and J. J. Burton, Nature Physical Science 243, 100 (1973).

24. C. L. Briant and J. J. Burton, to be published in J. Chem. Phys.

25. J. K. Lee, J. A. Barker, and F. F. Abraham, J. Chem. Phys. 58, 3166 (1973).

26. R. C. Tolman, J. Chem. Phys. 17, 333 (1949).

PREPARATION AND THERMOSTABILITY OF SUPPORTED METAL CATALYSTS

J. W. GEUS

Division of Inorganic Chemistry

University of Utrecht, The Netherlands

ABSTRACT

The different mechanisms that have been proposed for the sintering of supported metal particles are surveyed. It turns out that the mechanism of sintering depends on a number of properties that can vary appreciably in supported metal catalysts. Before the mechanism of sintering can be decided on, these properties must be known at least qualitatively.

The structure and stability of metal surfaces, the equilibrium shape of metal particles, the stability of small clusters of metal atoms and the mobility of metal atoms on the metal surfaces are reviewed. The adherence of metal particles to the support determines the mobility and the shape of small particles. The interaction with the support appears to depend on the gas atmosphere. The oxide support oxidation at the interface increases the interaction strongly. The experimental evidence of the mobility of metal particles over non-metallic surfaces is surveyed.

When the interaction with the support is small and the surface of the support is smooth, sintering of small particles by surface migration, collision and coalescence is likely. The presence of water has effects that are difficult to predict. Either the interface is oxidized, and anchors the particles to the substrate, or the interaction with the substrate is decreased and the surface of the support is smoothed, which leads to rapid sintering.

INTRODUCTION

An active catalyst must meet the following requirements:
(i) a large surface area per unit volume and, hence, very small active particle size;
(ii) an active surface easily accessible from the gas phase, which implies the presence of large pores;
(iii) active particles deposited upon porous particles of about 100 μ (fluid-bed) or 5 mm (fixed-bed) having a large mechanical strength.

Metal catalysts used at high temperatures in reducing atmospheres cannot meet the above demands without addition of non-metallic compounds. As will be apparent below, the mobility of metal atoms over metal surfaces is high. Consequently, small metal particles in contact with each other rapidly decrease their large surface energy by coalescence. Figure 1 is an instance of the sintering of iron particles. Small needles of β-FeOOH have been reduced in hydrogen at 250°C. It can be seen that reduction of the needles for several hours leads to large metal particles of a low specific surface area.

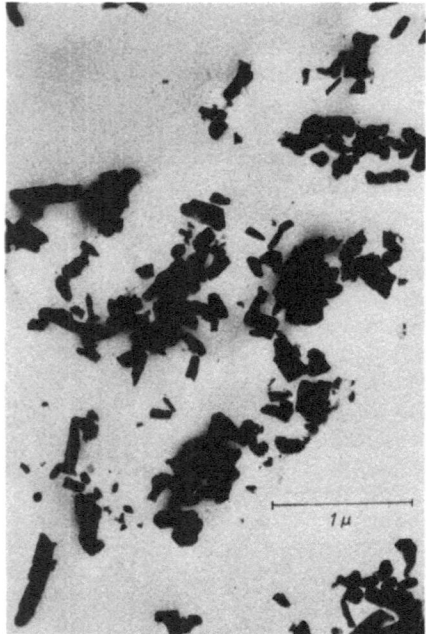 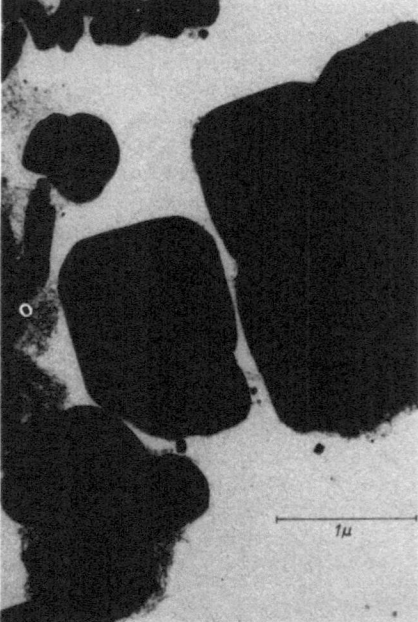

Figure 1.   Growth of iron particles. Left: β-FeOOH needles
            Right: after reduction at 250°C.

To produce thermostable metal catalysts, the metal particles
are dispersed on a thermostable non-metallic carrier (support).
The carrier must maintain the separation, and hence the small size,
of the metal particles. The carrier, which generally has not an
active surface itself, dilutes the catalyst. Nevertheless the support stabilizing small metal particles leads to a larger, thermostable, metal surface area per unit volume of catalyst. As the
weight proportion of metal-to-support is raised, the active surface
area per unit volume of catalyst passes through a maximum. It is
very difficult to prevent agglomeration and clustering of metal
particles when the support is highly loaded. The metal content
therefore is generally below the value of the above-mentioned maxi-
mem. Besides increasing and stabilizing the metal surface area, the
support may also affect the structure of the exposed metal surface.
As will be explained later, an effect on the structure of the metal
surface can change the intrinsic activity (activity per unit surface area) of the metal.

Support materials must display a large surface area that remains stable at elevated temperatures. The carrier must moreover
render it possible to produce mechanically strong pellets having
large pores. Compounds having a low tendency to sinter, such as
silica, alumina, carbon or titania are generally used as catalyst
carriers. Some clay minerals can also provide cheap, excellent
carriers.

To prepare supports of a large surface area two methods can be
utilized: 1) Restricting growth of particles from individual
molecules or ions, and 2) removing parts of the constituents of
larger crystals that maintain their external shape (Figure 2). An
example of the second method is the preparation of porous alumina
by dehydration of aluminum hydroxide. The first method can be

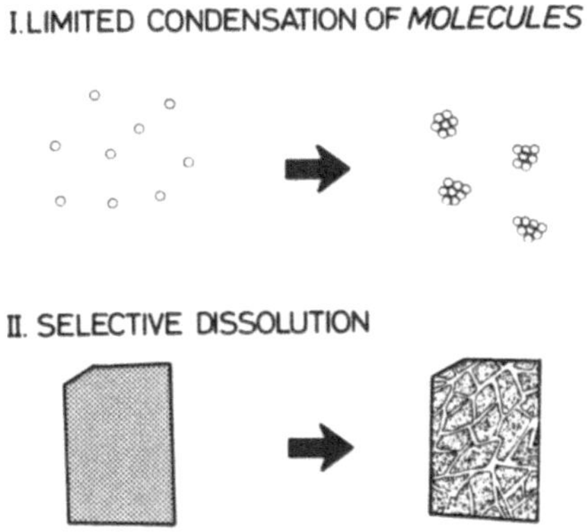

I. LIMITED CONDENSATION OF *MOLECULES*

II. SELECTIVE DISSOLUTION

Figure 2. Methods of preparation of a large surface area carriers.

used with precipitation from liquids under conditions of a rapid
nucleation and a very slow growth. Silica-alumina cracking catalysts are produced in this way. The method can also be used in the
preparation of small particles from the gaseous phase. A good
example is production of silica by flame hydrolysis (Teichner et
al., 1959, 1961a,b,c):

$$SiCl_4 + 2H_2O \rightarrow SiO_2 + 4HCl$$

As the concentrations in the gas phase are low, very small silica
particles can be produced by this method. Figure 3 gives a micrograph of a silica carrier produced by DEGUSSA (West Germany). The
carrier consists of a mixture of non-porous particles of about
300 Å and highly clustered particles of only about 20 Å.

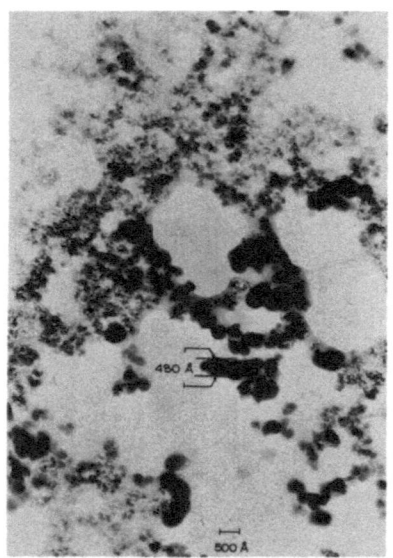

Figure 3. Electron micrograph of a silica carrier produced by
           flame hydrolysis.

Though with some noble metals, metallic particles can be deposited sirectly upon the support, smaller particles that are distributed more homogeneously are generally obtained by first applying a non-metallic compound onto the support. Deposition is
carried out mostly from an aqueous solution. The loaded carrier is
dried, in some methods without being separated from the liquid, in
others after separation. In a separate step the catalytic precursor is reduced to the corresponding metal or alloy. Formation of

the catalyst particles of the desired shape and size can be done
both either before or after deposition of the catalytic precursor.
Generally formation is done before reducing the catalyst.

    Application of the precursor can be accomplished by impregna-
ting the support with a solution of a salt of the metal to be deposited and drying the slurry. Unless the metal content is low,
impregnation and drying seldom leads to a homogeneous distribution
of the precursor over the support surface. The metal salt crystal-
lizes as evaporation proceeds. Despite intensive agitation of
the drying slurry, it is difficult to obtain evaporation homogeneous throughout the support Nor does precipitation of a catalytic
precursor in a suspension of the support lead to a homogeneous distribution of the precursor over the carrier. Precipitation takes
place where the precipitant enters the suspension. The usual addition of precipitant does not occur homogeneously throughout a suspension of the carried. Figure 4 shows an electron-micrograph of

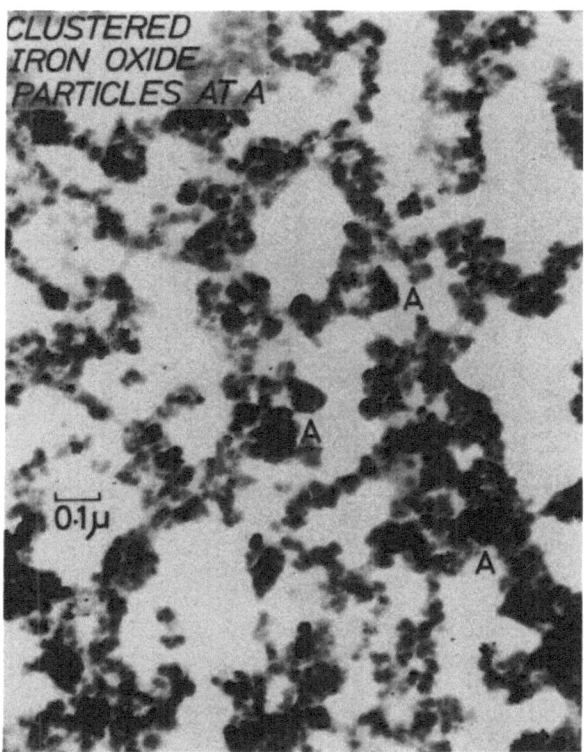

Figure 4.   Electron micrograph of inhomogenously precipitated iron
            (III) (hydr)oxide in a silica suspension.   Clustering
            indicated at A.

a sample where iron (III) ions were inhomogeneously precipitated
in a silica suspension. It can be seen that small iron particles
have clustered together; large portions of the silica surface have
remained bare. Reduction of this sample leads to large iron particles.

Very good dispersions of catalytically active precursors on
support surfaces can be obtained by precipitation from homogeneous
solution (Cartwright, Newman and Wilson, 1967; Geus, 1967, 1968).
In this method the precipitant is generated homogeneously throughout a suspension of the support in a solution of the active metal.
The homogeneous generation completely prevents local supersatura-
tions. If the precipitating compound nucleates more easily at the
surface of the carrier than in the pure solution, an exclusive
precipitation on the support surface is obtained (deposition-
precipitation). Preferred nucleation of a precipitating compound
at the surface of the support can be inferred from p_H-curves.
Geus, Hermans and de Rooij have studied the precipitation of nickel
from a homogeneous solution using the hydrolysis of urea to increase the hydroxyl ion concentration. In Figure 5, some of their
data are represented. The increase in the p_H-value was recorded
during a continuous generation of hydroxyl ions. From Figure 5 it
can be seen that the consumption of hydroxyl ions by precipitating

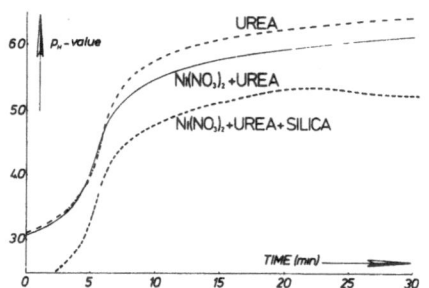

Figure 5.   Change of p_H-values of nickel nitrate solution with and
            without suspended silica during homogenous increase of
            hydroxy ion concentration. Temperature 90°C.

nickel hydroxide started at a markedly lower p_H-value when silica
was suspended in the solution. As a result, the nickel hydroxide
was completely deposited onto the carrier. The thin layer of
nickel hydroxide is much more difficult to reduce than bulk nickel
hydroxide as will be shown later.

In Figure 6 electronmicrographs of a nickel-on-silica catalyst
prepared by precipitation from homogeneous solution are represented.
It can be seen that the nickel hydroxide has been deposited as a
continuous layer on the silica. Dehydration at about 500°C did not

change the appearance of the thin layer. On reduction the continuous layer was broken up into very small nickel particles, as is
apparent from the photograph to the right of Figure 6.

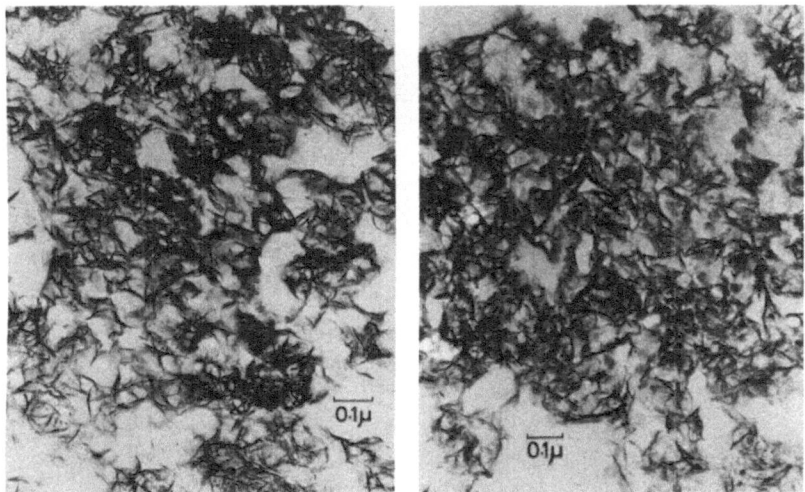

Figure 6.   Electron micrographs of silica homogenously covered by
            nickel (hydr)oride before reduction (left) and nickel
            metal particles after reduction (right).

In Table I some properties of nickel catalysts prepared by
Eurlings and Geus according to the method described above are
collected. The size of the nickel particles was determined by
X-ray line broadening. The values found ($\bar{d}_w$) are in agreement with
the dimensions observed in the micrographs. The exposed nickel
surface area was measured by hydrogen adsorption. The mean dimension of the nickel particles calculated from the exposed nickel
surface area ($\bar{d}_{vs}$) is markedly larger than $\bar{d}_w$. The difference is
due to relatively large contact areas of the nickel particles with
the support, which are inaccessible fro adsorving hydrogen molecules.

SINTERING OF SUPPORTED METAL CATALYSTS

a. Survey of Mechanisms

When the dispersion of metal particles over the carrier surface
has been done adequately, most metal particles are not in contact.
With non-contacting metal particles two sintering mechanisms can be
considered:

(i) transport of individual metal atoms or metal containing molecules from smaller to larger particles
(ii) surface migration of metal particles that coalesce on collision.

TABLE I

Exposed Nickel Surface Area and Nickel

Particle Size of Nickel-on-Silica Catalysts.[a]

Nickel Content %	Nickel Surface Area m^2/gNi	Particle Size $\bar{d}_{vs}$ [b] A	$\bar{d}_w$ [c] A
13.1	213	32	15
34.2	117	59	-
65.8	106	64	30
49.5	186	36	23
64.1	82	82	40

[a] Catalysts prepared by deposition-precipitation.

[b] Calculated from S, the nickel surface area, and

V, the nickel volume, with

$$\bar{d}_{vs} = 6V/S$$

[c] As determined from X-ray line broadening.

In mechanism (i) individual metal atoms are dissociating from
the particles to the surface of the support or to the gas phase.
The bonding energy of metal atoms to metal particles depends on the
size of the particles. Consequently the concentration of metal
atoms around smaller particles is higher than that around larger
particles. The difference in concentration leads to a transport
of atoms from smaller to larger particles. For solid particles in
a liquid phase the same phenomenon is rather well known "Ostwald
ripening" (Ostwald, 1900; Hulett, 1901, 1904; Greenwood, 1956).
Dissociation of atoms from particles can proceed more easily, when
the bond strength of the atoms to the surface of the support is
appreciable. The bonding energy must, however, not be so large
that migration of the atoms over the surface and capture by the
larger particles be impeded.
    Besides dissociation of atoms from metal particles, reaction
of metal atoms with reactive gases to form relatively volatile compounds must be envisaged. This mechanism can be important when the
metal reacts to form a volatile compound; instances are nickel carbonyl and platinum oxide. Also reactive gases can promote coales-

cence of particles only under conditions that the bulk metals are
stable. Neither the metal compound nor the (surface) compound with
the support can be more stable than the bulk metal. According to
the mechanism (ii) at least a fraction of the metal particles
should be mobile on the surface of the support.

When two particles collide, they can coalesce by surface or
volume diffusion. Pashley (1965), using Kingery and Berg's (1955)
relation has estimated that for spheres having diameters of about
0.05 μ surface diffusion predominates over volume diffusion. The
rate of coalescence of two contacting small metal particles hence
will be governed by the mobility of metal atoms over metal surfaces.

Chakraverty (1967a,b) derived theoretical expressions for the
grain size distribution in discontinuous thin metal films as a
function of time based on dissociation and surface diffusion of
metal atoms. This author considered separately cases where the
dissociation and where the surface diffusion were rate-determining.
Flynn and Wanke (1947a,b) assuming that the dissociation of metal
atoms is rate-determining at temperatures where supported metal
particles markedly sinter, developed a consistent model of catalyst
coalescence.

To account for their results (Skofronik and Phillips, 1967)
with discontinuous gold films deposited on amophous carbon and
"silicon monoxide" substrates, Phillips, Desloge and Skofronick
(1968) derived expressions for the mean size and the dispersion of
the particle size distribution. Assuming migration of at least a
fraction of the gold particles over the substrate and rapid coalescence of colliding particles they obtained the best agreement with
the experimental data. Also starting from a fraction of mobile
metal particles, Ruckenstein and Pulvermacher (1973) (Pulvermacher
and Ruckenstein, 1974) developed a mathematical model for the particle size distribution of supported particles as a function of
sintering time. According to these authors either the surface diffusion or the coalescence of metal particles is rate-determining.
Both mechanisms have been rather well investigated theoretically.
To decide which mechanism is operative in supported metal catalysts, more experimental data must be taken into consideration
then has been done so far. It will be apparent below, that depending on the metal (or alloy)-support combination, the impurities
present and the composition and pressure of the gas atmosphere, the
rates of sintering of supported metal systems vary dramatically.
As in catalytic reactions a large variety of metals and supports
are used and as supported metals are pretreated, used and regenerated in different atmospheres, it is not likely that one mechanism
can account for the sintering of supported metals in all cases. To
be able to make a well based decision about the sintering mechanism,
a large number of properties must be considered. The following
factors must be known at least qualitatively:
(a) the structure and stability of metal surfaces.
(b) the equilibrium shape of metal particles.

(c) the stability of metal particles against dissociation as a function of particle size and shape. As supported metal catalysts can contain clusters of a few tens of metal atoms, very small particles must be considered.

(d) the mobility of metal atoms over and out of metal surfaces having a different structure.

(e) the interaction of metal particles and individual metal atoms with carrier surfaces.

(f) the mobility of metal particles and individual metal atoms over carrier surface.

We shall first deal with the properties (a) to (d) whicn are pertaining to pure metals. Then (e) and (f) will be considered.

b. No marked effect of support

It has been known that metal catalysts accelerate chemical reactions by formation of a chemical bond between the metal surface atoms and at least one of the reacting species. A high catalytic activity requires a surface complex having a stability that lies within rather narrow limits. A small variation in the bonding energy of the intermediate complex will strongly affect the rate of the reaction. The chemical reactivity of a metal surface is consequently of paramount importance to its catalytic activity. That the surface structure affects the chemical reactivity has been demonstrated abundantly for the interaction of oxygen with metal surfaces (Bénard, 1962; Cathcart et al., 1967; Geus, 1970a). Sosnowsky (1959) has shown that the catalytic activity also depends on the structure of the metal surface. He studied the activity of the decomposition of formic acid on different crystallographic planes of silver. It is, however, difficult to establish beyond doubt that the activity depends on the structure of the surface of the catalyst. A catalytic reaction may proceed predominantly at a very small, differently structured fraction of the exposed metal surface, which very effectively catalyze the reaction. Preparation of metal specimens exposing exclusively an uncontaminated surface of a uniform structure is generally not possible.

The structure of clean metal surface can be ordered or disordered with respect to the bulk structure. Field-ion emission (FIM) and low-energy electron diffraction (LEED) have shown that the close-packed surfaces of most metals have the bulk structure. LEED data show that parallel to the surface the lattice spacings do not deviate markedly from the bulk spacing (for a review, see Ehrlich 1971). The lattice dimensions normal to the surface are more difficult to assess by LEED. Some metals (e.g., platinum and gold) display close-packed surfaces being differently ordered. It is still debated whether the deviation from the bulk structure is an intrinsic surface property or due to the presence of impurities (Ehrlich, 1971; Lambert et al., 1971; Nielsen, 1973; Dobson and Dennis, 1973).

Vicinal surfaces contain flat close-packed surfaces separated
by steps (Frank, 1958; Winterbottom, 1973). As argued by Houston
and Park (1970, 1971, Houston et al, 1973), LEED is not well suited
to investigate atomic details of vicinal surfaces. Theory predicts
a roughening of straight steps at temperatures well below the
melting point of the metal; kinks and adatoms should be formed at
a relatively low temperature (Burton et al, 1951; Karge et al,
1967). In view of Houston and Park's remarks, LEED patterns from
vicinal surfaces are not conclusive as to thermal roughening. FIM
shows a thermal disordering of all surfaces except the closely
packed planes. Figure 7 shows results obtained by Holscher (1967).
The photograph on the left shows a completely ordered tungsten surface that has been obtained by field-desorption of some layers of
metal atoms at a very low temperature. A thermally annealed tip is
shown in the picture at the right. In agreement with the theory,
all stepped surfaces appear to be completely disordered. Only
packed planes remain ordered.

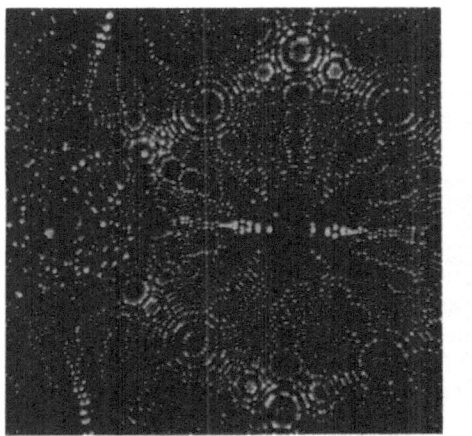 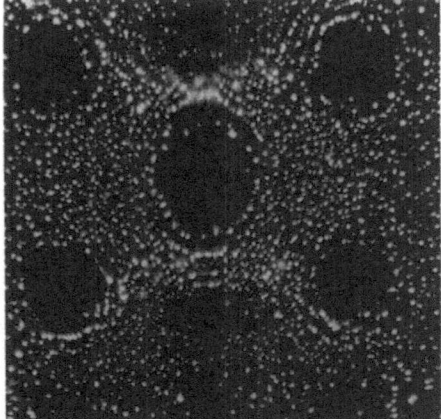

Figure 7. Field-ion emission patterns of a field-desorbed (left)
           and a thermally annealed (right) tungsten tip (after
           Holscher 1967).

To describe the surfaces of supported metal particles adequately, we must know the fractions of close-packed and stepped
crystallographic planes at the particle surface. The nature and
the extent of the crystallographic planes present on the surface of
solid particles depend on the shape of the particles. We shall
therefore investigate the equilibrium shape of metal particles that

are not markedly affected by interaction with non-metallic materials.
The effect of the support will be considered later.

The thermodynamical background for the problem has been provided by Herring (1953). Using a pair-wise interaction model
Drechsler and Nicholas (1967), and Nicholas (1968) have calculated
the surface energy for different crystallographic planes. They
calculated the number of intermetallic bonds broken in the formation of the surface without considering a redistribution of valence
electrons. Potential functions adjusted to fit with experimental
parameters such as sublimation energy, lattice, constant, and com-
pressibility were used.

As indicated in Figure 8, the range of the pair-wise interaction energy determines the equilibrium shape of metal crystallites.

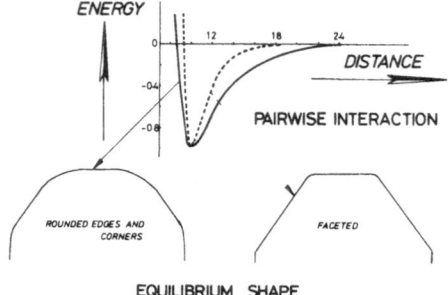

Figure 8.   Effect of range of interaction pairs of metal atoms on
            the equilibrium shape of metal particles.

If the interaction is short-range, the surface energy will vary
appreciably with the structure of the surface. In such a case the
most closely packed surfaces have a surface energy much lower than
that of less closely packed ones. The equilibrium shape will then
contain almost exclusively the closest packed planes, though this
can lead to a large deviation from a spherical shape. When, on the
other hand, the interaction is more long-range, a shape approximating a sphere is energetically more favourable. The above calculations led to a variation of the surface energy with the crystallographic orientation varying for different metals from 5 to 11%.

The experimental determination of the equilibrium shape of
metal crystallites is difficult. The profiles of field-emission
tips, or small metal particles on non-metallic substrates and of
bubbles of inert gases in metals have been used. Profiles of
field-emission tips show small flat surfaces surrounded by rather
extended curved surfaces. As stated by Müller (1953), it may,
however, be questioned whether the shape of a field-emission tip is
a true equilibrium shape. Whereas the cleanliness of the surface
of field-emission tips can be easily ascertained, that of metal

particles is much more difficult to control. Small metal particles
must be used in order to obtain an equilibrium shape in a reasonable time, as remarked by Herring (1953). With small metal particles, however, interaction with the substrate may affect the
equilibrium shape. Sundquist (1964a) carefully investigated the
equilibrium shapes of gold, silver, copper, nickel and γ-iron particles present on ceramic substrates (mostly beryllium oxide). He
found that the equilibrium shapes display flat planes with rounded
corners and edges. The flat planes are larger than those observed
on field-emission tips. Equilibrated noble gas bubbles in metals
are completely facetted. (Nelson et al, 1965, Miller et al, 1969).
     Sundquist also observed that the presence of foreign atoms
or molecules strongly affects the shape of the metal particles.
Earlier Müller and Gomer observed facetting of field-emission tips
by adsorbed impurities. Metallic impurities, on the other hand,
strongly reduce the anisotropy surface energy in many cases. The
reduction in the variation of the surface energy is apparent from
an almost completely rounded shape (Sundquist, 1964b).
     Whether Sundquist's data have been obtained on contaminated
particles is still debated. Pilliar and Nutting (1967) using
alumina as a substrate confirmed Sundquist's results that were obtained with beryllium oxide. Winterbottom (1967) could also repro-
duce Sundquist's results with silver particles obtained by annealing of a silver film deposited in a dirty vacuum of 10^{-6} torr. A
silver film deposited in ultrahigh vacuum (10^{-9} torr) failed, to
break up into small silver particles. Though Sundquist deposited
his metals in a bad vacuum, we believe that his metal particles
had reasonably clean surfaces. This is based on the fact that
Sundquist annealed his specimens in hydrogen after having exposed
them to air. From Auger analysis it is known that oxidation followed by reduction effectively removes carbon impurities from
metal surfaces. We shall return to Winterbottom's results later.
     As mentioned above, the stability of metal atoms in small
clusters can play an important role in the sintering of extremely
small supported metal particles. The average number of intermetallic bonds per metal atom and, hence, the bonding energy of metal
atoms to metal particles falls, as the size of metal particles decreases. Comparison of the dissociation energy of diatomic metal
molecules with the intermetallic bonding energy of bulk metals can
give information about the effect of the number of neighbouring
atoms on the pair-wise interaction energy. Verhaegen et al. (1960)
have reviewed the published data of the dissociation energies of
diatomic metal molecules. The dissociation energies appear to be
smaller than the heats of sublimation of the corresponding metals
by a factor varying from 1.5 to 2.1. Exceptions are the metals of
Group II of the Periodic System; these metals (e.g., Zn and Hg)
have dissociation energies that are from 5 to 13 smaller than the
corresponding heats of sublimation. To make a rough comparison of
the intermetallic bond strength in bulk metals, we only consider

bonds between nearest neighbours. With 12 nearest neighbours the heat of sublimation corresponds to breaking of 6 intermetallic bonds per atom. Hence the bond strength in diatomic molecules is 3 to 4 times the intermetallic bonding energy. Metals of Group II have a dimeric bond strength that is 0.5 to 1.0 times the bulk intermetallic bonding energy. The above data indicate that the stability of small clusters of metal atoms is large. Though with very small clusters the heat of desorption of metal atoms will be smaller owing to the decrease in the average number of neighbours, this trend is counteracted by the increase in intermetallic bond strength. Since most catalytically active metals have negligible vapour pressures at temperatures below about 800 K, small clusters of metal atoms are not apt to dissociate at the temperatures used in catalytic reactions. The stability does not pertain to small clusters of the metals of Group II, that will dissociate more rapidly. Hence the mechanism according to which supported metal particles are sintering by the dissociation and migration of individual metal atoms is not likely. Unless the bonding energy to the support is appreciable, the surface concentration of individual metal atoms is generally too small to lead to a marked rate of sintering. With reactive gases, however, the amount of metal containing molecules can be much larger.

The rate with which the equilibrium shape is approached is governed by the mobility of metal atoms on the metal surfaces. Field-emission techniques have rendered possible the investigation of the mobility of metal atoms over surfaces the structure of which was accurately known (Ehrlich and Hudda, 1966). Also much work on surface self-diffusion of metals by other methods has been published (Bonzel and Gjostein, 1969; Gjostein, 1970; Neumann and Neumann, 1972; Rhead, 1975).

Using the pair-wise interaction model mentioned above to estimate the activation energy for surface self-diffusion over different crystallographic surfaces, activation energies appreciably depending on the structure of the surface are obtained. The large variation in surface mobility is experimentally confirmed. In Table II results obtained by Ayrault and Ehrlich (1974) on rhodium surface are collected. As can be expected the activation energy for migration over the smooth (111) plane is low. The (110) plane, where the adatoms migrate along the ridges in the surfaces, exhibits a higher activation energy. The (100) plane shows an activation energy of 20 kcal mole^{-1}. Ayrault and Ehrlich observed the metal adatoms to be mobile on the (111) plane already at room temperature.

The b.c.c. metal tungsten shows the same trend. The results are collected in Table III (Ehrlich and Hudda, 1966: Bassett and Parsley, 1970). Again the activation energies are low, but in tungsten the (110), (211) and (321) planes exhibit activation energies that are about equal. The broken bond model suggests a value for the close-packed (110) surfaces that is lower than those

TABLE II

Diffusion of rhodium atoms over rhodium surfaces

Ayrault and Ehrlich (1974)

plane	Do $cm^2\ sec^{-1}$	E_A kcal mole^{-1}
(111)	2×10^{-4}	3.6
(311)	2×10^{-3}	12.4
(110)	3×10^{-1}	13.9
(331)	1×10^{-2}	14.8
(100)	1×10^{-3}	20.2

TABLE III

Diffusion of tungsten atoms over tungsten surfaces

plane	Ehrlich and Hudda (1966) Do $cm^2\ sec^{-1}$	E_A kcal mole^{-1}	Bassett and Parsley (1970) Do $cm^2\ sec^{-1}$	E_A kcal mole^{-1}
(110)	2.6×10^{-3}	21.2 ± 1.1	2.1×10^{-3}	19.9
(211)	3.0×10^{-8}	12.3 ± 0.9	3.8×10^{-7}	13.0
(321)	3.7×10^{-4}	20.1 ± 1.8	1.2×10^{-3}	19.4

Graham and Ehrlich (1973)

plane		Do $cm^2\ sec^{-1}$	E_A kcal mole^{-1}
(211)	single atoms	10^{-3}	18
	pairs	10^{-12}	7

for the other planes. The small difference in activation energy might be due to a repulsion energy between nearest neighbours (Gilman, 1962).

The data of Table III indicate that the surface diffusion is a complicated phenomenon. The first experiments pointed to a value for the (211) plane that is appreciably lower than that of the (110) plane. Later on, Graham and Ehrlich (1973) demonstrated that the low experimental value was due to the fact that pairs of atoms present in neighbouring ledges migrate with a much lower activation

energy than single atoms. More experimental data are required to
explain satisfactorily the peculiar behaviour of migrating pairs of
adatoms.

The above results have been obtained with metal atoms vapour-
deposited on the field-emission tips. When a metal particle assumes its equilibrium shape or when two contacting metal particles
sinter, metal atoms must migrate out of metal surfaces. On a
close-packed surface, formation of an adatom and a vacancy requires
much energy. The activation energy to dissociate an atom from a
step to an adsorbed position at the step leaving a kink behind is
much lower. This relatively low energy causes the steps to be
roughened already at temperatures well below the melting of the
metal. Therefore, a change in shape of metal crystallites is
chiefly due to metal atoms dissociating from the steps. The dis-
sociated atoms migrate fast over smooth crystallographic surfaces
but cross the steps much more slowly.

The mobility of metal atoms out of and over stepped surfaces
can be determined from field-electron emission experiments. Indeed the rate of change in the shape of the tip is used to measure the surface mobility (Melmed, 1967; Bettler and Barnes, 1968).
In Table IV the surface diffusion coefficients obtained from these
experiments are listed. Comparison of the values found for close-
packed surfaces of rhodium and tungsten (Table III) with the data
of Table IV shows that migration over stepped surfaces is much
more difficult. Nevertheless, metals generally used as catalysts
such as nickel, platinum, and copper, display a mobility over
stepped surfaces that is large at relatively low temperatures.

TABLE IV

Surface Self-Diffusion over Stepped Surfaces

metal	structure	cohesive energy kcal mole^{-1}	E_A kcal mole^{-1}	temp.range °K	author(s)
W	b.c.c.	200	64	2000–2600	
Re	h.c.p.	180	50	1500–2300	Bettler &
Ir	f.c.c.	160	53	1700–2100	Barnes
Rh	f.c.c.	150	41.5	1200–1500	(1968)
Pt	f.c.c.	150	29.7	500–850	
Ni	f.c.c.	100	20.6	510–570	Melmed (1967)
Cu	f.c.c.	81	9	215–400	Melmed (1965)

The above surface mobilities suggest that contacting metal
particles rapidly coalesce as long as they are small. A fast coalescence has been observed with gold and silver particles deposited

on a molybdenite substrate. Pashley and coworkers (Pashley et al., 1964; Pashley and Stowell, 1966) studied molybdenite substrates exposed to incident metal atoms in the electron microscope. They found that small particles in contact with each other coalesce very rapidly (liquid-like coalescence). The Moire patterns displayed by the particles on the monocrystalline molybdenite show that the particles are solid. Pashley accounted for this rapid coalescence of the small particles by surface migration. Moreover it has been observed that immediately after coalescence the resulting particle has a rounded shape. On prolonged exposure to incident metal atom beam, the particle assumes again a facetted shape.

A detailed explanation of the liquid-like coalescence has been developed by Geus (1971). Owing to the rapid migration over flat surfaces, the nucleation on flat surfaces cannot keep up with the supply of metal atoms. The newly arriving atoms are therefore accommodated into steps at the rounded edges and corners, which leads to a facetted shape. Nucleation of an atomic layer at a flat surface proceeds only if the stepped parts of the surface have been filled almost completely. Some FIM results obtained by Tsong (1972, 1973, Johnson and White, 1973) may be relevant for the nucleation on close-packed metal surfaces. Tsong obtained the evidence that the binding energy of a tungsten atom in a cluster of tungsten atoms on the close-packed tungsten (110) plane is surprisingly low, viz. 6 to 13 kcal/mole. These values are much lower than is to be expected from a pair-wise interaction. Tsong's results consequently point to a difficult nucleation on close-packed surfaces. It can be questioned, however, whether results obtained with tungsten, which is b.c.c., can be applied to f.c.c. metals.

Impurities not only strongly affect the shape of metal particles but also the surface mobility as is apparent from the data in Table V. A monolayer of carbon almost completely impedes the migration of tungsten atoms over tungsten surfaces, whereas a small amount of carbon has no effect. Oxygen also decreases the mobility of tungsten atoms. As indicated in Figure 9, carbon and/or oxygen may block the dissociation of metal atoms from steps or the migration of metal atoms over the steps. Two monolayers of nickel atoms strongly increase the surface mobility of tungsten. As the melting point of a tungsten-nickel alloy is appreciably lower than that of tungsten, an increase in the mobility of tungsten results. Rhead and coworkers (Rhead, 1975) have observed spectacular increases in the surface diffusion coefficients of gold, silver and copper by adsorption of metals such as lead and bismuth. Also adsorption of chlorine and sulphur led to very large surface diffusion coefficients. At present, no detailed explanation of these effects has been offered.

The fact that carbon decreases the surface mobility of metal atoms can explain Winterbottom's (1967) results with silver films. When the presence of carbon leads to a low surface mobility, the

TABLE V

Effect of Adsorption on Surface Self-Diffusion of Tungsten

adsorbate	coverage monolayers	activation energy kcal mole^{-1}	
		Pichaud & Drechsler (1972, 1973)	Bettler et al (1974)
Carbon	0.1	71.3	
	0.5-1.0	200.1	
	low	-	69
	high	-	195.5
Silicon	low	-	69
	high	-	161
Oxygen	0.5	96.6	-
	1.5	103.5	-
Nickel	<1.0	71.3	-
	>2.0	20.7	-

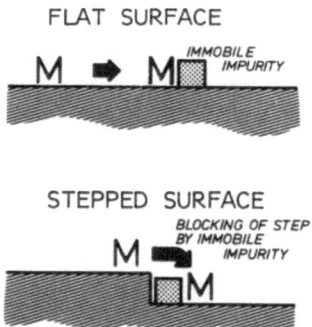

Figure 9.   Effect of impurities on the surface self-diffusion of
            metals (schematic).

silver particles coalesce only to a limited extent. As a result a
film containing many small crystallites sharing grain boundaries is
obtained. Grain boundary grooving by the high-temperature anneal
results in many small silver particles. The much larger mobility

of silver atoms in ultrahigh vacuum, gives rise to coalescence of silver crystallites to large irregularly shaped silver bodies. The flat surfaces of the large silver particles being extended leads to a very slow establishment of the equilibrium shape. The increased interaction with the beryllia substrate will be discussed in the next section.

c. Effects of Support

Interaction of metal particles with surfaces of carriers such as silica or alumina is determining the stability of the catalysts and the shape of supported metal particles. To investigate the interfacial energy of a metal and a support two types of experiments were performed. The contact angle of equilibrated solid particles or liquid drops has been measured, and the adhesion of vapour-deposited metal films to various substrates has been determined. Some results are listed in Table VI. The interaction of liquid and solid metals is about the same. Pilliar and Nutting found that the metal particles have the closest-packed (111) surface in contact with the substrate. When it is assumed that the metal surface atoms contacting the substrate only are engaged in the bond to the substrate, the interaction energy per metal atom can be calculated. For Ni/BeO it is 4 to 5 kcal mole^{-1}. Energies of this order of magnitude are characteristic of a physical interaction. Table VI shows that the interaction of metals with the oxides of thorium and beryllium is also physical. Tikkanen and coworkers (1962) found that the interaction of cobalt and nickel with the oxides of calcium and magnesium is small. From this, one may conclude that also the individual metal atoms are physi-sorbed on oxide substrates. This can be concluded from initial sticking coefficients and critical condensation rates of metal atoms incident on ionic substrates (Geus, 1971). The individual metal atoms desorb rapidly. The activation energy for surface migration is about 0.1 to 0.2 of the energy of adsorption (Ehrlich, 1959). Since the energy of adsorption is low, metal atoms are highly mobile. As it has been said above, the metal atoms that collide during migration on the surface, form clusters that do not dissociate when the temperature is not too high. As the number of metal atoms in contact with the substrate increases with the size of a cluster, the bonding energy to the surface rises too. Consequently at a high incidence rate, the sticking coefficient rises quickly, since the metal atoms are bonded to clusters of metal atoms. From the extensive work on growth of vapour-deposited metal films, it can be concluded that individual metal atoms are bonded weakly to ionic surfaces and therefore are highly mobile.

The weak interaction of metal atoms with ionic surfaces permits the inference that the desorption of metal atoms from supported metal particles is not assisted by an appreciable bonding energy to the substrate. The stability of supported metal parti-

TABLE VI

Interfacial energies of some metal/ceramic systems

System	Interface energy erg cm^{-2}	Remarks	Reference
Ni/BeO	187	molten metal	Humenik & Kingery (1954)
Ni/Al$_2$O$_3$	525–540	molten metal	Humenik & Kingery (1954)
Ni/Al$_2$O$_3$	645	solid metal	Pilliar & Nutting (1967)
Fe/ThO$_2$	900	molten metal	Kingery (1954)
Fe/Al$_2$O$_3$	800	solid metal	Pilliar & Nutting (1967)
Au/Al$_2$O$_3$	435	solid metal	Pilliar & Nutting (1967)

TABLE VII

Interaction between glass drops and solid metallic substrates as measured by contact angle (degrees) at 1000°C in various atmospheres

Pask & Fulrath (1962)

Metal	Vacuum	A, He N$_2$,CO$_2$	O$_2$	H$_2$	H$_2$O	CO
Au	62	62–65	35	65	65	65
Pt	22	17–24	15	52	60	60
Fe	55		12–24			

cles against dissociation is thus of the same order of magnitude as that of isolated particles. Since the interaction is small, a marked effect of the support on the shape of the particle cannot be expected.

The experiments with molten metals have demonstrated that with oxide substrates the interfacial interaction strongly depends on the degree of oxidation of the metallic phase. When metal drops or metal particles were annealed in an atmosphere that could oxidize the interface, or when oxygen from the substrate could oxidize the

interface, the interaction was observed to be much stronger.
Williams and Greenough (1967) found that aluminum wets alumina at
1200°C, whereas liquid nickel does not. Accordingly, Tikkanen
(1962) found that the wettability between cobalt and magnesium
oxide, and nickel and calcium oxide can be considerably improved
by addition of cobalt oxide to magnesium oxide and nickel oxide to
calcium oxide. Pask and Fulrath (1962) working with glass drops
on metal surfaces obtained results that confirm the above. Their
results are summarized in Table VII. The relatively large contact
angle indicates that the interaction of gold with glass is small.
Heating in oxygen leads to formation of gold ions in the interface,
that raise the interaction. It is well-known that platinum is
covered by a monolayer of oxygen atoms after exposure to air. The
ionic platinum species leads to an appreciable interaction at the
interface as can be seen from the low contact angle. At high temperatures in vacuum or an inert atmosphere, the interface remains
oxidized and the contact angle low. As appears from the small interaction, hydrogen or carbon monoxide reduce the interface at high
temperatures. The effect of water is more difficult to explain.
Presumably, water reacts with residual carbon in the system pro-
ducing carbon monoxide and hydrogen that reduces the interface.
The experiments with iron are interesting. Since iron oxide dissolves in molten glass, an oxide-free interface results and a small
interaction results. If the interfacial oxide is continuously re-
plenished by heating in oxygen, a smaller angle is observed.

Benjamin and Weaver (1960) investigated the adhesion of metal
films to glass substrates. Films deposited in an inert atmosphere
adhered very weakly to the substrate, whereas films deposited in
an oxidizing environment showed a strong adhesion to the substrate.
The authors, moreover, established that oxidation at the metal-
substrate interface can proceed also after deposition of the film.
The latter interfacial oxidation leads to a strong adherence.

Winterbottom's (1967) experiments with silver films have already been mentioned. This author deposited films both in the
relatively bad vacuum of an oil-diffusion pumped untrapped system
and in ultrahigh vacuum. He observed that films deposited in
ultrahigh vacuum after being exposed to air and annealed in an
inert atmosphere displayed a strong adherence to the beryllium
oxide substrate, whereas films deposited in a bad vacuum adhered
weakly. The difference in adherence may be due to the fact that
films deposited in ultrahigh vacuum are oxidized at the interface
by the substrate during annealing. Czanderna (1964) showed that a
silver surface exposed to oxygen at room temperature rapidly adsorbs some oxygen. During annealing the adsorbed oxygen, which is
not removed by annealing in helium, argon or vacuum, can oxidize
the metal at the interface by migration over the silver surface.
The films deposited under bad vacuum conditions contain some carbonaceous material. During annealing the adsorbed oxygen will be
removed by reaction with the carbonaceous material to carbon mon-

oxide and carbon dioxide. When the interface is completely reduced, the interaction of the silver particles with the beryllia
substrate is small.

Inasmuch as the surface migration of metal particles on non-
metallic surfaces is of paramount importance to the stability of
supported metal particles, we must deal with the surface mobility
in some detail. The surface mobility of metal particles depends on
the variation in interaction with the surface of the support. It
has been found that the activation energy of individual adsorbed
atom surface migration is generally 0.1 to 0.2 the adsorption energy
(Ehrlich, 1959).

Masson, Métois and Kern (1968a,b, 1970a,b, 1971, Métois et al.
1971) obtained the most direct evidence for the surface mobility
of small metal particles. They studied the movement of gold and
aluminum particles on KCl crystal cleavage planes. A (100) cleavage plane of potassium chloride one-half of which was screened by
a razor blade was exposed to a flux of gold or aluminum atoms resulting in production of metal particles on the exposed half of
the substrate only. Except from having a (111) plane parallel to
the substrate, the particles were randomly oriented. The partly
covered substrates were kept at temperatures up to about 250°C for
times up to 300 sec and in some experiments up to 1800 sec. During
this time some Au particles appeared on the shielded portion of
the crystal. That the effect on the redistribution of the particles over the substrate is due to migration of particles was
demonstrated by covering one-half of the particles before heating
by a carbon layer (Masson, Métois and Kern, 1971). The carbon
prevented migration of the particles. After the treatment at high
temperatures the distribution of the carbon covered substrate had
not changed. On the uncovered part particles had migrated into
the bare substrate surface. At 98°C the gold crystallites appeared to be quite mobile; within 60 sec they covered a mean distance of 220 Å. From these experiments Masson, Métois and Kern
(1970b, 1971) calculated the diffusion coefficients of particles
over the substrate. They were of the order of 10^{-12} to 10^{-13}
$cm^2 sec^{-1}$. The activation energy for surface diffusion depended on
the size of the particles. Gold particles of 24 Å diameter had an
activation energy from 14 to 18 kcal/mole; for particles of 31 Å
diameter the activation energy was 26 to 30 kacl/mole. Métois,
Gauch, Masson and Kern (1972a) also investigated the mobility of
aluminum particles on potassium chloride cleavage planes. Aluminum
particles were slightly less mobile; they migrated at higher temperatures. The activation energy varied with temperature.

The strength of physical interaction depends on the coordination number of adsorbed atoms with the atoms in the surface of the
adsorbent. A higher coordination number of the atoms of a metal
particle with those of the·substrate surface consequently must lead
to a lower mobility of the particle. Masson, Métois and Kern
(1970b, 1971) observed this decrease in mobility. When gold parti-

cles having a (111) plane parallel to the substrate were rotated to
a position with the <110> direction of the metal parallel to the
<100> direction of the substrate, their mobility strongly decreased.
Gold particles having a (100) plane in contact with the substrate
did not migrate at all under the experimental conditions used by
the authors.

Surfaces of supports are generally rough and not flat. Métois'
(1973) results can be used to predict the effect of surface roughness on the surface mobility of metal particles. Using the same
technique as Masson et al., Métois investigated the migration of
gold particles along and across monoatomic steps on potassium
chloride cleavage planes. As has been said above the strength of
physical adsorption depends on the coordination number with the
surface atoms of the adsorbent. Since at the steps on the surface
both atoms and small metal particles are better surrounded by adsorbent atoms, they will be adsorbed more strongly at steps. The
stronger adsorption leads to the decoration of steps, which results
from increased density of metal particles at the steps. Decoration
is obtained by exposing a surface containing steps to a flux of
metal vapour atoms. Figure 10 shows an instance of the decoration
of steps by Au on a sodium chloride cleavage plane. The occurrence
of decoration demonstrates that the individual metal atoms are mobile over the surface of alkali halides. Métois established that
the mobility of gold particles along steps is lower than that over
flat parts of the potassium chloride surface. He found also that
gold particles are unable to cross the monoatomic steps. From
Métois' results it can be concluded that even monoatomic irregu-
larities on surfaces of supports can impede surface migration of
metal particles.

To explain their results Kern, Masson and Métois (1971) considered two mechanisms of particle migration. According to the
first mechanism, migration is due to changes in the shape of the
particles, governed by migration of metal atoms on the surface of
the particle (Figure 11). As the surface migration of metal atoms
does not depend on the size of metal particles, the activation
energy for migration of particles by this mechanism should not depend on the particle size. Since Masson et al. (1971) observed
the activation energy for gold particles to increase with the size,
they rejected this mechanism.

According to the second mechanism the surface migration is due
to gliding of particles over the substrate already discussed.
Owing to the momentary unbalance of the thermal vibrations, the
particles behave like colloidal particles that can move in two dimensions. Spherical gold particles containing 500, 1,000, and
10,000 atoms have diameters of 25, 32 and 69 Å, respectively. The
average velocity $\bar{v}$ of their Brownian motion is

$$\bar{v} = (\frac{kT}{2\pi M})^{\frac{1}{2}}$$

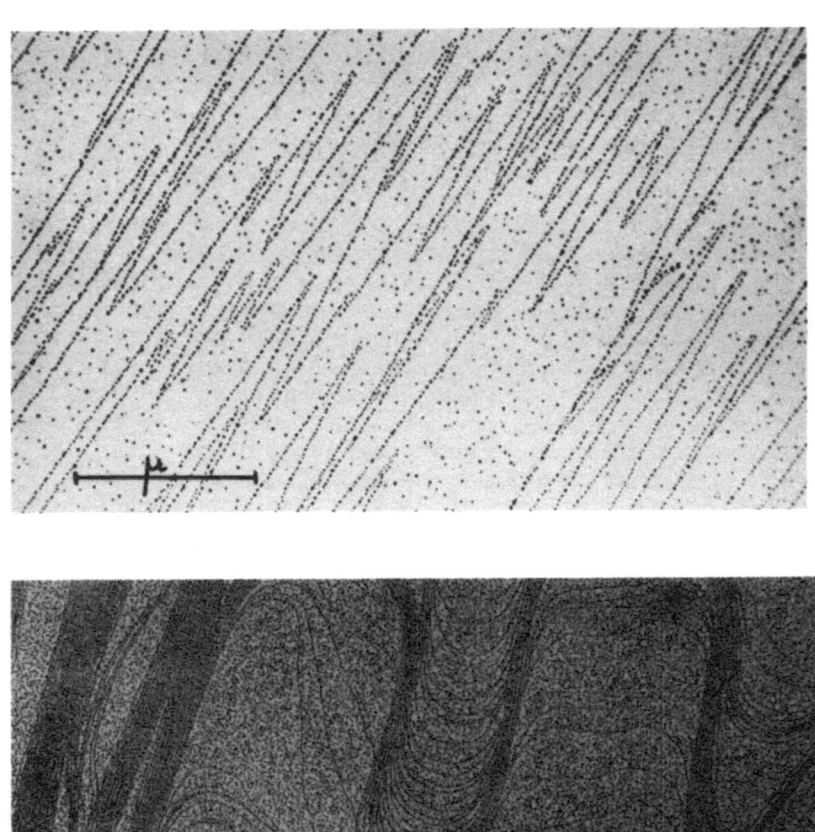

Figure 10.   Decoration of cleavage planes of NaCl by gold
             particles.

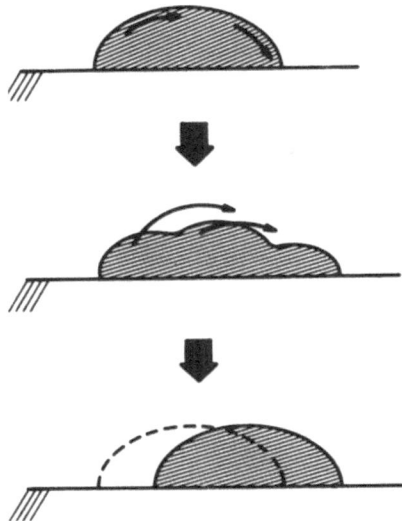

Figure 11. Surface migration of a particle by shape change
(schematic).

According to this equation the mean velocity of these particles at
300°K should be 200, 142 and 45 cm sec^{-1}. For gold particles of
the average diameter of 31 Å on potassium chloride Masson et al.
(1971) found a mean velocity at 367 K of 1.5×10^{-8} cm sec^{-1} and at
377 K of 4×10^{-8} cm sec^{-1}. The motion of the particles was thus
strongly impeded by interaction with the substrate.
     To account for these results Kern et al. (1971) developed an
island model of the interface, schematically represented in Figure
12. The interface is coherent only at the island as indicated in
the figure. Accordingly the activation energy for gliding of
metal crystallites over the substrate is the energy required to
disorder (to melt) the coherent islands. Since the number of islands is proportional to the interfacial area, the activation
energy increases with the crystallite size. The pre-exponential
factor as observed experimentally agrees with the value predicted
for this process. With aluminum particles Métois et al. (1972)
found that the activation energy did not depend on the size of the
particles at temperatures above 190°C. The authors surmounted
this difficulty by assuming that at higher temperatures the particles glide over a viscous layer. Using Eyring's (1941) theory
of viscosity, they explained the activation energy being independent of the particle size. However, there is no experimental evidence of the existence of the islands. Kern et al. invoked the
presence of coherent islands, because the critical shear stress as
evident from their results is much lower than that of a coherent

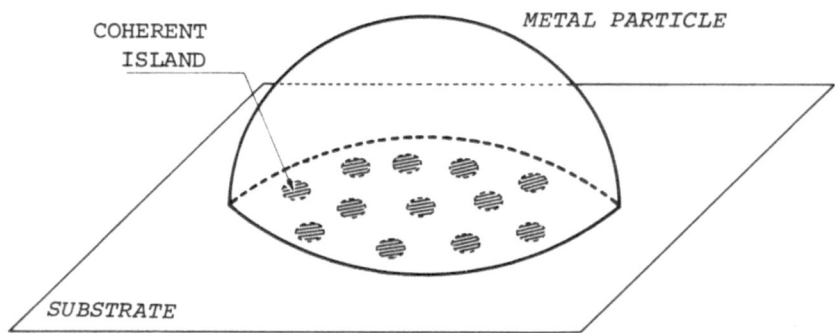

Figure 12.   Metal particle with coherent "islands" at the inter-
             face (schematic) according to Kern, Masson and
             Métois (1971).

dislocations-free interface. We believe that the interface between
a metal and an alkali halide is far from being coherent also when
the solids are ordered. A lower critical shear stress therefore
does not necessarily mean a partially disordered interfacial layer.
       The above data obtained with particles weakly interacting with
the substrate surfaces point to a rather high surface mobility.
Chemical interaction at the interface, such as oxidation with oxide
substrates, can strongly increase the adherence. We therefore expect that the particles will show a much lower mobility, when chemical interaction at the interface takes place. Experiments by
Bachmann, Sawyer and Siegel (1965) demonstrate the effect of oxidation at the interface on the mobility. These authors vapour-
deposited copper particles on "silicon monoxide" substrates under
conditions that oxidation was not completely prevented. They observed that keeping the covered substrates at elevated temperatures did not result in growth of the size of the metal particles.
On the other hand copper particles deposited in the same apparatus
on carbon substrates, sintered fast at the same temperatures.
 Since oxidation does not affect the interaction with the non-
oxidized carbon, the mobility on the carbon substrates is high.
That the difference in mobility on "silicon monoxide" and carbon
is not due to a difference in roughness of the substrates, can be
concluded from the results of Skofronik and Phillips (1967). These
authors deposited gold particles on the same substrates. As said
above, the gold interface with oxide surfaces can only be oxidized
by heating at high oxygen pressures. Though Phillips, Desloge and

Skofronik (1968) found the mobility on "silicon monoxide" to be slightly smaller than on carbon, the gold particles displayed a considerable mobility on "silicon monoxide".

As has been shown above, chemical reaction at the interface can increase the adhesion of metals to support and stabilize small metal particles. However, Sears and Hudson (1963) pointed out that an adsorbed layer can also decrease the metal/substrate interaction due to adsorbed water. During reduction of the catalysts water is formed, while many catalytic reactions are carried out in the presence of water or steam. Depending on the ratio of the partial pressures of water and hydrogen and on the nature of the metal, water can oxidize the metal at the interface. Water also increases the adherence of metals to oxide substrates. Caswell (1961) depositing tin on quartz substrates, found decreased mobility when water was present. With noble metals, such as platinum, silver, and gold, which are not oxidized by water, the effect on the mobility can, however, be completely different.

Water adsorbed on the surface of a support may increase mobility of non-oxidized metal particles. As the polarizability of water is lower than that of the ions of the usual supports, an adsorbed layer of water would decrease the physical interaction of metals with the supports. Stirland and Campbel (1966) found that the steps on a rock salt crystal surface exposed to air and briefly evacuated at 20°C were not decorated.

They started with a "silicon monoxide" substrate that was covered by water. Besides the water-covered substrate (I), they prepared two water-free substrates by depositing a fresh layer of "silicon monoxide" in ultra-high vacuum (II) and (III). Next, gold, or silver was vapour-deposited on the three substrates. To protect two of the substrates from the atmosphere, the water-covered substrate (I) and one of the water-free substrates (II) was covered with "silicon monoxide". The third substrate (III) remained unprotected. Investigation of the metal particles in the electron microscope after exposure of the substrates to air showed that the particles of substrates (I) and (III) coalesced, whereas those of substrate (II) had remained small. Further investigation revealed that the coarsening of the metal particles was brought about exclusively by exposure to water vapour. Inasmuch as the metel particles gain mobility in the presence of water vapour, both before and after their deposition, water molecules must be able to penetrate between the metal particles and the substrate.

As mentioned above, desorption of metal atoms from metal particles to the gas phase will occur markedly at rather high temperatures only. Since the interaction of metals with the non-metallic surfaces of supports is small, dissociation of metal atoms to an adsorbed mobile state at the support surface is not likely. Since adsorbed metal atoms are highly mobil on the support surfaces, metals that have a small but marked vapour pressure at the temperature where the catalyst is pretreated, used or regenerated, can

sinter with a measurable rate by this mechanism. Generally, however, Ostwald ripening can only be expected when reactive gases
capable of forming volatile compounds or species strongly interacting the support are present. A good instance is the sintering
of very small nickel particles in CO atmosphere. At carbon monoxide
pressures at which the bulk nickel is stable, nickel is transported
as nickel carbonyl from small to larger nickel particles. Wösten
(1971) has mentioned the sintering of silica-supported nickel particles by carbon monoxide.

The evidence available about platinum-on-alumina catalysts indicates that sintering proceeds much more rapidly in an oxygen than
in the hydrogen atmosphere. In oxygen Maat and Moscou (1965) observed a decrease in the exposed metal surface area by a factor of
15 after 17 hr. at 1053 K, and Hermann et al. (1961) by a factor of
12.5 after 14 hr at 898 K. In hydrogen Gruber (1962), found a decrease by only a factor of 2.5 after 1200 hr at 773 K and Hughes
et al. (1962) by a factor of 3 after 1000 hr at 811 K. Thermal
treatment in oxygen thus causes a more rapid sintering than in hydrogen. Ruckenstein and Pulvermacher (1973) assumed that in the
presence of oxygen, the rate of sintering controlled by the coalescence of mobile particles and in reducing atmospheres, migration of
the particles should be the rate-determining step.

However, one should consider the possibility of the formation of
platinum oxide, PtO_2, that dissociates on the surface of the support
and migrates to the larger platinum particles. Applying the modified Kelvin relation (Flynn and Wanke, 1974a) to Schafer and Tebben
(1960) experimental data, we obtain the oxygan pressure as a function of partical radius as listed in Table VIII with an oxygen
pressure of 0.2 atm. The surface energy for platinum was taken to
be 3120 erg cm^{-2} and the interfacial energy about 1800 erg cm^{-2}.
From the pressures given in Table VIII one can obtain the maximum
rates of desorption which are listed in TABLE IX. Though the
application of Kelvin equation to the particles of average diameter
of 20 Å can be questioned, we believe that the data of Table IX are
of the correct order of magnitude. A justification can be found in
the results of Sambles et al. (1970). These authors observed the

TABLE VIII

Platinum dioxide Pressures for Different Platinum Particles

Temp.		$\log(P_{PtO_2})$		
K	bulk	r = 10 Å	r = 20 Å	r = 40 Å
673	-12.77	-8.68	-11.04	-12.22
773	-11.04	-7.63	- 9.69	-10.71
873	- 9.71	-7.37	- 8.59	- 9.51

TABLE IX

Maximum Loss Rate of Platinum Atoms as PtO_2 Molecules

Temp.	Loss rate (atoms sec^{-1})		
K	$r = 10$ Å	$r = 20$ Å	$r = 40$ Å
673	0.12	0.002	0.0005
773	1.23	0.04	0.02
873	2.11	0.51	0.25
No.atoms/particle	372	2974	23794

Kelvin equation to be obeyed for liquid lead and solid silver particles of diameters down to 50 Å. Smaller particles evaporated more rapidly than predicted from Kelvin relation.

Therefore we believe that the data at hand indicate that desorption of platinum atoms as oxide from the platinum particles explains better the experimental data than a slow coalescence of platinum particles assumed by Pulvermacher and Ruckenstein. Still the low mobility of platinum particles in a hydrogen atmosphere remains to be explained. To understand the thermal stability of very small platinum particles in a reducing atmosphere, we refer to the experiments of Klemm and coworkers (Bronger and Klemm, 1962; Schultz et al. 1968). These authors found that a powder mixture of platinum and a base metal oxide in a hydrogen or ammonia atmosphere at 900 to 1200°C yielded platinum alloys with the base metal. The alloys Pt_3Al, $Pt_{13}Al_3$, Pt_3Mg, Pt_2Ca and Pt_3Cr were prepared by this method. Other metals of the platinum group reacted similarly. Though platinum catalysts are reduced at lower temperatures, we believe that a slight reduction of the alumina support may take place.

Finally the possibility of reduction of some ions of the support to metal atoms cannot be discarded. These atoms alloying with metal in the particles may help to bind them to the support.

The above survey shows that by trial and error many good combinations of strongly interacting metals and supports have already been developed. To meet the many demands heterogeneous catalysts have to satisfy the theories discussed above could be used as guidelines.

REFERENCES

Ayrault, G. and Ehrlich, G. (1974) J. Chem. Phys. 57, 1788.
Bachmann, L., Sawyer, D. L. and Siegel, B. M. (1965) J. Appl. Phys. 36, 304.
Bachmann, L. and Hilbrand, H. (1966) in "Basic Problems in Thin Film Physics" (R. Niedermayer and H. Mayer eds.) p. 77, Van den Hoeck and Ruprecht, Göttingen.
Bassett, D. W. and Parsley, M. J. (1970) J. Phys. D 3, 707.

Bénard, J. (1962) "Oxidation des métaux" Gauthier-Villars Paris.

Benjamin, P. and Weaver, C. (1960) Proc. Roy. Soc. (London) A 254, 177.

Bettler, P. C. and Barnes, G. (1968) Surface Sci. 10, 165.

Bettler, P. C., Bennum, D. H. and Case, C. M. (1974) Surface Sci. 44, 360.

Bonzel, H. P. and Gjostein, N. A. (1969) in "Molecular Processes on Solid Surfaces" (E. Drauglis, R. D. Gretz and R. J. Jaffee, eds) p. 533 McGraw-Hill, New York.

Bronger, W. and Klemm, W. (1962) Z. anorg. allgem. Chem. 319, 58.

Burton, W. K., Cabrera, N. and Frank, F. C. (1951) Phil. Trans. Roy. Soc. (London) 243, 299.

Cartwright, P. F. S., Newman, E. J. and Wilson, D. W. (1967) The Analyst 92 663.

Cathcart, J. V., Petersen, G. F. and Sparks, C. J. (1967) in "Surfaces and Interfaces I" (J. J. Burke, N. L. Reed and V. Weiss eds) p. 333 Syracuse University Press, Syracuse (N. Y.).

Caswell, H. L. (1961) J. Appl. Phys. 32, 105.

Chakraverty, B. K. (1967a) J. Phys. Chem. Solids 28, 2401.

Chakraverty, B. K. (1967b) J. Phys. Chem. Solids 28, 2413.

Coenen, J. W. E. (1958) Thesis, Technological University of Delft, The Netherlands.

Czanderna, A. W. (1964) J. Phys. Chem. 68, 1765.

Dobson, P. J. and Dennis, P. N. J. (1973) Surface Sci. 36, 781.

Drechsler, M. and Nicholas, J. F. (1967) J. Phys. Chem. Solids 28, 2609.

Ehrlich, G. (1959) in "Structure and Properties of Thim Films" (C. A. Neugebauer, J. B. Newkirk and D. A. Vermilyea eds) p. 423 John Wiley, New York.

Ehrlich, G. and Hudda, F. G. (1966) J. Chem. Phys. 44, 1039.

Ehrlich, G. (1971) in "Interatomic Potentials Simulation Lattice Defects, Battelle Inst. Mater. Sci. Colloq., 6th 1971" (P. C. Gehlen, J. R. Beeler, Jr. and R. J. Jaffee, eds) p. 573 Plenum Publishing Corp., New York (Publ. 1972).

Eyring, H. Glasstone, S. and Laidler, K. J. (1941) "The Theory of Rate Processes" p. 482 McGraw-Hill, New York.

Flynn, P. C. and Wanke, S. E. (1974a) J. Catal. 34, 390.

Flynn, P. C. and Wanke, S. E. (1947b) J. Catal. 34, 400.

Frank, F. C. (1958) in "Growth and Perfection of Crystals" (R. H. Doremus et al. eds) p. 411 John Wiley, New York.

Geus, J. W. (1967) Dutch Patent Application 6, 705, 259; Chem. Abstr. 72 36, 325 (1970)

Geus, J. W. (1968) Dutch Patent Application 6, 813, 236.

Geus, J. W. (1970a) in "Physical and Chemical Aspects of Adsorbents and Catalysts" (B. G. Linsen, ed) p. 529, Academic Press, London and New York.

Geus, J. W. (1971) in "Chemisorption and reactions on Metallic Films" (J. R. Anderson ed) p. 129. Academic Press, London and New York.

Gilman, J. J. (1962) Met. Soc. Conf. 20, 541.

Gjostein, N. A. in "Physicochemical Measurements in Metals Research Part 2" (R. A. Rapp ed) p. 405 Interscience Publishers, New York.

Graham, W. R. and Ehrlich, G. (1973) Phys. Rev. Letters 31, 1407.

Greenwood, G. W. (1956) Acta Met. 4, 243.

Gruber, H. (1962) J. Phys. Chem. 66, 48.

Herring, C. (1953) in "Structure and Properties of Solid Surfaces" (R. Gomer and C. S. Smith eds) p. 5 University of Chicago Press, Chicago.

Herrmann, R. A., Adler, S. F., Goldstein, M. S. and DeBaun, R. M. (1961) J. Phys. Chem. 65, 2189.

Holscher, A. A. (1967) Thesis, University of Leiden, The Netherlands.

Hougen, O. A. (1961) Ind. Eng. Chem. 53, 509.

Houston, J. E. and Park, R. L. (1970) Surface Sci. 21, 209.

Houston, J. E. and Park, R. L. (1971) Surface Sci. 26, 269.

Houston, J. R., Laramore, G. E. and Park, R. L. (1973) Surface Sci. 34, 477.

Hughes, T. R., Houston, R. J. and Sieg, R. P. (1962) Ind. Eng. Chem. (Proc. Des. Devel.) 1, 96.

Hulett, G. A. (1901) Z. Phys. Chem. 37, 385.

Hulett, G. A. (1904) Z. Phys. Chem. 47, 357.

Humenik, Jr., M. and Kingery, W. D. (1954) J. Am. Ceram. Soc. 37, 18.

Jaeger, H., Mercer, P. D. and Sherwood, R. G. (1968) Surface Sci. 11, 265.

Jesser, W. A. and Kuhlmann-Wilsdorf, D. (1967) J. Appl. Phys. 38, 5128.

Johnson, R. A. and White, P. J. (1973) Phys. Rev. B 7, 4016.

Karge, H., Heyer, H. and Pound, G. M. (1967) Z. Phys. Chem. (Frankfurt) 53, 294.

Kern, R., Masson, A. and Métois, J. J. (1971) Surface Sci. 27, 483.

Kingery, W. D. (1954) J. Am. Ceram. Soc. 37, 42.

Kingery, W. D. and Berg, M. (1955) J. Appl. Phys. 26, 1205.

Lambert, R. M., Weinberg, W. H., Comrie, C. M. and Linnett, J. W. (1971) Surface Sci. 36, 653.

Linsen, B. G. (1964) Thesis, Technological University of Delft, The Netherlands.

Maat, H. J. and Moscou, L. (1965) in "Proceedings of the Third International Congress on Catalysis" (W. M. H. Sachtler, G. C. A. Schuit and P. Zwietering eds) Vol. II p. 1277 North Holland Amsterdam.

Masson, A., Métois, J. J. and Kern, R. (1968a) J. Crystal Growth 3/4, 196.

Masson, A., Métois, J. J. and Kern, R. (1968b) C. R. Acad. Sci. Paris 267, 64.

Masson, A., Métois, J. J. and Kern, R. (1970a) C. R. Acad. Sci.

Paris, 271, 235.

Masson, A., Métois, J. J. and Kern, R. (1970b) C. R. Acad. Sci. Paris B 271, 298.

Masson, A., Métois, J. J. and Kern, R. (1971) Surface Sci. 27, 463.

Melmed, A. J. (1965) J. Chem. Phys. 43, 3057.

Melmed, A. J. (1967) J. Appl. Phys. 38, 1885.

Métois, J. J., Gauch, M. and Masson, A. (1971) C. R. Acad. Sci. Paris B 272, 958.

Métois, J. J., Gauch, M., Masson, A. and Kern, R. (1972a) Thin Solid Films 11, 205.

Métois, J. J. Gauch, M. Masson, A. and Kern, R. (1972b) Surface Sci. 30, 43.

Métois, J. J. (1973) Surface Sci. 36, 269.

Miller, W. A., Carpenter, G. J. C. and Chadwick, G. A. (1969) Phil. Mag. 19, 305.

Müller, E. W. (1953) in "Structure and Properties of Solid Surfaces" (R. Gomer and C. S. Smith eds) p. 73 University of Chicago Press, Chicago.

Nelson, R. S., Mazey, D. J. and Barnes, R. S. (1965) Phil. Mag. 11, 91.

Neumann, G. and Neumann, G. M. (1972) "Surface Self-Diffusion of Metals" Diffusion Information Center, Solothurn.

Nielsen, P. E. H. (1973) Surface Sci. 36, 778.

Nicholas, J. F. (1968) Austr. J. Phys. 21, 21.

Ostwald, W. (1900) Z. Phys. Chem. 34, 495.

Pashley, D. W. (1956) Adv. Phys. 5, 173.

Pashley, D. W. (1965) Adv. Phys. 14, 327.

Pashley, D. W., Stowell, M. J., Jacobs, M. H. and Law, T. J. (1964) Phil. Mag. 10, 127.

Pashley, D. W. and Stowell, M. J. (1966) J. Vac. Sci. Technol. 3, 156.

Pask, J. A. and Fulrath, R. M. (1962) J. Am. Ceram. Soc. 45, 592.

Phillips, W. B., Desloge, E. A. and Skofronik, J. G. (1968) J. Appl. Phys. 39, 3210.

Pichaud, M. and Drechsler, M. (1972) Surface Sci. 32, 341.

Pichaud, M. and Drechsler, M. (1973) Surface Sci. 36, 813.

Pilliar, R. M. and Nutting, J. (1967) Phil. Mag. 16, 181.

Pulvermacher, B. and Ruckenstein, E. (1974) J. Catal. 35, 115.

Rhead, G. E. (1975) in "The Solid-Vacuum Interface, Proceedings of the Third Symposium on Surface Physics" (G. A. Bootsma and J. W. Geus eds) p. 207 North Holland Amsterdam.

Ruckenstein, E. and Pulvermacher, B. (1973) J. Catal. 29, 224.

Saleh, J. M., Wells, B. R. and Roberts, M. W. (1964) Trans. Faraday Soc. 60, 1365.

Sambles, J. R., Skinner, L. M. and Lisgarten, N. D. (1970) Proc. Roy. Soc. (London) 318, 507.

Schaeffer, H. and Tebben, A. (1960) Z. Anorg. u. Allgem. Chem. 304, 317.

Schulz, H., Ritapol, K., Bronger, W. and Klemm, W. (1968) Z. anorg.

allgem. Chem. 357, 299.
Sears, G. W. and Hudson, J. B. (1963) J. Chem. Phys. 39, 2380.
Skofronik, J. G. and Phillips, W. B. (1967) J. Appl. Phys. 38, 4791.
Sosnovsky, H. M. C. (1959) J. Phys. Chem. Solids 10, 304.
Stirland, D. J. and Campbell, D. S. (1966) J. Fac. Sci. Technol. 43, 182.
Sundquist, B. E. (1964a) Acta Met. 12, 67.
Sundquist, B. E. (1964b) Acta Met. 12, 585.
Teichner et al. (1959) Caillat, R., Cuer, J. P., Juillet, F., Pointud, R., Prettre, M. and Teichner, S. J., Bull. Soc. Soc. Chim. France 1959, 152.
Teichner et al. (1961a) Cuer, J. P., Elston, J. and Teichner, S. J., Bull. Soc. Chim. France 1961, 81.
Teichner et al. (1961b) Cuer, J. P., Elston, J. and Teichner, S. J., Bull. Soc. Chim. France 1961, 89.
Teichner et al. (1961c) Cuer, J. P., Elston, J. and Teichner, S. J., Bull. Soc. Chim. France 1961, 94.
Tikkanen, M. H., Rosell, B. O. and Wiberg, O. (1962) Powder Metallurgy 1962 no. 10, 49.
Tsong, T. T. (1972) Phys. Rev. B 6, 417.
Tsong, T. T. (1973) Phys. Rev. B 7, 4016.
Verhaegen, G., Stafford, F. E., Goldfinger, P. and Ackerman, M. (1962) Trans. Faraday Soc. 58, 1926.
Williams, D. I. T. and Greenough, A. P. (1967) Powder Metallurgy 10, 318.
Winterbottom, W. L. (1967) Acta Met. 15, 303.
Winterbottom, W. L. (1973) in "Structure and Properties of Metal Surfaces" (S. Shinodaira et al. eds) p. 36 Maruzen Co. Tokyo.
Wösten, W. J. (1971) in "Proceedings of the Fourth International Congress on Catalysis Moscow USSR 23-29 June 1968" II p. 310 Akadémiai Kiado, Budapest.

SINTERING AND CATALYSIS

M. ASTIER, S.J. TEICHNER and P. VERGNON

Université Claude Bernard (Lyon I)

69621 - Villeurbanne, France

Catalysts, commercial or laboratory made, show a decline in
activity when they are used in reactions. This behaviour, detrimental to the industrial efficiency and often not well understood
by the laboratory chemist, may result, in principle, from many reasons. One of the most suspected is the presence in the reactants
of poisons which cover permanently the catalyst and compete therefore for the chemisorption of reagents. But also, the reaction products undergo a further evolution, like polymerization or cracking,
and their deposit on the surface of the catalyst is comparable to
poisoning. These phenomena, however, are to be distinguished from
the sintering which is an additional, if not the main, cause of the
loss in activity. The sintering may result in the decrease of the
active area and in the change of the structure of the surface. For
instance, the porous oxide catalyst or the porous support of the
active metal may exhibit a narrowing or closing of its pores. The
permeability of the catalytic bed decreases together with the loss
of accessible area. The porous, high surface area oxide catalyst
may also change its surface structure by surface recrystallization,
and by formation or vanishing of surface defects, whose role may be
important in catalysis.

In the case of an active metal like platinum deposited on a
high surface area carrier, if the carrier itself is preserved from
any modification, the metal which initially is in a highly dispersed state, may sinter and this results in the growth of its crystallites and in the change of its surface structure.

These phenomena, which are here reunited in a general term of
sintering are responsible for the decrease of the catalytic activi-

ty and also, very often, for a change in the reaction selectivity.
Moreover, in the case of poisoning of a catalyst a reactivation process is required in order to remove the materials (poisons, coke,
tar) covering the active surface. This reactivation may result in
a sintering of porous catalysts or carriers or in a grow of metal
particles.

The understanding of the influence of these various factors
on the catalytic activity and selectivity is still very poor. Besides, a clear definition of the catalytic activity is still lacking.
Generally based on kinetic results (e.g. number of moles of the
reactant transformed per unit time, per unit mass of the catalyst
or of the active metal in the catalyst or per unit area of the catalyst or active metal in the catalyst) the activity may also imply
the activation energy and the preexponential factor, because its
change with the temperature may not be simple.

On the other side the catalysts are active through their surface properties (particular crystal surface planes, or surface defects or surface coordination of active atoms or ions). These surface properties impose a particular catalytic activity and selectivity which may undergo a change if these properties are modified,
for instance through the recrystallization process.

In the search for correlations between the catalytic activity
and the sintering, considered in the most general sense, it is necessary to determine if a reaction under study is a structure sensitive reaction (or demanding reaction) or a structure insensitive
reaction (or facile reaction). This very interesting concept, suggested by BOUDART (1) for reactions catalyzed by supported metals
(mainly platinum), was recently applied also to oxide catalysts (2).
The determination of this characteristic of the reaction (demanding
or facile) requires a series of catalysts of the same type, differing
only by the size of particles of the active component, like platinum
crystallites. The increase of their size may be achieved by a progressive heating of a given platinum catalyst or by a progressive
increase of the platinum content in a series of catalysts.

In this way, BOUDART showed that for platinum on alumina catalysts the catalytic activity, expressed as the number of molecules
of the reagent, converted per unit time per surface platinum atom
(turnover number) at a given temperature, decreases for certain
reactions (like hydrogenolysis of neopentane) when the size of platinum crystallites increases. On the other hand for facile reactions
like hydrogenation of benzene, the turnover number remains constant
when the particle diameter increases.

The number of surface defects or coordinatively unsaturated
sites increases when the particles diameter decreases (3). Certain
reactions therefore require these defects, whereas other are produced on any type of active surface.

Two types of a decrease of the catalytic activity must be therefore distinguished when the active surface decreases or the catalyst "sinter" through the decrease of the surface area. The first
decrease, for demanding reactions, is characterized by a simultaneous decrease of the turnover number, whereas the second type of
decrease, for facile reactions, characterized by a constant turnover number or a constant quality of the catalytic surface, better
represents the decrease of the active surface.

It is of interest to consider in the light of the work of
SOMORJAI (4), what kind of surface may offer platinum and which
reactions are liable to be structure sensitive. Ultra-high vacuum
molecular beam techniques coupled with LEED and Auger-Electron
Spectroscopy were used for the study of catalytic reactions on a
series of low and high Miller index platinum single crystal planes.
The low Miller index plane was the hexagonal close packed plane(111)
in which each atom has six nearest neighbors. The high Miller index
planes, (997) and (553) are actually terraces of (111) orientation,
on the average nine atoms wide for (997) and five atoms wide for
(553) planes, limited by ordered monoatomic height steps of (111)
planes (fig. 1).

The rate of reactions like dehydrocyclization of n-heptane into
toluene and even the hydrogen-deuterium exchange is much higher on
stepped surfaces than on (111) surface. Investigation of the hydrogen-deuterium exchange reaction has moreover indicated that the steps
present on high index surfaces are necessary for the dissociation
and subsequent recombination of hydrogen (5). Therefore atomic steps
on the platinum surface must play a controlling role in dissociating
the diatomic molecules.

It is obvious that recrystallization of platinum crystallites
on supported catalysts decreases the probability of the presence of
stepped surfaces and may be, through this process, responsible for
the decrease of the catalytic activity.

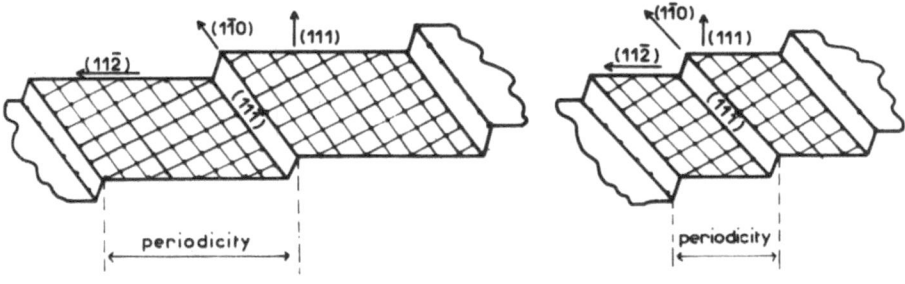

Figure 1 - Terraces on platinum crystals (Ref. 4).

The problem which now arises is how to measure the dispersion of the metal in supported catalysts. For platinum, BENSON and BOUDART (6) propose to saturate first the surface of platinum with oxygen and then to titrate oxygen with hydrogen according to reaction:

$$PtO + \frac{3}{2} H_2 \rightarrow PtH + H_2O$$

However this stoichiometry may be questionable according to MEARS and HANSFORD (7) who propose the reaction:

$$PtO + 2 H_2 \rightarrow PtH_2 + H_2O$$

Indeed, the stoichiometry of the interaction between hydrogen and oxidized platinum surface may vary with the catalyst thermal treatment and moreover the amount of oxygen, fixed on the surface in the first place also depends on this treatment (8). It seems now that a mere adsorption of hydrogen (at 200°C) on evacuated platinum catalyst would be the best method of determining the dispersion of the metal. A pulsed chromatographic technique gives results in a good agreement with those obtained by the selective chemisorption of carbon monoxide, the small-angle X-Ray scattering and the direct electron microscope observations (9).

In the case of supported copper, the decomposition at 90°C of nitrous oxide with oxidation of copper according to the stoichiometry:

$$2 Cu^+_{(surface)} + N_2O \rightarrow Cu_2O_{(surfact)} + N_2$$

gives good determination (10,11) of the dispersion of copper.

A good example of a structure insensitive reaction on copper is the dehydrogenation of butanol-2 (fig. 2) (12). For all catalysts, with various dispersions of copper, supported by alumina, the rate of reaction at 178°C per gramme of copper in the catalyst, is directly proportional to the number of surface copper atoms per gramme of copper in the catalyst, determined by N_2O decomposition. The slope of the straight line gives directly the turnover number. This example of a facile reaction is one of the best defined. Indeed, the activation energy of the reaction is the same (12.5 kcal/mole) for all catalysts and the preexponential factor is proportional to the number of surface copper atoms. It is shown below that a simple comparison of reaction rates at a given temperature for various particle sizes of the catalyst, taken as a measure of the catalytic activity, may lead to erroneous correlations between the catalytic activity and the dispersion of the catalyst.

Finally, it must be pointed out that in the determination of

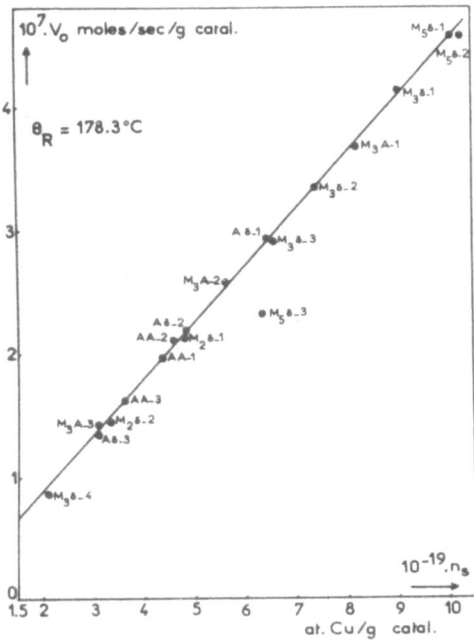

Figure 2 – Correlation between the rate of dehydrogenation of butanol-2 (at 178.3°C) per gramme of catalyst and the number of surface copper atoms (n_s) per gramme of catalyst.

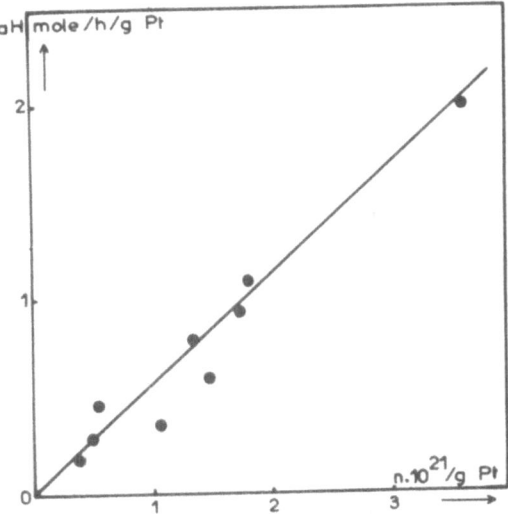

Figure 3 – Activity in the hydrogenation of benzene as a function of the number of surface Pt atoms per gramme of platinum (Ref.13).

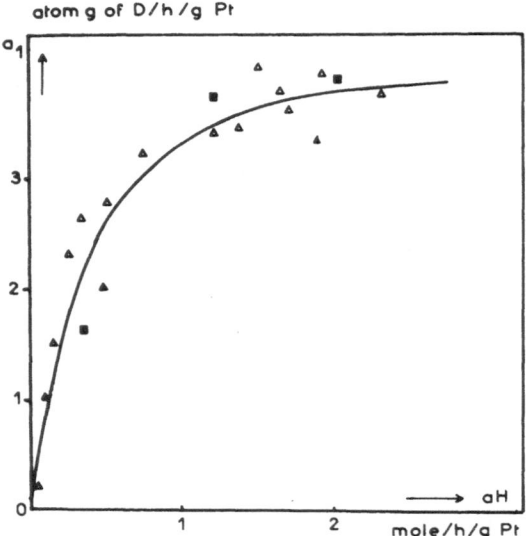

Figure 4 - Activity in the monoexchange D_2-C_6H_6 as a function of the activity in the hydrogenation of benzene (Ref.13).

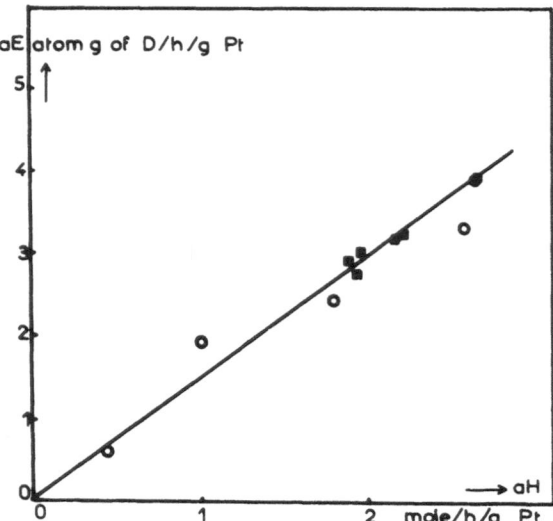

Figure 5 - Activity in the monoexchange D_2-C_6H_6 as a function of the activity in the hydrogenation of benzene of Pt-alumina catalyst without sulfur (Ref.13).

the nature of the reaction, facile or demanding, spurious results
may be obtained if some poisoning effect is not avoided. The best
example of this behaviour is taken from the work of BARBIER and
MAUREL (13). The catalytic activity in the hydrogenation of benzene
at 85°C is proportional to the number of platinum surface atoms dispersed on alumina (fig. 3). The dispersion of platinum determined
by O_2-H_2 titration was decreased in a given catalyst by a progressive sintering under hydrogen between 300° and 450°C. This reaction
is therefore facile and may be used as a reference test. In other
words the catalytic activity of the hydrogenation of benzene may be
used as a measure of the active surface of platinum.

If now the catalytic activity in the monoexchange between benzene and deuterium at 85°C is plotted as a function of the catalytic
activity in the hydrogenation of benzene (fig.4) the resulting curve
shows that the monoexchange is apparently a demanding reaction. The
alumina carrier contains sulfates which are reduced to sulfur during
progressive sintering in hydrogen. If platinum is deposited on a
different alumina carrier, with a very low sulfur content, both acti-
vities remain proportional when the dispersion of platinum decreases
(fig. 5) by sintering under hydrogen.

This work was extended to show the influence of various poisons
on the preceeding and other reactions on platinum and thus the concept of the demanding and facile poisoning was developed by the
authors (13).

Suprisingly enough, the catalytic activity of oxides was never
considered as being dependent on the state of dispersion of the
oxide, whereas this type of correlation was envisaged for metals by
virtue of demanding reactions.

The case of titanium dioxide (anatase), catalyst of the carbon
monoxide oxidation, is very interesting in this respect because it
shows the necessity of a better definition of the catalytic activity,
expressed till now as a simple reaction rate at a given temperature,
for a series of catalysts differing only by dispersion.

Titania homodispersed non porous particles, of a diameter in
the range of 60Å to 2000Å, are commonly obtained in a hydrogen-oxygen
flame reactor (14). The morphology of particles may be also controlled by the temperature of the flame and some other parameters. Two
families of particles were prepared, of polyhedral and spherical
shape differing only in diameter in the previous range. It has been
shown that the mechanism of the carbon monoxide oxidation on these
particles depends on the nature of the activation pretreatment at
500°C i) in vacuum, ii) in hydrogen or iii) in carbon monoxide (15).

For vacuum activation the overall order of reaction is 1/2 and

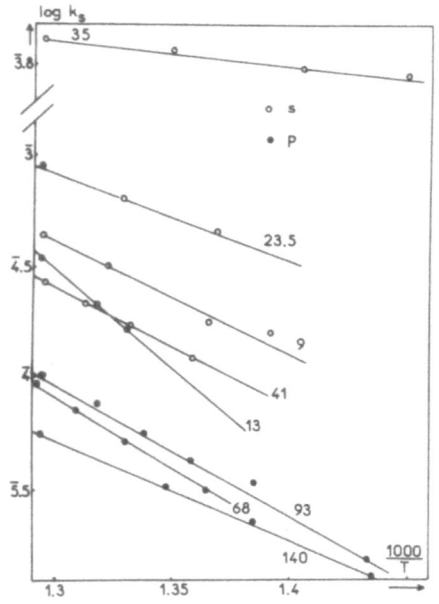

Figure 6 - Arrhenius plot of rate
constant R_s for CO oxidation on
anatase samples: o-spherical particles, •-polyhedral particles (the
numbers by the lines refer to the
surface area of the sample, in
m^2/g).

and the Arrhenius plot of the rate constant (per m^2 of the oxide) is
shown in Figure 6 for titania spherical and polyhedral particles of
various diameter which are indicated in the diagram by their surface
areas. No clear correlation can be observed between the reaction
rate constant at a given temperature, which is a kind of catalytic
activity, and the particle diameter or surface area. This is because activation energy and the preexponential factor vary for all
samples. However, if the logarithm of preexponential factor A is
plotted against activation energy E as in fig. 7, polyhedral and
spherical particle data fall on two separate lines, exhibiting well
known compensation effect, (16) . Therefore the initial classifica-
tion of samples, based on their morphology, is found again here. For
polyhedral, but not for spherical particles, a linear relationship
is found moreover between the activation energy and the surface area.
In summary, the catalytic activity of anatase activated in vacuum,
depicted by the rate constant referred to 1 m^2 of the surface, varies
with the state of dispersion of the solid (surface area or particle
diameter) and with its morphology (polyhedral or spherical particles).

For hydrogen activated titania particles, where the overall order of reaction is unity, the compensation law is again observed and
polyhedral and spherical particles give separate plots of the type
given in figure 7.

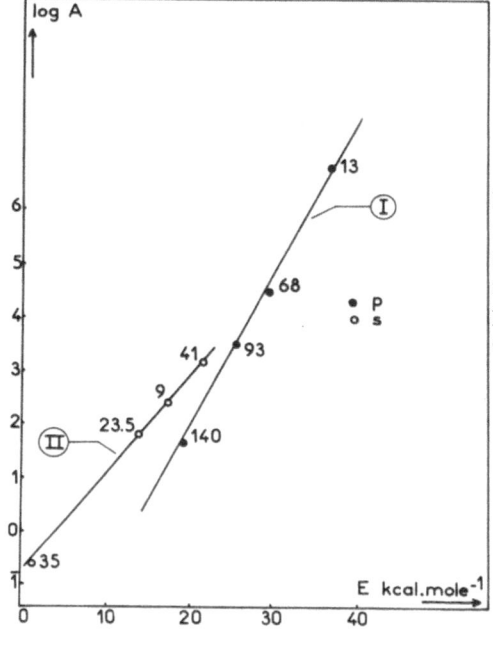

Figure 7 - Compensation effect for polyhedral particles (I) and spherical particles (II). The numbers close to the experimental points refer to the surface area of samples (m^2/g).

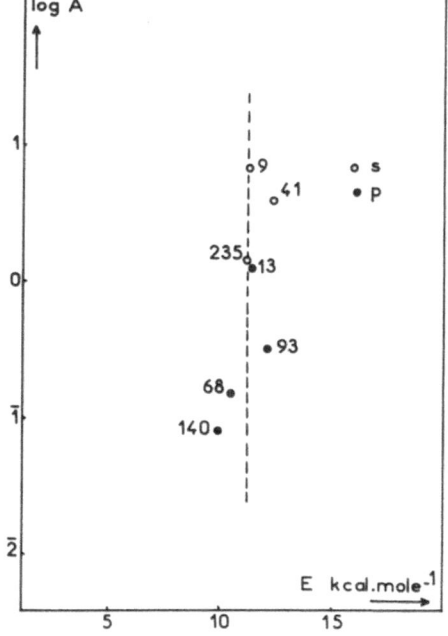

Figure 8 - Non-compensation effect for polyhedral (●) and spherical (o) particles for CO activated anatase.

Finally, for carbon monoxide activated titania particles where the overall order of reaction is also unity, a non-compensation effect is observed because the diagram log A = f(E) exhibits a mean straight line (fig. 8), perpendicular to the abscissa, with E = 11.5 ± 1.5 kcal/mole.

If the compensation effect, for vacuum or hydrogen activated particles, is correlated with the energetic heterogeneity of the surface of catalysts, the identity of slopes i) for polyhedra ii) for spheres in the compensation diagram (fig.7), after vacuum or hydrogen activation, seems to show that the same law of energy distribution for the active sites applies for both activation treatments, having in common the creation of anionic vacancies (17). The absence of the compensation effect in the case of the activation by CO seems to indicate that the active sites in this case may be of a different nature. Indeed, after this activation treatment almost no anionic vacancies are formed, but interstitial Ti^{3+} ions (17). If these interstitials are the sites for the adsorption of carbon monoxide, which reacts, by Rideal mechanism, with oxygen from the gas phase, the constant activation energy, irrespective of the size and morphology of particles, seems to suggest that these sites are isoenergetic. On the other side, the anionic vacancies, developed during vacuum or hydrogen activation treatment, which are sites for oxygen adsorption before its reaction with CO (Langmuir-Hinshelwood mechanism) are responsible for the energetic heterogeneity during the reaction.

In summing up, the oxidation of carbon monoxide on vacuum or hyrogen activated anatase would be a demanding reaction, but the same catalytic reaction would be facile on the same anatase samples activated in carbon monoxide. But even in this last case, despite of constant activation energy, there is no simple correlation between the preexponential factor, calculated per unit area of the anatase particles and the surface area of particles, polyhedral or spherical. Clearly, the efficiency of a square meter of the anatase in the oxidation of CO does not remain constant for particles of various diameters. The turnover number or the number of molecules reacting per unit time on one active site cannot be established here, because the number of active sites per unit surface is not known. Although the number of Ti^{3+} ions, in nodal or interstitial positions, per unit surface, determined by E.S.R. (2) is constant for all samples, probably not all these ions are sites for the catalytic reaction. A simple definition of the catalytic activity, which would be useful in the search for correlations between this activity and the dispersion of the oxide is therefore difficult to establish, in particular for spherical and polyhedral particles of anatase activated in vacuum or in hydrogen. These findings seem even to bring into question any attempted correlations between the "catalytic activity" and some properties of the oxide catalysts (18).

It is noteworthy that the studies concerning the sintering of
catalysts were mainly performed on supported metals, like platinum,
where it seems that the nature of the active site is more obvious
than for oxides.

The sintering of platinum by the increase of the platinum islands on the support, with the simultaneous reduction of the active
surface area, may be beneficial to the catalytic activity and not
always detrimental. This was shown in the case of the hydrogenation
of cyclopropane (19).   Indeed, in this reaction, the self-poisoning
process through the formation of some carbonaceous residues formed
from cyclopropane (reagent) and propane (reaction product) is delayed if the platinum-on-alumina catalyst is presintered in H_2 at
$600^{\circ}C$ for 6 h. The choice of a test reaction in the sintering studies of catalysts must be therefore dictated by the absence of self-
poisoning.

Another unusual result concerns the progressive poisoning by
arsine of a platinum film supported by mylar foil (20). It was found
indeed that the surface area of the film as a whole and not only the
active surface area decreases linearly with poisoning by arsine due
to the heat developed by the formation of $PtAs_2$.

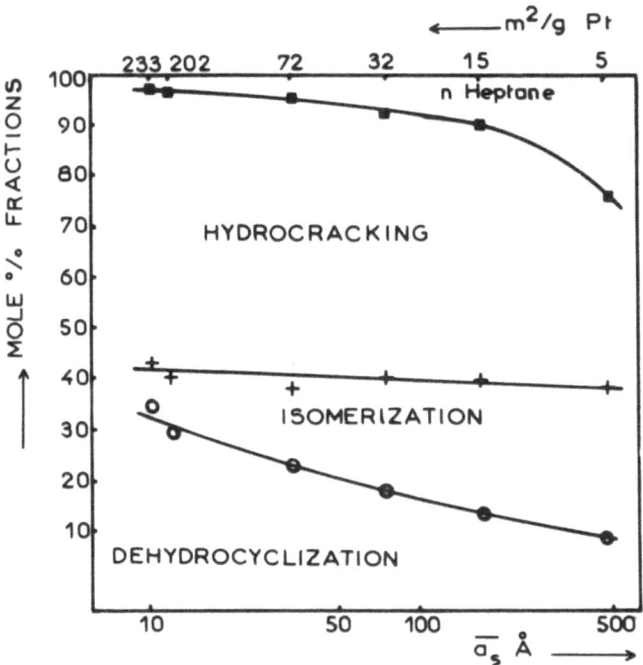

Figure 9. Product distribution of n-heptane reforming (Ref. 21).

As a general rule there is a requirement of a fine dispersion
of platinum on support in order to observe a good efficiency in the
catalytic reaction. For instance platinum-on-alumina reforming catalyst is primarily designed to produce aromatics from paraffins.
MAAT and MOSCOU (21) determined the selectivity in the reforming of
n-heptane as a function of the particle size of platinum on alumina
catalysts in the progressively sintered samples at 780°C. Fig. 9
gives the selectivity in dehydrocyclization, isomerization and hydro-
cracking versus the average particle size of Pt crystallites, assumed
to be cubes, determined by hydrogen chemisorption. As the aromatics
formed by dehydrocyclization contribute to the octane number more
considerably than the isomerized paraffins, it is obvious that an
increase in platinum size of the catalyst results in lower octane
number. Therefore a highly selective platinum reforming catalyst,
with a maximum dehydrocyclization effect, must have platinum crystallites in the range of 10 - 20Å.

The isothermal sintering of platinum, initially highly dispersed
on alumina, followed by hydrogen adsorption, was found by MAAT and
MOSCOU (21) and previously by HERRMANN and coworkers (22) to be a
second order reaction with respect to the sintering time, according
to the integral equation:

$$\frac{1}{S} = \frac{1}{S_o} + kt$$

where k is the sintering rate constant and S platinum surface area
determined by hydrogen chemisorption.

However, analyzing the available in the literature experimental
results RUCKENSTEIN and PULVERMACHER (23)came to the conclusion that
in reality the surface area decrease of platinum follows the rate
equation:

$$\frac{dS}{dt} = - kS^n$$

where the range of the exponent n is between 2 and 8. For temperatures higher than the Tamman temperature (about 550°C for Pt) the
crystallites of platinum are believed to migrate as a whole on the
surface of the support. Now, if the interaction between particles
which come in contact is strong they will form a single unit within
a time which is short compared to the time in which the migration
process takes place. The rate of sintering is then diffusion (migration) controlled and the model gives for the exponent n a value
higher than 4. If the merging into a single unit of two particles
in contact is assumed to be slow, compared to the diffusion (migration process), the rate of sintering is controlled by the sintering
(coalescence) process itself and the exponent n has a value smaller
than 3. The mobility of the crystallites does not depend exclusively

on temperature. The rate of sintering depends also on the method of
preparation of the metal - supported catalyst and upon the chemical
atmosphere (24) in which the decay takes place because these factors
affect the strength of the interaction between the crystallites and
support and, consequently, the diffusion (migration) coefficients of
the crystallites upon the support and the rate constant of the merging process. The chemical atmosphere affects the process because,
due to chemisorption, the surface diffusion and the surface free
energies are modified. The wetting angle of the metal crystallite
on the support depends on the surface free energies. If the wetting
angle is decreased, due to chemisorption, the surface area of contact
between crystallite and support is increased and the mobility of the
crystallites (diffusion coefficient of crystallites) upon the support is decreased. Also the merging of two colliding particles into
one depends upon the surface diffusion of the metal and the surface
free energy and consequently upon the chemistry of the atmosphere.

It should be mentioned here that in the studies of the sintering of catalysts the experiments and the models based on the surface
area decrease would be more appropriate than those based on the
shrinkage of compacts, which is usually the case in the study of the
sintering of materials. Indeed, surface properties are more representative of the catalytic properties than bulk properties.

For the nonsupported catalysts, like oxides, the available models are those of spheres (25) or of overlapping cubes (26) and of a
polycrystalline body with pores (27,28). The relative surface area
change ($\Delta S/S_0$) in a spherical powder compact (26) depends upon the
average coordination number of the particles and hence the bulk decreases occur actually prior to the occurrence of any shrinkage(28).
This tends to show that the transport mechanism is surface diffusion
(26). However $\Delta S/S_0$ does not exceed 10 - 12% therefore this model
(26) represents only the initial sintering of cubes or spheres. The
time exponent for the relative surface decrease is 1/3 or 2/7 for
spheres and 1/4 for overlapping cubes (26).

A particularly desirable material for experiments in sintering
of unsupported catalysts are titania anatase particles in the form
of homodispersed spheres or polyhedra, mentioned previously in the
description of carbon monoxide oxidation catalysis.

The shrinkage of compacts made out of this material was studied
in our laboratory by an isothermal technique (29). The compacts of
anatase spheres, whose surface area of 11 m^2/g corresponds to the
average diameter of 1430Å, show a linear dependence of shrinkage
with time (fig. 10) during which no grain growth occurs. This behaviour was explained on the basis of secant spheres model without
a neck having a negative curvature (28) (fig.11). Indeed, electron
microscope examinations gave evidence that for such small spheres

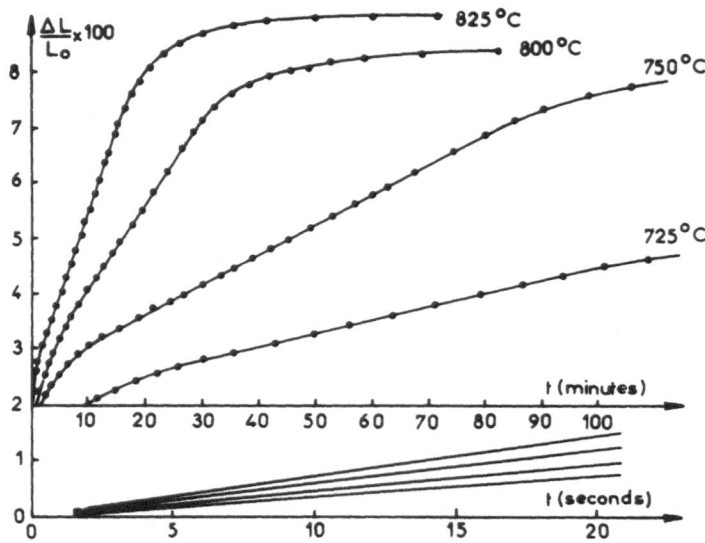

Figure 10 - Shrinkage isotherms for anatase compacts of sperical particles (11 m²/g).

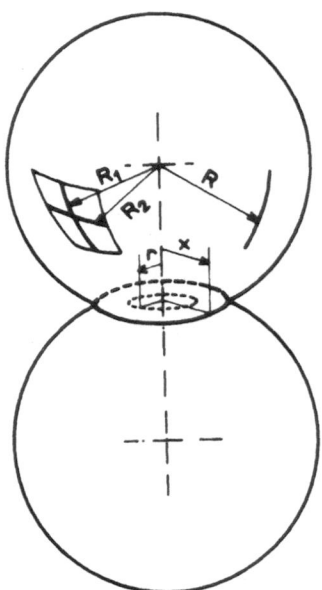

Figure 11 - Secant spheres model for submicronic particles without formation of a neck.

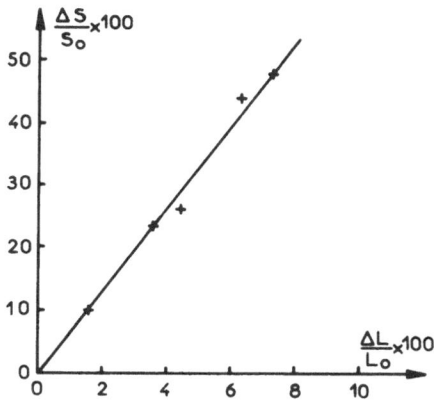

Figure 12 – Fractional Change in total surface area as a function of fractional shrinkage yield for spherical particles of anatase (diameter 1430Å).

the picture for coalescence is that represented in figure 11. The determination of the surface area S of samples in the course of sintering shows that the variation $\Delta S/S_o$ (up to 50%) is directly proportional to the fractional shrinkage $\Delta L/L_o$ for this sample (fig.12) or to the time. The value of the proportionality coefficient (6.6) in figure 12 corresponds to the mean number of interparticular contacts. This figure shows also that the rate of the surface area decrease is a zero order reaction with respect to the sintering time ($dS/dt = -K$), instead of order 2 as for merging process (23) or order 1/3 or 2/7 for the initial stage of sintering of spheres where $\Delta S/S_o$ does not exceed 10 - 12% (26). This result also demonstrates that the shrinkage is exclusively caused by the decrease of the distance between the particle centers as a result of the extension of the junction area. The activation energy of 75 kcal/mole was attributed to the formation and migration energies of equilibrated vacancies (30).

The isothermal shrinkage of compacts made with polyhedral particles of 115Å mean diameter, with a surface area of 140 m^2/g, is shown on figure 13. Again a linear behaviour is observed and no grain growth is recorded during the linear period. The activation energy of 23 kcal/mole corresponds here to the energy of migration of vacancies trapped in a large excess (above the thermodynamic equilibrium concentration) in the junction zone (30). The loss of the surface area before the shrinkage starts is already of 18%. This almost instantaneous loss of the surface area corresponds to the formation of junction zones between partciles of the compact. This

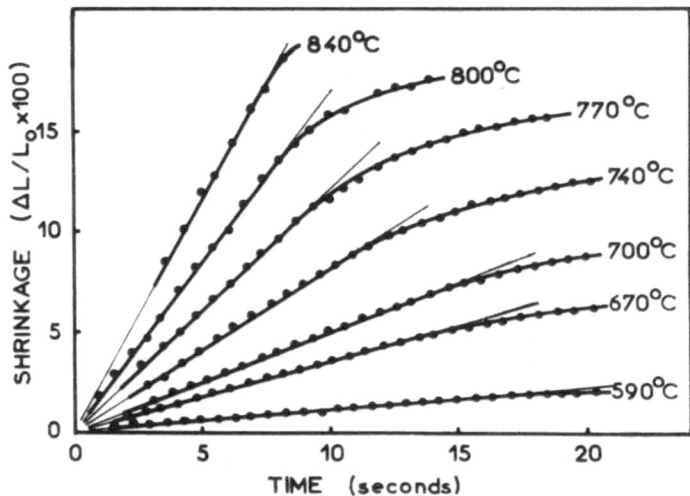

Figure 13 – Shrinkage for anatase compacts of polyhedral particles
(140 m^2/g).

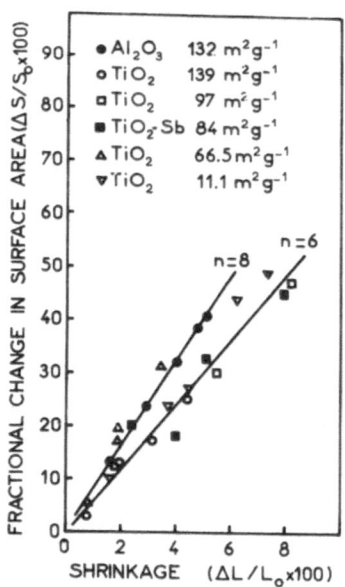

Figure 14 – Fractional change in surface area as a function of
fractional shrinkage yield for polyhedral and spherical particles
of anatase (diameter 115Å) and other materials.

phenomenon is important for small particles and becomes negligible
for particles of 1430Å (no instantaneous loss of surface area).
Indeed, for particles of 200Å diameter, at a distance of 20Å from
the point of contact, their surfaces are only separated by 4Å, which
is of the order of lattice distance in the solid. The fast formation of junctions between particles, before the shrinkage starts,
involves trapping of a high excess of vacancies. The shrinkage of
the compact results then from a decrease of the distance between the
centers of particles due to annihilation of trapped vacancies in
the junction zone and to the transport of matter through a diffusional mechanism in the grain boundaries just formed. But the interparticular voids form pores and the compact may be considered as an
aggregate of grains, similar to polycrystalline body with pores,
each pore being connected to another one by a grain boundary. The
proposed model (28) gives also a linear relationship between the
fractional shrinkage ($\Delta L/L_o$) and the time (fig. 13).

Here again the fractional change of the surface are ($\Delta S/S_o$)
(where S_o is the surface area, extrapolated, after the initial, almost instantaneous decrease, before the shrinkage starts) is proportional to the fractional shrinkage (fig. 14) or to the time. This
behaviour seems to be quite general for submicronic particles of titania and also alumina. Again, this decrease of the surface area is
a zero order reaction with respect to the sintering time.

Real catalysts exhibit high surface areas, are in a highly divided state or are porous. In the first case (individual particles)
the previous model, taking into account the formation of junction
zones with trapped vacancies, would explain the initial surface
loss. The elimination of the excess vacancies which results in the
linear decrease with time of the surface area (zero order) should
be explained on the basis of this model.

After this period the solid catalyst is similar to a porous
body with interconnected pores and Nabarro-Herring mechanism of
diffusional creep (31) should be applicable, as it was shown previously (28). It is our hope that the experiments, in progress, on
the activity of the CO oxidation on anatase particles, gradually
sintered and activated in CO, will cast some light on the expected
correlation between catalytic activity and sintering, where the
catalytic quality of the surface is not neglected to the profit of
the extent of the surface.

REFERENCES

(1) M. BOUDART, Advan. Catal., 1969, 20, 153.

(2) J.M. HERRMANN, P. VERGNON and S.J. TEICHNER, J. Catal., 1975, 37, 57.

(3) O.M. POLTORAK and V.S. BORONIN, Russ. J. Phys. Chem., 1966, 40, 1436.
R. VAN HARDEVELD and F. HARTOG, Surface Sci., 1969, 15, 189.

(4) G.A. SOMORJAI, Catal. Review, 1972, 7, 87.

(5) S.I. BERNASEK and G.A. SOMORJAI, Surface Sci., in press.
J. Chem. Phys., in press.

(6) J.E. BENSON and M. BOUDART, J. Catal., 1965, 4, 704.

(7) D.E. MEARS and R.C. HANSFORD, J. Catal., 1967, 9, 125.

(8) G.R. WILSON and W.K. HALL, J. Catal., 1970, 17, 190.

(9) P.A. COMPAGNON, C. HOANG-VAN and S.J. TEICHNER, Bull. Soc. Chim. Fr., 1974, 2311 and 2317.

(10) J.J.F. SCHOLFEN and J.A. KONVALINKA, Trans. Farad. Soc., 1969, 65, 2465.

(11) G. PAJONK, M.B. TAGHAVI and S.J. TEICHNER, Bull. Soc. Chim. Fr., 1975, in press.

(12) B. ECHEVIN and S.J. TEICHNER, Bull. Soc. Chim. Fr., 1975, in press.

(13) J. BARBIER and R. MAUREL, Private Communication.
J. BARBIER, Thesis, Poitiers 1975.

(14) J. LONG and S.J. TEICHNER, Rev. Int. Hautes Temper. et Refract. 1965, 2, 47.
M. FORMENTI, F. JUILLET, P. MERIAUDEAU, S.J. TEICHNER AND P. VERGNON, J. Colloid Interface Sci., 1972, 39, 79.
F. JUILLET, F. LECOMTE, H. MOZZANEGA, S.J. TEICHNER, A. THEVE-NET and P. VERGNON, Farad. Symp. Chem. Soc., 1973, 7, 57.

(15) M.T. VAINCHTOCK, P. VERGNON, F. JUILLET and S.J. TEICHNER, Bull. Soc. Chim. Fr., 1972, 2806 and 2812.

(16) F.H. CONSTABLE, Proc. Roy. Soc. London, Ser. A, 1925, 108, 355.
E. CREMER, Advan. Catal., 1950, 2, 251.

(17) J.M. HERRMANN, P. VERGNON and S.J. TEICHNER, Bull. Soc. Chim. Fr., 1972, 3034.

(18) S.J. TEICHNER, "Catalysis, Progress in Research," R.L. Burwell and F. Basolo Ed., Plenum Press, London 1973, p. 91.

(19) L.L. HEGEDERS and E.E. PETERSEN, J. Catal., 1973, 28, 150.

(20) R.D. CLAY and E.E. PETERSEN, J. Catal., 1970, 16, 32.

(21) H.J. MAAT and L. MOSCOU, Proc. 3rd Intern. Congr. Catalysis, II, 1277, Amsterdam (1965).

(22) R.A. HERRMANN, S.F. ADLER, M.S. GOLDSTEIN and R.M. de BAUN, J. Phys. Chem., 1961, 65, 2189.

(23) E. RUCKENSTEIN and B. PULVERMACHER, A.I. Ch. E. Journal, 1973, 19, 356, 1286.

(24) T. BAIRD, Z. PAAL and S.J. THOMSON, J. Chem. Soc. Faraday. Trans., 1, 1973, 69, 50.

(25) G.C. KUCZYNSKI, Trans. Amer. Inst. Min. Metall. Engr., 1949, 185, 169.

(26) S. PROCHAZKA and R.L. COBLE, J. Phys. Sint., 1970 2, (1), 1.
1970 2, (2), 1,15.
(27) R.L. COBLE, J. Appl. Phys., 1961 32, 787 and 793.
G.C. KUCZYNSKI, "Sintering and Related Phenomena", Mater. Sc.
Res. vol. 6, Plenum Press, London 1973, p. 217.
(28) P. VERGNON, M. ASTIER and S.J. TEICHNER, ibid. p. 301.
M. ASTIER and P. VERGNON, Rev. Int. Hautes Temper. et Refract.,
1972, 9, 265.
(29) P. VERGNON, M. ASTIER and S.J. TEICHNER, Rev. Int. Hautes
Temper. et Refract., 1972, 9, 271.
(30) P. VERGNON, M. ASTIER and S.J. TEICHNER, 2nd Intern. Conf.
Electochem. Soc. Inc. (Princeton, N.J.), 1974, p. 299.
(31) C. HERRING, J. Appl. Phys., 1950, 21, 437
F.R.N. NABARRO, Rep. Conf. Strength, Sol. Phys. Soc., London
1948, p. 75.

CRYSTALLITE SINTERING AND GROWTH IN SUPPORTED CATALYSTS

Paul Wynblatt and Tae-Moon Ahn*

Scientific Research Staff, Ford Motor Company
Dearborn, Michigan 48121

1. INTRODUCTION

Supported metal catalysts, consisting of metal crystallites dispersed on the surface of non-metallic substrates, are widely used both in the chemical and automotive industries. The activity of these catalysts tends to decrease during service, as a result of various chemical and physical phenomena. The present review will focus on the phenomenon of supported metal particle growth, or sintering, which leads to a decrease in the total active metal surface area and a concomitant decrease in total activity.

Wynblatt and Gjostein[1] have recently proposed a general modeling scheme for the various stages in the formation and growth of supported metal particles. In the present paper we shall not reiterate considerations which apply to the formation of three-dimensional particles from an initially monomeric dispersion of metal atoms. Rather, we shall presume that small three-dimensional particles exist on the support and proceed to consider the mechanisms which can control the growth of these particles. Also, we shall confine attention primarily to the case of platinum supported on alumina since this has been one of the most widely studied systems.

The paper will be divided into two major sections. In the first, we shall review the two major concepts which have been employed to model the growth of particles in supported catalysts,

*Permanent Address: Henry Krumb School of Mines, Columbia University, New York, N. Y. 10027

namely, growth by particle migration, collision and coalescence, and growth by interparticle transport. In addition, some recent modifications of classical ripening theories will be presented. The second section will be devoted to the interpretation of various experimental results derived from model experiments of particle growth on flat substrates as well as experiments of particle growth on microporous supports.

2.   PARTICLE GROWTH MODELS

2.1 Growth by Particle Migration, Collision and Coalescence

The possibility that metal particles supported on non-metallic substrates can grow by particle migration followed by interparticle collisions and coalescence has been recognized for some time. Skofronick and Phillips[2] and Phillips et al.[3] interpreted their experiments on the growth of gold particles supported on carbon and silicon monoxide substrates on the basis of such a process and developed a biparticle collision model for particle growth. More recently, Kern and co-workers[4,5] conducted elegant experiments on KCℓ-supported gold particles which leave little doubt about the ability of metal particles to migrate on substrates. The most comprehensive (albeit phenomenological) treatment of this subject has been put forward by Ruckenstein and Pulvermacher[6,7] (R&P). They have developed a binary collision model for particles migrating on a planar substrate in which the rate controlling step for particle growth can be either particle migration or particle coalescence. We give here a brief summary of their approach.

Let n_k be the number of metal particles (per unit area of support) containing k units. Collisions between particles containing i and k-i units will result in an increase of n_k whereas collisions between particles containing k units and any other size will result in a decrease of n_k. Thus,

$$\frac{dn_k}{dt} = \frac{1}{2} \sum_{i+j=k} K_{ij}\, n_i\, n_j - n_k \sum_{i=1}^{\infty} K_{ik}\, n_i \tag{1}$$

where K_{ij} are the rate constants for bi-particle collisions and depend on the nature of the rate controlling process. By solving the diffusion equation in cylindrical coordinates R&P find two limiting cases of interest: growth controlled by particle migration, and growth controlled by particle coalescence. In both these cases, their solutions for the rate of change of the exposed surface area of the particles per unit area of support, S, lead to expressions of the form:

$$\left[\frac{S_o}{S}\right]^n = 1 + (\text{const})\, t \tag{2}$$

where S_o is the initial value of S and n is an integer. The values of the exponent n and the constant depend explicitly on certain assumptions. For example, in the case of particle migration controlled growth, the exponent n is given by:

$$n = 3 + m \qquad (3a)$$

where m is an integer defined by:

$$D_{pi} = C \left(\frac{1}{R_i} \right)^m , \qquad (3b)$$

D_{pi} and R_i are the diffusivity and radius respectively of a particle containing i units and C is a constant. It is worth noting that the average particle radius, $\overline{R}$, can be related to the surface area of particles by:

$$\overline{R} \approx \frac{const.}{S} \qquad (4a)$$

so that Eq. (2) can also be written:

$$\left[\frac{\overline{R}}{\overline{R}_o} \right]^n \approx 1 + (const.) t . \qquad (4b)$$

Thus, the R&P formalism leads to a family of rate expressions, of the type described by Eqs. (2) or (4b), in which the exponent n can adopt a wide range of values depending on certain model assumptions. In addition R&P[7] have derived particle size distributions for these processes and have shown that, when appropriately normalized, the distributions obtained in the case of particle migration controlled growth become time independent (or quasi-stationary) after short transient periods. However, the normalized particle size distribution functions in the case of coalescence controlled growth do not appear to converge to quasi-stationary distributions, although their rate of change becomes slow after long periods of time. Finally, P&R[8] have indicated how their models may be compared with different types of particle growth measurements, such as: electron microscopy, X-ray line broadening, small angle scattering, magnetic measurements and static chemisorption, each of which averages particle size in a somewhat different fashion.

R&P have not attempted to formulate physical models either for the particle migration process or for the coalescence process of a pair of contacting particles. Thus, for example, the interpretation of experiments showing a third power law for the relative decrease in surface area (i.e. n = 3) might be interpreted as particle migration controlled growth; however, that would require particle diffusivity to be independent of particle radius [see Eq. (3a)],

a requirement of dubious physical significance. However, a number
of physical models for particle migration have been developed and
some of these might be usefully incorporated into the R&P model.

In their experiments on particle growth, Phillips et al.[3]
assumed that particle migration was the rate controlling process
and that the particle velocity could be treated as in the case of a
two-dimensional gas, i.e.

$$\bar{v}_p = \left(\frac{\Pi kT}{2M_p} \right)^{1/2} \text{ and } D_p \propto \left(\frac{1}{R} \right)^{3/2} \tag{5}$$

where M_p is the particle mass, k is the Boltzmann constant, T is
the absolute temperature and D_p the particle diffusivity. In their
model this yields a growth law of the type shown in Eq. (4b), but
with an exponent n of 7/2 rather than the 9/2 exponent which would
be predicted from the R&P model.

Kern et al.[5] found experimentally that particle diffusivity
obeyed an Arrhenius type relation in which both constants were
particle size dependent. As a result, they developed a particle
migration model based on interfacial viscous shear which satisfied
their experimental observations. Unfortunately, their expression
for particle diffusivity does not have the simple form of Eq. (3b)
and cannot therefore be combined readily with the R&P model.

More recently, Wynblatt and Gjostein[1] have shown how migration
and coalescence models based on surface self diffusion might be
adapted and combined with the R&P formalism. A brief summary of
their approach is given here. Consider first the coalescence of
two identical spheres by surface diffusion, a process which has been
treated elegantly by Nichols and Mullins.[9,10] The relaxation time
for complete coalescence may be written:

$$\tau_{cc} = 0.89 \frac{R_o^4}{B} \quad , \quad B = \frac{D_s N_o \gamma_m \Omega^2}{kT} \tag{6}$$

where R_o is the radius of the spheres, D_s is the surface self-diffusion coefficient over the particles, N_o is the average density of
surface sites on the particle, γ_m is the surface energy and Ω is
the atomic volume of the particle constituent. The second column of
Table 1 illustrates the change with particle size of the relaxation
time for coalescence of Pt particles at a temperature of 700°C,
(with $D_s = 2.8 \times 10^{-9}$ cm^2/sec,[11] $N_o = 10^{15}$ sites/cm^2,
$\gamma_m = 2100$ ergs/cm^2,[12] $\Omega = 1.5 \times 10^{-23}$ cm^3; i.e. $B = 9.9 \times 10^{-24}$
c$_m^4$/sec).

Table 1

R_o (Å)	τ_{cc} (sec)	D_p (cm^2/sec)	t_p (sec)
25	3.5×10^{-4}	1.3×10^{-13}	12
100	8.9×10^{-2}	5.0×10^{-16}	5×10^4
250	3.5	1.3×10^{-17}	1.2×10^7
2500	3.5×10^4	1.3×10^{-21}	1.2×10^{13}

We consider next the process of particle migration. It is assumed that monomers (adatoms) diffusing over the particle surface will occasionally accumulate on one side of the particle, by random fluctuations, causing the particle to execute Brownian motion over the substrate. As an approximation to particle diffusivity, D_p, we employ a result obtained by Gruber[13] for the migration of bubbles in solids by surface diffusion:

$$D_p = 0.3 \, D_s \left(\frac{a_o}{R_o} \right)^4 \qquad (7)$$

where a_o is the atomic diameter. Columns 3 and 4 of Table 1 illustrate the change with particle size of D_p and of the average time for a platinum particle to travel a distance of 10 R_o, $t_p = (10 \, R_o)^2/4 \, D_p$, at 700°C ($a_o = 2.77$Å).

The results in Table 1 show clearly that the times for coalescence are considerably shorter than the times required for a particle to migrate over significant distances. Thus, if surface diffusion is invoked as the mechanism for both migration and coalescence of particles, as has been done here, then particle migration will obviously be the rate controlling process for growth in a migration-coalescence sequence. Furthermore, for interparticle distances of interest in catalyst systems, typically 20 R_o to 200 R_o for R_o in the range 25Å to 2500Å, we can conclude that growth by particle migration and collision will be an improtant process only for particles smaller than about 50Å.

2.2 Growth by Interparticle Transport

Wynblatt and Gjostein[1,14] have recently derived expressions for particle growth by interparticle transport in which two modes of transport have been considered: diffusion through the vapor phase and diffusion over the substrate surface. The approach used was based on the classical treatments of Ostwald ripening by Wagner[15] and by Lifshitz and Slyozov,[16] and extensions thereof by Chakraverty.[17] We present here a summary of that approach.

Consider a distribution of partially wetting particles in the form of spherical segments, making a contact angle θ with the

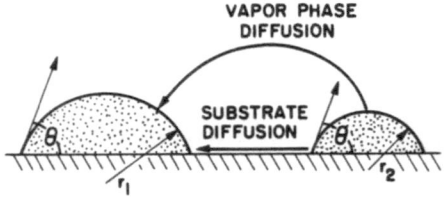

FIG. 1   Schematic of partially-wetting, isotropic particles on a
           substrate.

substrate, as illustrated schematically in Fig. 1. The angle θ is
defined by:

$$\gamma_s = \gamma_{ms} + \gamma_m \cos(\theta) \qquad (8)$$

where γ_s and γ_m are the surface energies of substrate and particle,
respectively, and γ_{ms} is the particle-substrate interfacial energy.
The chemical potential, μ, of any particle within the distribution
will depend on its radius of curvature R according to the Gibbs-
Thompson relation:

$$\mu = \mu_o + 2\gamma_m \Omega/R \quad . \qquad (9)$$

Thus, the larger particles with lower chemical potential will grow
at the expense of smaller particles with higher chemical potential,
the driving force for growth being the reduction in the total surface energy of the system. Under those conditions, some particles
grow and others dissolve away leading to an overall increase in the
average particle size of the distribution.
      a) Growth by Substrate Diffusion. By way of example we now
present a summary of the ripening model for the case of diffusion
over the substrate surface.

      Figure 2 shows schematically the changes in the concentration
of diffusing entities on and about a particle, as well as some of
the important energy transitions for such an entity. In the case of
supported platinum particles it is assumed that under reducing
conditions the diffusing entities are platinum monomers whereas in
the presence of oxygen it is assumed that the diffusing entities
are PtO_2 molecules. The net flux of diffusing entities onto a
particle is given by:

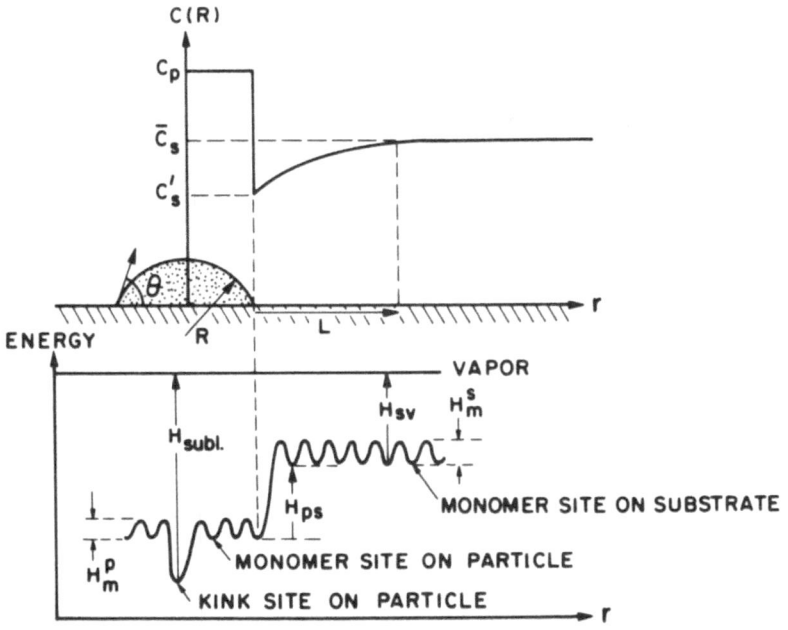

FIG. 2   Schematic of monomer concentration and energies on and
           around a particle.

$$J = \frac{[2\Pi R \sin(\theta) a \beta'] \; [2\Pi D_1/\ln (L/R \sin \theta)]}{[2\Pi R \sin(\theta) a \beta']+[2\Pi D_1/\ln (L/R \sin \theta)]}(\bar{c}_s - c_p\beta) \qquad (10a)$$

where

$$\beta' = \nu_s \exp\{- H_m^s/kT\} \quad , \qquad (10b)$$

$$\beta = (\nu_p/\nu_s) \exp\{- H_{ps}/kT\} \quad , \qquad (10c)$$

$\bar{c}_s$ is the average concentration of diffusing species on the substrate away from the immediate vicinity of the particle (i.e. the
far field concentration), c_p is the concentration of diffusing
species on the particle surface (both c's being expressed as a
number/unit area), ν_s and ν_p are the vibration frequencies of the
diffusing species on the substrate and particle, respectively, L is
the distance required for c to reach its far field value $\bar{c}_s$, D_1 is
the diffusivity of diffusing entities over the substrate, H_m^s is the
activation energy for migration over the substrate and H_{ps} is the
energy difference between a diffusing entity adsorbed on the particle and on the substrate. Equation (10a) shows that two limiting

cases generally arise in this type of problem.   If

$$2 \Pi R \sin(\theta) \, a \, \beta' \gg 2 \Pi D_1 / \ell n (L/R \sin \theta) \tag{11}$$

then the term $[2 \Pi R \sin(\theta) \, a \, \beta']$ drops out of the r.h.s. of Eq. (10a) and the overall process is "diffusion controlled", i.e. particle growth is controlled by the rate at which the diffusing species migrate over the substrate.   If, on the other hand, the inequality is reversed, then the process becomes "interface controlled", i.e. particle growth is limited by the rate of attachment or detachment of the diffusing species from particles.   By expressing the diffusivity over the substrate as:

$$D_1 = a^2 \nu_s \exp\{- H_m^s /kT\} \tag{12}$$

it can be shown that the condition (11) will be met as long as $R \gtrsim 10a$ (i.e. $R \gtrsim 25\text{Å}$).   Thus, we can conclude that in the case of interparticle transport by diffusion over the substrate particle growth will be diffusion controlled.   The flux expression Eq. (10a) can be manipulated by the techniques developed by Wagner[15] and Lifshitz and Slyozov[16] to yield the kinetic laws governing growth. For the case of growth controlled by diffusion over the substrate:

$$\left[\frac{R^*}{R_o^*} \right]^4 = 1 + \frac{K_D}{R_o^{*4}} \, t \tag{13a}$$

where

$$K_D = \frac{27 \, D_1 \, c_s^{eq} \, \gamma_m \, \Omega^2}{64 \, \log(L/R \sin \theta) \, \alpha_1 \, kT} \quad , \tag{13b}$$

$$\alpha_1 = (2 - 3 \cos \theta + \cos^3 \theta)/4 \quad , \tag{13c}$$

R^* is the radius of the particle in equilibrium with the far field concentration $\bar{c}_s$, R_o^* is the initial value of R^* and c_s^{eq} is the concentration of diffusing entities on the substrate which would be in equilibrium with an infinite sized particle.   This type of model also leads to a quasi-stationary distribution when particle size is normalized as R/R^*.   Thus it is possible to obtain a simple relationship between R^* and the average particle radius of the distribution, $\bar{R}$.   In this particular case $\bar{R} \approx 1.03 R^*$, so that Eq. (13a) may be rewritten as:

$$\left[\frac{\bar{R}}{\bar{R}_o} \right]^4 = 1 + \left(\frac{1.03}{\bar{R}_o} \right)^4 K_D t \quad . \tag{14}$$

It can be seen, therefore, that the general form for the growth laws obtained from this type of model is indistinguishable from the form of Eq. (4b) derived for particle growth by migration and coalescence.

Estimation of the rate constant K_D reduces to an evaluation of $D_1 c_s^{eq}$. This term will differ depending on whether we consider particle growth to be occurring under reducing or oxidizing conditions, as has been outlined previously.[18] For reducing conditions:

$$K_D \approx \frac{54}{64} \frac{\gamma_m \Omega^2 \nu_s}{\alpha_1 kT} \exp\{-(H_{subl.} + H_m^S - H_{sv})/kT\} \qquad (15)$$

where $H_{subl.}$ is the sublimation energy of Pt (135 k cal/mole[19]), H_{sv} is the heat of adsorption of a Pt monomer on the substrate (see Fig. 2) and can roughly be estimated[1] from the work of adhesion of Pt on $A\ell_2O_3$ determined (under oxidizing conditions) by McLean and Hondros.[12] H_m^S is not known, but must lie in the range $0 < H_m^S < H_v^S$. An estimated upper bound for K_D has been obtained on this basis[18] and shows that this mechanism leads to negligible growth rates, primarily as a result of the very large value of $H_{subl.}$ of Pt.

For oxidizing conditions:

$$K_D \approx \frac{27 a^2 \gamma_m \Omega^2}{64 \alpha_1 kT} \frac{P_{O_2} K_{eq}}{(2\Pi M kT)^{1/2}} \exp\{-(H_m'^S - H_{sv}')/kT\} \qquad (16)$$

where P_{O_2} is the oxygen partial pressure of the surrounding atmosphere, M is the mass of a PtO_2 molecule, the primed quantities are the ones appropriate for a PtO_2 molecule (assumed to be the diffusing entity) and K_{eq} is the equilibrium constant for the reaction:

$$Pt_{(s)} + O_{2(g)} = PtO_{2(g)} \quad . \qquad (17)$$

Although K_{eq} is known[20], there is no independent information available at this time for the quantity $(H_m'^S - H_{sv}')$. However, since the activation energy associated with K_{eq}, ΔH_{ox} (which represents the energy required for the removal of a Pt atom from a crystallite, in the form of a PtO_2 molecule) is ~ 42 kcal/mole and therefore much smaller than $H_{subl.}$, we can conclude that the rate of particle growth under oxidizing conditions will be considerably larger than under reducing conditions. This was indeed borne out in previous estimates[18] and will be discussed further in Section 3.

Finally, it is worth mentioning a model for particle growth by substrate diffusion proposed recently by Flynn and Wanke.[21,22] Their basic expression for the net number of atoms gained by a particle, per unit time, is basically similar to Eq. (10a), in the limit of interface control. We rewrite Eq. (10a) in that limit after substitution for β and β' from Eqs. (10b) and (10c) as:

$$J = 2\Pi R \sin(\theta) \, a \, \nu_s \, \exp\{-H_m^s/kT\} \, \bar{c}_s$$

$$- 2\Pi R \sin(\theta) \, a \, \nu_s \, \exp\{-H_m^s/kT\} \, c_p \, (\nu_p/\nu_s) \, \exp\{-H_{ps}/kT\} \, . \quad (18)$$

They then assume that $[R \, c_p]$ in the second term on the r.h.s. of Eq. (18) (which represents atom loss from a particle) is independent of R, and rewrite the second term as: $[A \exp\{-H_{ps}/kT\}]$. The assumption involved here is valid over a restricted range of R. In addition, they multiply the first term on the r.h.s. (representing atom gain by the particle) by a "sticking coefficient", α, which is used as an adjustable parameter. Up to this point, the changes introduced do not affect the basic model significantly, and if their formulation were combined with the ripening model discussed above it would yield a growth law of the form of Eq. (14) but with an exponent n of 3. However, Flynn and Wanke proceed to treat the rate constant for atom loss, A, as an adjustable parameter independent of the rate of atom gain; this decoupling of the atom gain and loss processes does not have any physical justification. As a result, they obtain growth laws where the exponent n adopts various values and changes with time, thus violating the more rigorous solution for their model. It should be mentioned, however, that even the classical ripening model can yield effectively varying n values, in a plot of $\log(\bar{R}/R_o)$ vs. $\log t$, during the period over which the time dependent term is not large compared with 1.

    b) <u>Growth by Vapor Phase Diffusion</u>. Only the main results of the ripening model under these conditions will be presented. Here, growth is controlled by the "interface" process, because of the high rate of vapor phase transport, and yields a second power growth law:

$$\left[\frac{\bar{R}}{R_o}\right]^2 = 1 + \frac{K_v t}{R_o^2} \, . \quad (19)$$

Under reducing conditions K_v is negligibly small for platinum, but under oxidizing conditions we obtain:

$$K_v = \frac{128 \, \alpha_2 \, \gamma_m \, \Omega^2}{192 \, \alpha_1 \, kT} \cdot \frac{P_{O_2} \, K_{eq}}{(2\Pi \, M \, kT)^{1/2}} \cdot \quad (19a)$$

All quantities which enter into K_v can be readily estimated so that direct comparison with experiment can be made in this case.

2.3 Nucleation Inhibited Growth

The two general models discussed above yield particle growth laws with well defined exponents, n, which can attain values as large as 7. However, some experiments on particle growth have yielded values of n as large as $\sim$ 12 (i.e. a very drastic deceleration of the growth process with time), as well as time dependent exponents.[23,24] As a result, Wynblatt et al.[1,14,18] have recently attempted to develop a formalism capable of predicting this type of behavior. In this approach, it has been proposed that the growth of particles is inhibited by a nucleation process. This type of concept has also been employed by Chen and Cost[25] and by Willertz and Shewmon[26] to rationalize low rates of bubble migration in irradiated metals.

It is well known from crystal growth theory[27] that the continued growth of faceted crystals (i.e. crystals bounded by singular surfaces) requires repeated nucleation events, leading to rates of growth that are lower than those of unfaceted crystals. Furthermore, the presence of faceted particles can be observed in published photomicrographs of microporous catalysts[21] as well as model catalysts consisting of particles supported on flat substrates.[23] We give here a brief summary of this model.

Consider an array of faceted particles lying on a substrate as shown schematically in Fig. 3a. Since $(\gamma_{ms} - \gamma_s) \approx 0$ for the case of platinum on alumina, we can approximate the chemical potential of any given particle in terms of its height H[28]:

$$\mu = \mu_o + 2\,\gamma_m\,\Omega/H \quad . \tag{20}$$

Thus, the driving force for growth in this case is analogous to that of the array of particles in the shape of spherical segments. As a faceted particle grows, incoming monomers (or other diffusing entities) will tend to settle at energetically favorable sites such as kink and ledge sites; eventually all such sites will be consumed, leading to a perfect particle. At this point, the continued growth of the particle will have to await the nucleation of a new atom layer before large numbers of monomers can once again be accommodated on the particle. After nucleation, a whole new layer of growth can occur, as shown in Fig. 3b, before another nucleation event is required. Under those conditions, the rate of change in height of a growing particle may be approximated by:

$$\frac{dH}{dt} = S_p \, \dot{N} \, h \tag{21}$$

where S_p is the exposed surface area of the particle, $\dot{N}$ is the nucleation rate (nuclei per unit time per unit area) on the particle surface and h is the height of a monatomic ledge or nucleus (see Fig. 3b). Equation (21) may be rewritten[14,29]:

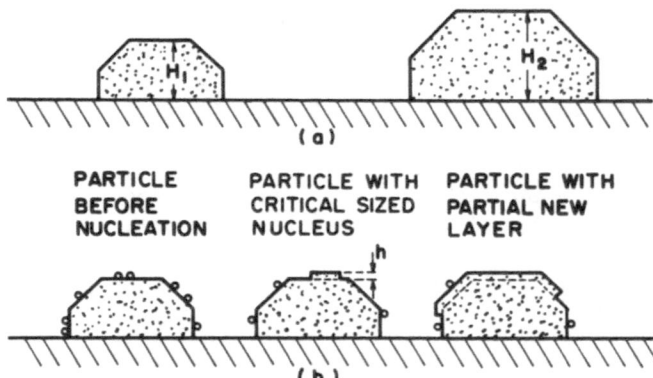

FIG. 3   Schematic of (a) partially-wetting faceted particles on a
substrate and (b) nucleation process on a faceted particle.

$$\frac{dH}{dt} = B \left[\frac{H^3 (H - H^*)}{H^*} \right]^{1/2} \exp \left\{ \frac{2 \gamma_m \Omega}{kT \, H^*} - \frac{\Pi h \gamma_m}{2 \, kT} \left(\frac{\epsilon}{h \gamma_m} \right)^2 \frac{H \, H^*}{(H - H^*)} \right\} \quad (22a)$$

where

$$B = \frac{4\Pi \, \alpha_2 \, D_s \Omega}{a_o^5} \left[\frac{2 \gamma_m h}{kT} \right]^{1/2} \quad , \quad (22b)$$

$$\alpha_2 = (1 - \cos \theta)/2 \quad , \quad (22c)$$

H^* is the height of the particle in equilibrium with the far field
concentration of diffusing entities on the substrate and ϵ is the
energy per unit length of a surface ledge of height h.

The nucleation barrier to growth described above does not apply
to dissolving particles (i.e. particles for which $H < H^*$) since
these always possess atoms in high energy sites that are easily removed. Thus, the expressions governing the dissolution of small
particles will be the same as in the case of non-inhibited growth or
classical ripening. For example, the rate of dissolution of small
particles controlled by substrate diffusion under oxidizing conditions can be described by an expression derived from Eq. (10):

$$\frac{dH}{dt} \approx \frac{a^2 \gamma_m \Omega^2}{\alpha_1 \, kT} \frac{P_{O_2} \, K_{eq}}{(2\Pi \, M \, kT)^{1/2}} \exp\{-(H_m'^s - H_{sv}')/kT\} \frac{1}{H^2} \left(\frac{1}{H^*} - \frac{1}{H} \right) (23a)$$

or in the case of vapor phase transport controlled by the interface process as:

$$\frac{dH}{dt} = \frac{2\,\alpha_2\,\gamma_m\,\Omega^2}{\alpha_1\,kT}\ \frac{P_{O_2}\,K_{eq}}{(2\Pi\,M\,kT)^{1/2}}\ \left(\frac{1}{H^*} - \frac{1}{H}\right). \qquad (23b)$$

By combining Eqs. (22a) and (23) with the classical ripening formalism[15,17] it is possible to obtain either numerical solutions for the average particle height as a function of time, $\bar{H}(t)$[14], or closed form asymptotic solutions under somewhat different assumptions, as has been shown by Ahn and Tien.[30] In contrast to the classical ripening theories, the numerical solutions lead to time dependent effective exponents n which can attain large values, while the asymptotic solution can be expressed as:

$$t = \frac{k\,T\,C_1}{2\,\gamma_m\,\Omega} \int_{\bar{H}(t=0)}^{\bar{H}(t)} \bar{H}(t)^{-3/2} \exp\left\{\frac{C_2\,\Pi\,\varepsilon^2\,\bar{H}(t)}{2\,h\,\gamma_m\,kT} - \frac{2\,\gamma_m\,\Omega}{kT\,\bar{H}(t)}\right\} d\bar{H}(t) \qquad (24)$$

where C_1 and C_2 are constants. By appropriate expansion of the exponential terms, Eq. (24) may be reduced to a power series in $\bar{H}(t)$. The dominant term in this power series will depend on the magnitudes of both $\bar{H}(t)$ and the constants in Eq. (24) and will consequently change with time. Thus, time dependent effective exponents and small growth rates (i.e. large n) also result in this treatment.

While the numerical solutions do not appear to produce quasi-stationary particle size distributions, the asymptotic solutions do yield such distributions. This result, however, might be limited to the particle growth regime assumed in the asymptotic treatment.

Finally, it should be mentioned that the concepts of nucleation inhibition can also be applied to growth by particle migration, collision and coalescence, as has been discussed previously in general terms.[1] There also, one would expect to obtain particle growth laws with time dependent exponents.

3. INTERPRETATION OF PARTICLE GROWTH EXPERIMENTS

3.1 Summary of Model Predictions

Before proceeding to a direct comparison with experiment, it is interesting to contrast certain aspects of the theoretical models. Table 2 contains some of the salient features of each model.

The first row of the Table lists the major models and the second row identifies the various mechanisms. The third row compares the exponent of the growth laws for the various cases and indicates the large extent of overlap which exists between the different models. Thus, it will often be insufficient to rely on the value of

TABLE 2

MODEL	R&P MIGRATION-COALESCENCE		CLASSICAL RIPENING Diffusion		NUCLEATION INHIBITED RIPENING Diffusion	
Mechanism	Particle Migration	Particle Coalescence	Through Vapor	Over Substrate	Through Vapor	Over Substrate
Exponent	$1 \leq n \leq 2$	$3 \leq n \leq 7$	$2 \leq n \leq 3$	$3 \leq n \leq 4$	$2 \leq n(t)$	$3 \leq n(t)$
Rate Constant	Proportional to ϕ_o [†]		Independent of ϕ_o		Independent of ϕ_o	
Particle Size Distribution	Time Dependent	Quasi-Stationary	Quasi-Stationary		Quasi-Stationary	

[†] ϕ_o = number of particles per unit area of substrate.

the exponent alone for the purpose of mechanism identification. It is also clear that time dependence of the exponent would not necessarily imply the operation of a nucleation inhibited ripening mechanism. Other complex mechanisms could possibly yield that type of behavior, such as for example, particle growth by migration based on Kern et al.'s[5] interfacial viscous shear model, or nucleation inhibited particle migration based on a surface diffusion model.

The fourth row of Table 2 compares one aspect of the rate constant which can give a clue to mechanism. A general feature of biparticle collision models is that the rate constant depends on the particle concentration, ϕ_o. In particular, the linear dependence of the rate constant on ϕ_o in the R&P models[6,7] has been used advantageously by Bett et al.[31] to identify the mechanism of growth in a system of carbon supported platinum particles. Such a test of mechanism, however, requires growth experiments to be carried out for several different metal loadings on the support.

R&P have emphasized the usefulness of the particle size distribution function and its characteristics (e.g. various moments) for the purposes of mechanism identification. Some of the characteristics of the distributions are indicated in the last row of Table 2 and the distributions are compared in Fig. 4. The figure shows that with the exception of coalescence control the other distributions do not differ radically. We feel that this similarity of the distributions, coupled with the extensive experimental effort which must be expended to obtain well characterized distributions, detracts from usefulness of this approach for the purposes of mechanism identification.

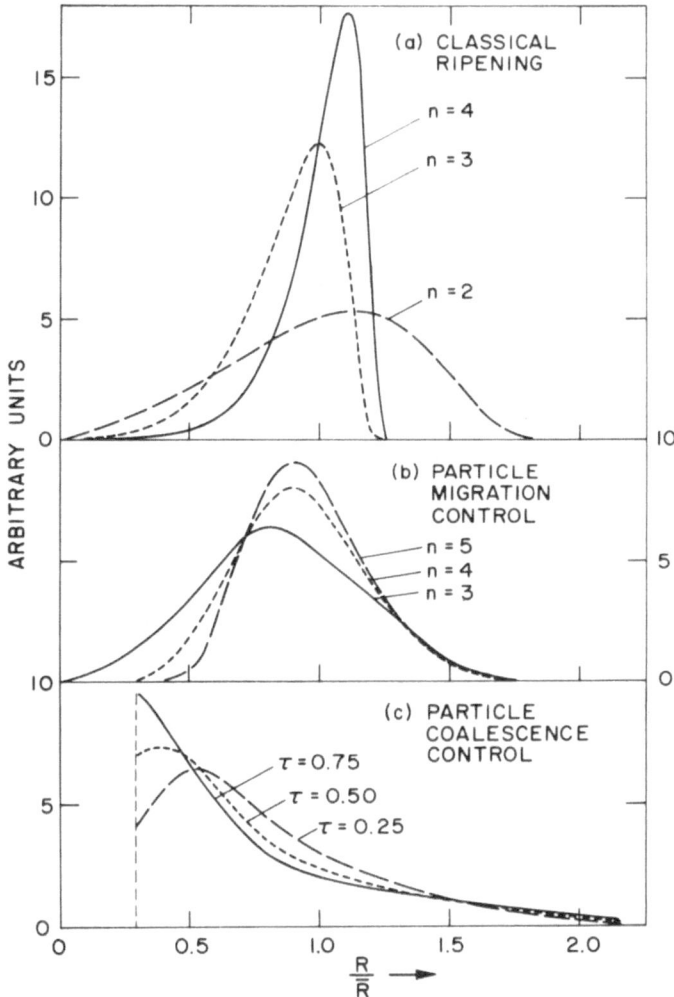

FIG. 4 Particle size distribution functions from various models
plotted as number of particles of size R/R̄ (arbitrary units)
vs. R/R̄. (a) and (b) are quasi-stationary distributions,
(c) shows the time evolution of a given distribution for
growth by coalescence control, where τ is a reduced time
scale.[8]

3.2   Growth of Particles on Flat Substrates

Experimental investigations of particle growth in systems consisting of metal particles supported on flat substrates should in
principle be particularly useful for mechanistic studies. First,
this type of experimental geometry conforms closely to the simple
framework of the theoretical models described in Section 2, and thus
avoids effects which might stem from the large substrate curvature
present in microporous catalysts. Second, these experimental systems
are far more amenable to study by transmission electron microscopy
than microporous catalysts where the particulate nature of the support can lead to difficulties in distinguishing between metal and
support particles.

There have been several experiments performed on the growth of
supported particles at low temperatures, in which the behavior of Cu,
Ag and Au particles supported on various substrates (silicon monoxide,
carbon and alkali halides) has been studied. That work has recently
been reviewed by Geus[32] and several interesting phenomena resulting
from the interaction of particles with oxygen and water vapor have
been catalogued. We shall confine our attention here to the case of
platinum supported on alumina.

Huang and Li[33] have studied the growth of large (initial size
1500 to 3000Å) platinum particles supported on flat single crystal
sapphire surfaces under oxidizing conditions. Their results yield a
particle growth law with an exponent n of 4, which is consistent with
a classical ripening mechanism controlled by interparticle diffusion
over the substrate. This conclusion is supported by the fact that
the rate constant for the process, K_D, was found to depend on the
crystallographic orientation of the substrate surface. Our estimates
of the rates of particle diffusivity (see Table 1) for such large
particles would seem to rule out a mechanism of growth based on particle migration and coalescence. These results do not show the effects of nucleation inhibition, even though their published SEM
photomicrograph does show some tendency to flat faces. However, the
mere presence of faceting in a particle does not preclude the existence of defects, such as dislocations, which can produce self-propagating surface steps and hence eliminate the need for the nucleation
of new steps. We shall return to the issue of defects in the discussion of other experiments.

Wynblatt and Gjostein[14,23] have studied the growth of smaller
platinum particles (initial size 50-80Å) on flat polycrystalline γ-
alumina substrates under oxidizing conditions. The growth kinetics
obtained do not fit any particular growth law and show wide variation

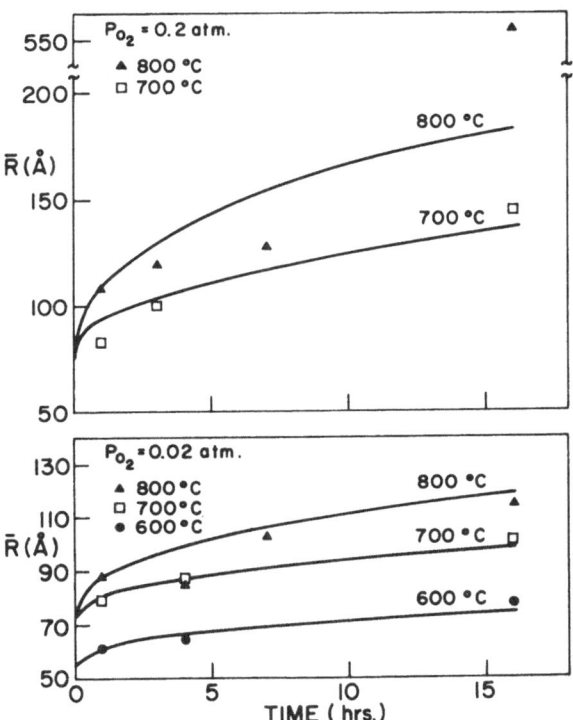

FIG. 5   Comparison of data from flat substrate experiments with
theoretical predictions from nucleation inhibited particle
growth[14] (see text). Note break in ordinate in upper
portion of the figure.

as well as time dependence of the exponent n.   Some data obtained at
temperatures ranging from 600 to 800°C and two oxygen partial pressures are shown in Fig. 5.

These results can be interpreted by means of the model based on
nucleation inhibited growth described in Section 2.3, coupled with a
substrate diffusion mechanism.   There are however two parameters in
the model which are not well known: $(H_m'^S - H_{sv}')$ and $\varepsilon/h\gamma$ [see Eqs.
(22) and (23)].   In order to obtain theoretical curves it has there-
fore been necessary to fit one set of data in Fig. 5, so as to establish these parameters (the data at 700°C and 0.02 atm O_2) and check
the model predictions by application of the same parameters to the
remaining data.   The curves shown in Fig. 5 are the result of preliminary numerical calculations obtained in this fashion and show a
reasonable fit with the data except for the set corresponding to
800°C and 0.2 atm O_2.

Study of the heat-treated samples, by transmission electron microscopy, indicates that beyond growth by about a factor of 2, some
particles in the system begin to grow "abnormally" fast, as evidenced
by the presence of particles that are considerably larger than the
average particle size.   This can be seen in the photomicrographs of
Fig. 6.   It is speculated that some (small) fraction of the particles
in the initial state contain defects, such as dislocations, and that
consequently the growth of those particles is not inhibited by the
constraints of ledge nucleation.   Thus, as growth proceeds, the defect containing particles are able to grow faster than, and eventually at the expense of, their more perfect neighbors.   When all of the
growing particles are of this dislocated type, the kinetics should
revert back to classical ripening.   It is believed that the set of
data in Fig. 5, for 800°C and 0.2 atm O_2, is showing such a transition from an inhibited to a non-inhibited mode of growth.

Within the framework proposed above one may rationalize the observations of Huang and Li[33] by suggesting that since the probability of finding a defect in a particle increases with particle size,
it is likely that the majority of particles in their experiments
(initial size 1500-3000Å) were dislocated.   Such a situation would
lead to their observation of non-inhibited growth kinetics.   Thus,
it seems that under oxidizing conditions, one can interpret the data
available at this time on the growth of platinum particles supported
on flat alumina substrates, by means of interparticle transport over
the substrate either by classical or inhibited ripening.

Under reducing conditions, the ripening rate constants for all
mechanisms are too small to account for any significant growth.   Consequently, it must be concluded that particle growth under those
conditions is controlled by particle migration.

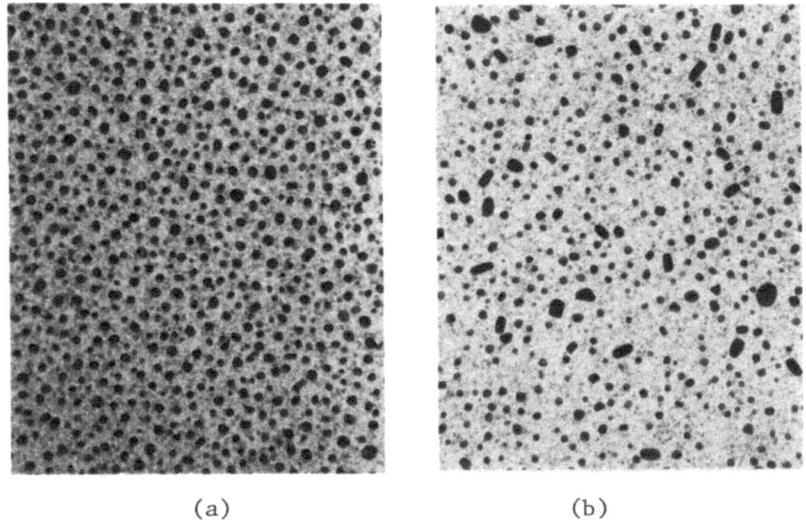

(a)   (b)

FIG. 6   Typical microstructures from flat substrate experiments.[14]
(a) 3 hours at 700°C in O_2 at 0.02 atm pressure, X 61000,
(b) 68 hours at 700°C in O_2 at 0.02 atm pressure; note
beginning of abnormal growth, X 44000.

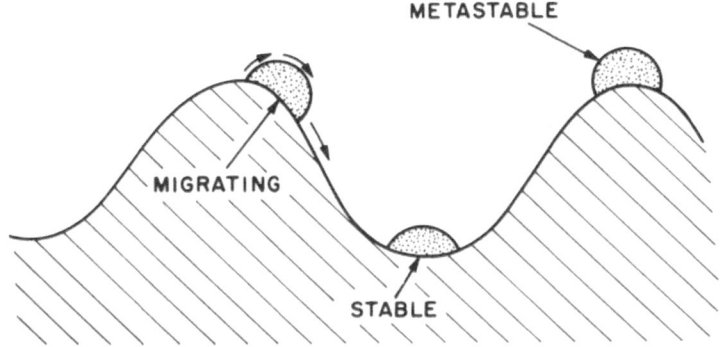

FIG. 7   Schematic of particles on curved substrate.

3.3 Growth of Particles in Microporous Supports

The models of particle growth described in Section 2 have all been derived under the assumption that particles are supported on flat substrates and are not therefore strictly applicable to particles dispersed throughout the pore system of a microporous substrate. Consequently, it is worthwhile to devote some space here to a discussion of possible effects on particle growth stemming from the complex nature of microporous substrates.

a) <u>Effects of Microporous Supports on Particle Migration</u>. All models of growth by particle migration, collision and coalescence assume that particles execute Brownian, or random motion over the substrate. However, in the presence of substrate curvature, the constraints on a partially wetting particle to attain local equilibrium with the substrate, at all points on its periphery, will lead to a curvature gradient along the exposed surface of the particle, as shown schematically in Fig. 7. This effect will tend to impose a driving force on particles to migrate preferentially towards regions of positive substrate curvature (valleys). The biased motion of the particles will therefore tend to produce higher rates of growth than would be expected from random motion. However, since particles will tend to be trapped in concave regions, the higher rates of growth should be observed only in the early stages of the process; in the later stages, growth should be slower than predicted by theory.

b) <u>Effects of Microporous Supports on Interparticle Transport</u>. The chemical potential of a particle with a spherically shaped exposed surface will depend only on the radius of curvature of that surface. Thus, the constraints of achieving a given contact angle with the substrate would lead to a higher chemical potential for a particle residing on a convex substrate than for a particle residing on a concave substrate (given equal volumes for the two particles) as shown schematically in Fig. 7. Thus there will exist a tendency for net transport from particles on "hills" to particles in "valleys". This effect may not influence particle growth kinetics considerably and is certainly amenable to theoretical and experimental modeling within certain limits.

A second effect on particle growth comes about from the pore size of the support which must impose some spatial constraint on particle size. This type of constraint applies to all mechanisms of particle growth and would tend to slow down the growth rate as the average particle size approaches the average pore size.

c) <u>Effects of Microporous Supports on Particle Size Measurement</u>. Metal surface area measurement by chemisorption has been used as the major technique to infer particle size in supported catalysts. In order to convert the metal surface area to particle size it has generally been assumed that particles adopt a cubic shape with five of

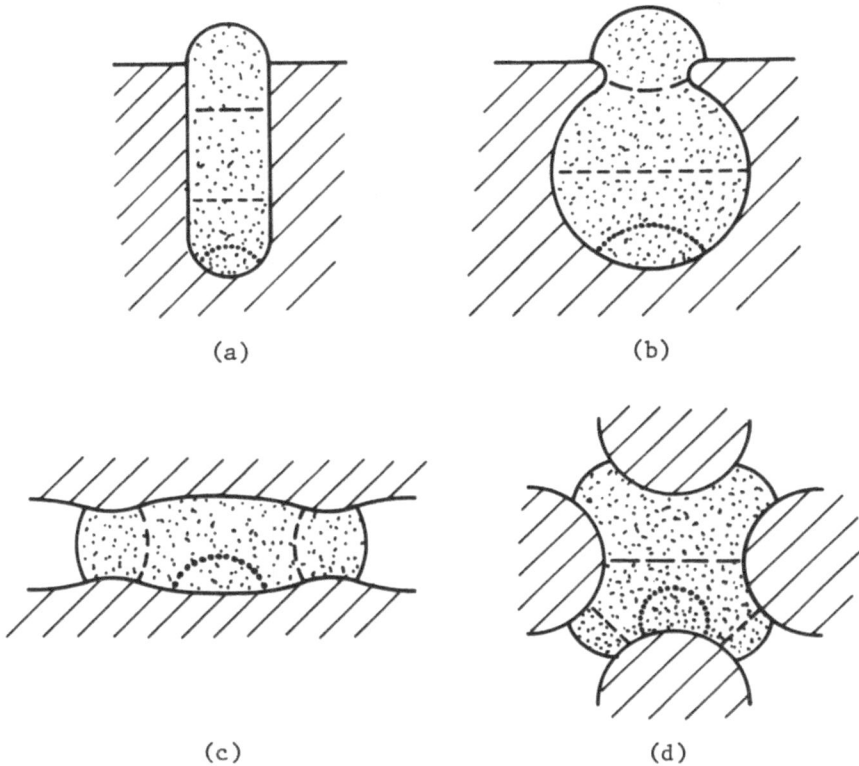

FIG. 8   Schematic of particle evolution in pores of various simple
         geometries. Dots show initial particle outline and dashed
         lines indicate intermediate particle outlines.

the six cube faces exposed to the gaseous environment. That this
approach tends to overestimate particle size in the later stages of
growth can be illustrated by considering Fig. 8. The figure indicates the possible evolution of particle shape in some simple pore
geometries. As a particle grows in a pore, an increasing fraction of
the particle surface will come into contact with the wall of the pore
to form a metal-substrate interface. Thus, the fraction of particle
surface area accessible to gas phase chemisorption will decrease
sharply and the apparent particle size inferred from that surface
area (assuming, say, cubes with 5/6 of their surface exposed) will
be grossly overestimated.

        Difficulties with transmission electron microscopy of small
particles embedded in the pore system of a microporous catalyst have
been alluded to in Section 3.2. Some of the problems which arise in

various possible microscopy techniques have recently been summarized by Sprys et al.[34]

It is worth pointing out that unless the microporous material is stabilized by pre-sintering at temperatures above the particle growth temperatures before impregnation of the metal component, sintering of the support during particle growth experiments will obscure particle growth phenomena. Finally, since introduction of the metal component into microporous catalysts is achieved by impregnation with various compounds, which are subsequently decomposed to metal, there can be a variety of contaminants left on either particles or substrates which could affect the growth process in different ways. Such effects have recently been reported by Baker et al.[35]

d) Particle Growth in Microporous Supports. In view of the complexity of systems composed of metal particles supported on microporous substrates, the physical and chemical differences among various experiments and the different techniques used to measure particle size, it is not surprising to find results in the literature which appear to fit any theoretical model. We can summarize some of the data on the growth of alumina supported platinum as follows.

Under oxidizing conditions, Maat and Moscou[36] find a particle growth law having an exponent n of 1; Herrmann et al.[37] also report a value of n=1 although a plot of their data show significant deviation from that value; the data of Wynblatt et al.[18] shows time dependent exponents ranging from 3 to 9; and Somorjai's results[24] also indicate time dependence of the exponent with values of n as large as 12. Under reducing conditions the values of the growth law exponent extracted from a number of sources[24,38,39] show a variation from 7 to 13.

It is felt that a detailed analysis of this type of data on the basis of the models of Section 2 is premature. However, in spite of the complex nature of supported catalysts, and the difficulties in formulating detailed interpretations of particle growth kinetics in such systems, it is clear that substantial progress has been made in understanding and predicting their behavior. The identification of the rate controlling mechanism of particle growth for any given metal particle-substrate system is possible in principle from experiments performed on model, or flat substrate systems. Armed with that information together with knowledge of the form of the rate constant for the operating mechanism, it should in turn be possible to predict the major qualitative trends in the behavior of the microporous system. Some of the principal trends to be expected in alumina supported platinum are summarized in the last section.

4. CONCLUDING REMARKS

The results of the few model experiments described in Section

3.2 indicate that, under oxidizing conditions, the growth of alumina supported platinum particles occurs either by a classical ripening mechanism or by nucleation inhibited ripening in which the mode of interparticle transport is by diffusion over the substrate surface. Thus, from the rate expressions describing these processes, Eqs. (22) and 23a), it is possible to draw certain qualitative conclusions about the behavior of the corresponding microporous system.

i) The particle growth rate will increase with increasing oxygen partial pressure.

ii) The effect of temperature on growth rate will occur primarily through a Boltzmann factor with activation energy of $\sim$ 30 k cal/mole.

iii) Increased wetting of the substrate by particles will slow down the rate of growth. This might be achieved in practice by doping of either particle or substrate. The models therefore point to possible ways of increasing the resistance to particle growth.

iv) It is advantageous to attain particles free of defects, such as dislocations, grain boundaries, etc. in order to promote the inhibited mode of growth. Thus, research into the structure of particles produced by different impregnation techniques may be useful.

v) Growth under oxidizing conditions is strongly dependent on the enthalpy change of the reaction:

$$M_{(s)} + x\, O_{2(g)} = MO_{2x(g)} \quad .$$

Thus, it is possible to predict an approximate stability series for supported noble metals as follows:

$$Os < Ru < Ir < Pt < Pd = Rh \quad .$$

vi) Under reducing conditions, Eq. (15) indicated that particle growth by interparticle transport is controlled in a major way by the heat of sublimation of the metal. Similarly, growth by particle migration based on surface diffusion would have an activation energy roughly proportional to the sublimation energy. On this basis, it is therefore possible to rank the stability of noble metals under reducing conditions as:

$$Pd < Rh = Pt < Ru < Ir < Os \quad .$$

ACKNOWLEDGMENT

One of us (T-M.A.) is grateful to the National Science Foundation for financial support under Grant No. NSF-GH-44311. We are indebted to Dr. N. Gjostein for his critical review of the manuscript.

REFERENCES

1. P.Wynblatt and N.A.Gjostein, "Progress in Solid State Chemistry", Vol. 9, 1974, in press.
2. J.G.Skofronick and W.B.Phillips, J.Appl.Phys. <u>39</u>, 304 (1965).
3. W.B.Phillips, E.A.Desloge and J.G.Skofronick, J. Appl. Phys. <u>39</u>, 3210 (1968).

4. A.Masson, J.J.Metois and R.Kern, Surface Sci. 27, 463 (1971).

5. R.Kern, A.Masson and J.J.Metois, Surface Sci. 27, 483 (1971).

6. E.Ruckenstein and B.Pulvermacher, AIChE J. 19, 356 (1973).

7. E.Ruckenstein and B.Pulvermacher, J. Catal. 29, 224 (1973).

8. B.Pulvermacher and E.Ruckenstein, J. Catal. 35, 115 (1974).

9. F.A.Nichols and W.W.Mullins, Trans. AIME 233, 1840 (1965).

10. F.A.Nichols, J. Appl. Phys. 37, 2805 (1966).

11. N.A.Gjostein, "Surfaces and Interfaces, I", p. 271, Eds. J.J.
    Burke, N.L.Reed and V.Weiss, Syracuse Univ. Press, N.Y. (1967).

12. M.McLean and E.D.Hondros, J. Mater. Sci. 6, 19 (1971).

13. E.E.Gruber, J. Appl. Phys. 38, 243 (1967).

14. P.Wynblatt and N.A.Gjostein, to be published.

15. C.Wagner, Z. Electrochem. 65, 581 (1961).

16. I.M.Lifshitz and V.V.Slyozov, J. Phys.Chem.Solids 19, 35 (1961).

17. B.K.Chakraverty, J. Phys. Chem. Solids 28, 2401 (1967).

18. P.Wynblatt, R.A.Dalla Betta and N.A.Gjostein, Proceedings of
    the Battelle Colloquium on The Physical Basis for Heterogeneous
    Catalysis, Gstaad 1974, in press.

19. C.Kittel, "Introduction to Solid State Physics", p. 99, John
    Wiley & Son Inc., New York.

20. H.Schafer and A.Tebben, Z. Anorg. Chem. 304, 317 (1960).

21. P.C.Flynn and S.E.Wanke, J. Catal. 34, 390 (1974).

22. P.C.Flynn and S.E.Wanke, J. Catal. 34, 400 (1974).

23. P.Wynblatt and N.A.Gjostein, Scripta Met. 9, 969 (1973).

24. G.A.Somorjai, Prog. Anal. Chem. 1, 101 (1968).

25. K.Y.Chen and J.R.Cost, J. Nucl. Mater. 52, 59 (1974).

26. L.E.Willertz and P.G.Shewman, Met. Trans. AIME 1, 2217 (1970).

27. W.K.Burton, N.Cabrera and F.C.Frank, Phil. Trans. Roy. Soc.
    A243, 299 (1950).

28. W.L.Winterbottom, "Surfaces and Interfaces, I", p. 133, Eds.
    J.J.Burke, N.L.Reed and V.Weiss, Syracuse Univ.Press,N.Y.(1967).

29. J.P.Hirth and G.M.Pound, "Progress in Materials Science",
    Vol. 11, Ed. B.Chalmers, Macmillan, N.Y. 1963.

30. T-M.Ahn and J.K.Tien, to be published.

31. J.A.Bett, K.Kinoshita and P.Stonehart, J.Catal. 35, 307 (1974).

32. J.W.Geus, "Chemisorption and Reactions on Metallic Films",
    Vol. 1, Ed. J.R.Anderson, Academic Press, N.Y. 1971.

33. F.H.Huang and C-Y.Li, Scripta Met. 7, 1239 (1973).

34. J.W.Sprys, L.Bartosiewicz, R.C.McCune and H.K.Plummer, J.
    Catal., in press.

35. R.T.K.Baker, C.Thomas and R.B.Thomas, J. Catal., in press.

36. H.J.Maat and L.Moscou, Proc. 3rd Int. Congr. Catal. II, p. 1277
    (1965).

37. R.Herrmann, S.F.Adler, M.S.Goldstein and R.M.Debaun, J. Phys.
    Chem. 65, 2189 (1961).

38. T.R.Hughes, R.J.Houston and R.P.Sieg, I&EC Proc. Des. Dev. 1,
    96 (1962).

39. H.L.Gruber, Anal. Chem. 34, 1828 (1962).

SINTERING OF SUPPORTED METAL CATALYSTS: APPLICATION OF A MECHANISTIC MODEL TO EXPERIMENTAL DATA

Sieghard E. Wanke

Department of Chemical Engineering

The University of Alberta, Edmonton, Canada

INTRODUCTION

When noble metals are used as catalysts, the metal is generally dispersed on a high surface area support in the form of small crystallites (< 3 nm in diameter). One of the functions of the support is to physically separate the metal crystallites and thereby hinder the agglomeration of the metal crystallites. This agglomeration would decrease the metal surface area and therefore cause a loss of catalytic activity. This process of metal particle growth is generally called catalyst sintering or aging. It is in this context that the term sintering will be used throughout this paper.

Although the use of high surface area supports to separate the metal crystallites results in thermally stable catalysts, growth of metal crystallites still occurs, especially if the catalysts are used at elevated temperatures. The mechanism of this sintering process is not known at the present, but recently two mechanistic models have been developed for the sintering of supported metal catalysts. The first model, developed by Pulvermacher and Ruckenstein (1,2), envisages sintering to occur by the migration of entire metal crystallites over the support surface followed by coalescence with other metal crystallites upon collision (this model will be referred to as the crystallite migration model). The second model, by Flynn and Wanke (3,4), envisages sintering to occur by dissociation of atomic or molecular species from metal crystallites, migration of these species across the support surface, and capture of these species upon collision with crystallites (this model will be referred to as the atomic migration model).

In this paper a slightly modified version of the atomic migration model (3,4) will be presented, and three sets of experimental data will be examined in light of this atomic migration model. The objective of this work is to determine whether or not the atomic migration model can adequately describe experimental sintering data. Adequate description of data by a model does not constitute proof of the mechanism used in the model, but it does lend support to the proposed mechanism.

An understanding of the mechanism of sintering would be a great asset in the design of supported metal catalysts with increased thermal stability. For example, if crystallite migration is the mechanism then strong metal-support interactions would immobilize the crystallites and thereby decrease the rate of sintering. On the other hand, strong metal-support interactions could increase the rate of sintering if atomic migration is the mechanism, since the loss of metal atoms from the crystallite to the support surface would be facilitated by strong metal-support interactions. A detailed comparison of the two models has been presented elsewhere (5).

DESCRIPTION OF MODEL

The atomic migration model presented in this paper is essentially the same as that presented previously (3). The sintering process is modelled as a three step process: one, escape of metal atoms (or molecules such as PtO in an oxygen atmosphere) from the metal crystallite to the support surface; two, migration of these species over the support surface; and three, capture of the migrating species by metal crystallites upon collision. The support surface is assumed to be energetically homogeneous.

The rate of loss of metal atoms undoubtedly is a highly activated process and qualitative arguments, based and strong metal-support interactions, have been presented previously (3) that could lower this activation energy. The quantitative expression for the rate of loss of metal atoms from the ith crystallites is based on the Kelvin equation and is given by

$$\frac{dL_i}{dt} = A \; r_i \; \exp[\alpha/r_i T - E/RT] \text{ in atoms/sec} \qquad (1)$$

where r_i is the particle radius (metal particles are assumed to be hemispheres), A is a constant, $\exp[\alpha/r_i T]$ is a term from the Kelvin equation (to be discussed below), $\exp[-E/RT]$ is the postulated Arrhenius temperature dependences, and the subscript i refers to the ith crystallite. It should be pointed out that in the previously presented model (3) the metal crystallites were assumed to be

cubes and that the rate of loss was assumed to be independent of
metal crystallite size.

The rate of loss as given by Equation 1 was arrived at as
follows. Consider a supported metal catalyst containing metal
crystallites all of the same radius r_k. At equilibrium there would
exist a certain concentration of metal atoms on the surface support. This concentration would be proportional to the two dimensional spreading pressure, ϕ, of these freely migrating species.
The value of ϕ would be related to the equilibrium spreading pressure ϕ_0 for large metal crystallites by the Kelvin equation, i.e.

$$\phi = \phi_o \exp[\frac{\alpha}{r_k T}] \tag{2}$$

where $\alpha = \bar{\sigma}V/R$, $\bar{\sigma}$ = average (unknown) value of the interfacial
energies between metal-support and metal-atmosphere, and V is the
molar volume of the metal. Since this system is at equilibrium,
and all crystallites are the same size then the rate of loss of
atoms from the crystallites has to be equal to the rate of gain.
The rate at which crystallites gain migrating atoms is proportional
to ϕ and r_k. Hence in the equilibrium situation

$$\frac{dL_k}{dt} = \frac{dG_k}{dt} = a\ r_k\ \phi = a\ \phi_o\ r_k\ \exp[\frac{\alpha}{r_k T}] \tag{3}$$

Since the rate of atom loss from crystallites is independent of
whether or not the system is at equilibrium as long as energy
transfer is not rate controlling and the Arrhenius temperature dependence is included in the proportionality constant, a, then
Equation 1 results. Equation 1 then gives the rate of loss of
metal atoms from an individual metal crystallite for the case where
crystallites of different sizes are present.

Once the rate of loss from individual crystallites is known
the rate of gain of each crystallite can be calculated if the surface concentration is known. In the present model the assumption
is made that the migration of atoms across the support surface and
the capture of migrating atoms by collision with metal crystallites
is rapid, i.e., the rate of loss is the rate determining step. In
this situation the concentration of atoms migrating on the support
is small at all times and the number of atoms lost by all crystallites is equal to the number of atoms gained in a specific time
interval. The redistribution of the lost atoms among the crystallites is according to their size, i.e., the number of atoms gained
by the ith crystallite is

$$\frac{dG_i}{dt} = \frac{(r_i + a_o)}{\sum\limits_{j=1}^{M} (r_j + a_o)} \left[\sum\limits_{j=1}^{M} \frac{dL_j}{dt} \right] \tag{4}$$

M is equal to the total number of metal particles (time dependent). The a_o is added to r_i to account for the size of the migrating species.

Equations 1 and 4 can readily be solved by a single-step finite difference technique to yield the number of atoms in each particle as a function of time. The solution involves the following steps:

1. generation of the initial particle size distribution, PSD. (In this work the PSD were obtained from experimental data as number of particles in a certain size interval. The size of each metal particle was calculated by the method previously described (4) so that each particle had a different size.)

2. conversion of particle size to number of atoms, N_i, for each particle. Since the particles were assumed to be hemispheres the conversion from r_i to N_i for Pt particles is given by

$$N_i = 138.7 \, r_i^3 \text{ atoms [with } r_i \text{ in nm]} \tag{5}$$

3. selection of parameter values (A, α, E, a_o) and size of time increment. A time increment of 360 sec and an a_o equal to 0.554 nm were used; the value of the other parameters was varied in order to fit the experimental data.

4. calculation of the number of atoms lost by each particle in the time increment according to Equation 1.

5. calculation of number of atoms gained by each particle in the time increment according to Equation 4.

6. calculation of the number of atoms in each particle at the end of the time increment by adding the difference of the number gained minus the number lost for each particle to the number of atoms in the particle at the beginning of the time increment. If the number of atoms in any particle was $\leq$ 58 (corresponds to $r_i \approx 0.75$ nm) then the particle was assumed to have disappeared and the number of atoms in the particle was added to the total number of atoms lost.

7. steps 4 to 6 are repeated for the next time increment.

This model and the above computational procedure was applied to three sets of experimental data and the results are described in the following section.

APPLICATION OF MODEL TO EXPERIMENTAL DATA

In order to test the proposed model it is necessary to have
initial PSD. Data of this type are scarce in the literature since
usually only average metal particle sizes or specific metal surface
areas are reported as a function of treatment conditions. The
results of Wynblatt and Gjostein (6,7) and Renouprez et al. (8) are
exceptions, since Wynblatt and Gjostein presented transmission
electron micrographs (TEM) of Pt/Al_2O_3 samples treated in two different atmospheres at 700°C for varying periods of time and Renouprez et al. presented PSD obtained by small angle X-ray scattering
(SAXS) and TEM for a 3.7% Pt/Al_2O_3 catalyst sintered in vacuum for
6 hr at 600, 700 and 800°C. The results for these three cases are
examined in terms of the model discussed above.

Cases I and II

The data of Wynblatt and Gjostein (6,7) were used for these
two cases; Case I refers to the treatment of Pt/Al_2O_3 at 700°C in a
2% O_2 in N_2 atmosphere and Case II refers to the treatment at 700°C
in air. The initial PSD for Case I, shown in Fig. 1, was obtained
by measuring the particle sizes from Fig. 2a of ref. 6 and Fig. 11a

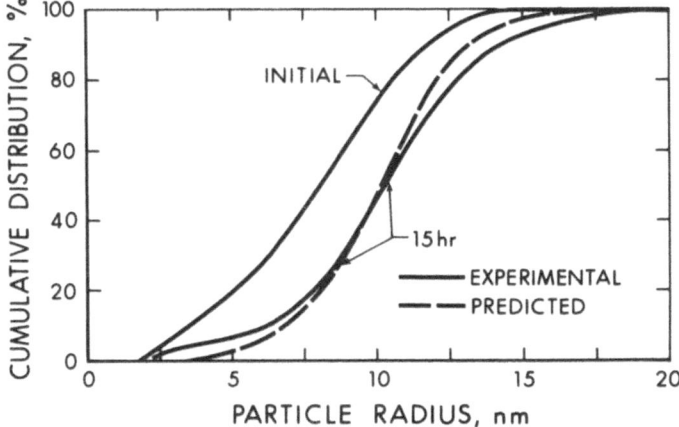

Figure 1. Comparison of experimental and predicted PSD for
     sintering at 700°C in 2% O_2 in N_2 (Aexp[-E/RT] = 2.55 x 10^{-2}
     (sec-nm)$^{-1}$; α = 1.46 x 10^4 (K-nm). Data from ref. 6 and 7).

of ref. 7. The initial PSD for Case II, shown in Fig. 2, was
obtained by measuring the particle sizes from Fig. 2d of ref. 6
(Fig. 2 in ref. 6 is incorrectly labelled (9)) and Fig. 11c of ref.
7. A total of 480 and 452 particles were measured for Cases I and
II, respectively.

Using these initial PSD various values of α and $A\exp(-E/RT)$
were used in the model to predict average particle radii, $\bar{r}$, as a
function time. (Varying A and E independently is meaningless in
this situation since all experiments were at the same temperature.)
The average particle radius used in this work is defined as

$$\bar{r} = \frac{1}{M} \sum_{i=1}^{M} r_i \tag{6}$$

The predicted values of $\bar{r}$ as a function of time were compared to
the experimental results reported in Fig. 1 of ref. 6. Figure 3
shows comparisons of predicted and experimental values of $\bar{r}$ for
Cases I and II. The values of α and $A\exp(-E/RT)$ used to obtain the
predicted results were 1.46×10^4 (K-nm) and 2.55×10^{-2} (sec-nm)$^{-1}$
for Case I and 1.95×10^3 (K-nm) and 12.8 (sec-nm)$^{-1}$ for Case II.
The agreement between experimental and predicted values of $\bar{r}$ is

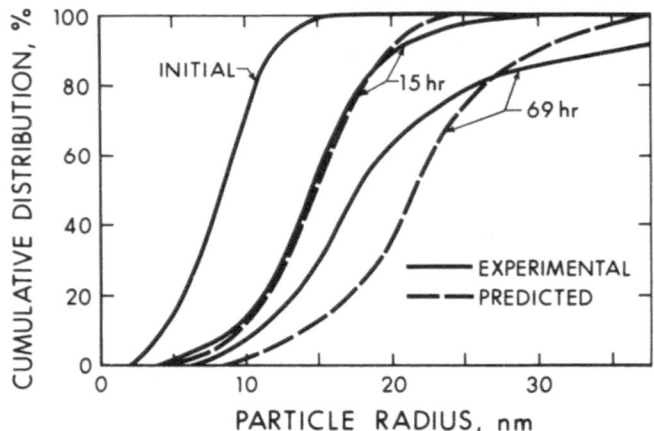

Figure 2.   Comparison of experimental and predicted PSD for
        sintering at 700°C in air ($A\exp[-E/RT]$ = 12.8 (sec-nm)$^{-1}$;
        $\alpha = 1.95 \times 10^3$ (K-nm). Data from ref. 6 and 7).

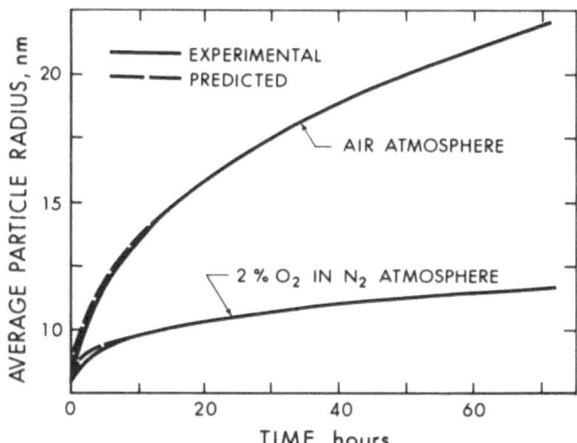

Figure 3. Comparison of experimental and predicted values of $\bar{r}$
    (Parameter values given in text. Data from ref. 6 and 7).

excellent for the above parameter values; small deviations occurring at low sintering times. Increasing the value of α results in more rapid particle growth at short times. Increasing $A\exp(-E/RT)$ results in increased sintering rates for all times.

In this work the agreement between experimental and predicted values of $\bar{r}$ was used as the criterion for determining the values of the parameters. A check in the validity of the predictions is to compare the predicted PSD at various sintering times with experimental PSD. Figs. 1 and 2 show such comparisons. The experimental PSD were measured from Figs. 2 and 11 of ref. 6 and 7. The agreement between predicted and measured PSD is not good, especially at large times. The predicted PSD are always narrower than the measured ones. Numerous reasons, based on deficiencies of the model (such as assuming homogeneous support surfaces, hemispherical particles, etc.) and the problem of measuring PSD (measuring in the excess of 1000 particles does not necessarily yield a representative PSD (8), and for the 69 hr treatment shown in Fig. 2 only 43 particles were measured) could cause the observed discrepancies.

Case III

Renouprez et al. (8) present PSD data for a 3.7% Pt/Al$_2$O$_3$ catalyst sintered in vacuum at 600, 700 and 800°C for 6 hr. They
measured PSD by SAXS and TEM. The initial PSD, estimated from Fig.
7 of ref. 8, are listed in Table 1 in columns 2 and 3. Renouprez
et al. presented PSD based on volume distribution; the PSD shown in
Table 2 are based on number distribution and they were obtained by
converting the volume PSD to a number PSD. The number of particles
in the various size ranges for the two initial PSD were adjusted so
that the dispersion of both initial PSD was 0.38. Dispersion is
defined as the ratio of surface metal to total metal atoms, and for
a hemispherical Pt particles the dispersion, D, is given by

$$D = 70.60 \sum_{i=1}^{M} r_i^2 \bigg/ \sum_{i=1}^{M} N_i \qquad \text{[with } r_i \text{ in nm]} \qquad (7)$$

where N$_i$ is given by Equation 5.

Various values of the parameters, α, A and E, were substituted
into the model together with the initial PSD listed in Table 1 in
order to find parameter values which yielded predicted dispersions
close to the experimental dispersions at the three sintering temperatures. Table 1 shows two sets of parameter values applied to
the initial SAXS PSD which result in predicted dispersions close to
the experimentally determined dispersions (columns 4, 5, 7, 8, 10
and 11 in Table 1). Since two sets of parameter values describe
the dispersion data equally well, it can be concluded that using
dispersion as a function of sintering temperature as the sole criterion is insufficient for determining a unique set of parameter
values. The second set of parameter values also resulted in good
predictions when applied to the initial PSD obtained by TEM (columns
6, 9 and 12 in Table 1).

Table 1 also shows the predicted PSD for the various parameter
values and initial PSD. It can be seen that increasing the value
of α, which requires an increase in the activation energy in order
to match predictions and experiment, results in a more rapid disappearance of small crystallites. This means that the predicted PSD
can be varied while keeping the dispersion constant. Hence it
should be possible to match experimental and predicted PSD at a
fixed dispersion. Unfortunately, reliable experimental PSD are
difficult to obtain. For example, Renouprez et al. (8) obtained the
same PSD by SAXS for sintering at 700 and 800°C. This is an unusual
result and in light of the PSD obtained by TEM it appears to be
incorrect.

Table 1. Initial and predicted PSD and dispersions for a 3.7% Pt/Al$_2$O$_3$ catalyst after sintering in vacuum at 600, 700 and 800°C for 6 hr (data from ref. 8).

PARTICLE RADIUS RANGE (nm)	NUMBER OF PARTICLES IN EACH SIZE RANGE										
	Initial PSD		Predicted Particles Size Distribution								
			600 °C			700 °C			800 °C		
	SAXS	TEM	SAXS		TEM	SAXS		TEM	SAXS		TEM
0.75–1.00	116	90	28[a]	10[b]	7[b]	3[a]	0[b]	1[b]	1[a]	0[b]	1[b]
1.00–1.25	43	70	29	26	36	10	5	7	1	1	0
1.25–1.50	20	26	28	32	42	18	18	21	3	1	3
1.50–1.75	5	8	13	15	16	15	16	20	7	5	6
1.75–2.00	2	6	3	4	7	7	8	9	8	10	9
2.00–2.25	2	7	3	2	7	2	2	6	4	4	6
2.25–2.50	1	1	1	1	2	2	3	7	2	2	6
2.50–2.75	1		1	1	1	1	1	1	2	2	7
2.75–3.00	1		1	1		2	1		1	2	1
3.00–3.25	1		1	1		1	1		1	1	
3.25–3.50									2	1	
TOTAL	192	208	108	93	118	61	55	72	32	29	39
PREDICTED DISPERSION	0.38	0.38	0.32	0.32	0.31	0.26	0.27	0.28	0.22	0.23	0.22
MEASURED DISPERSION[c]	0.38		0.33			0.27			0.22		

Parameter values used to obtain predicted PSD are:

[a] $\alpha = 2.146 \times 10^3$ (K-nm); $A = 2.0 \times 10^4$ (sec-nm)$^{-1}$; and E/R = 15000K

[b] $\alpha = 5.880 \times 10^3$ (K-nm); $A = 2.0 \times 10^6$ (sec-nm)$^{-1}$; and E/R = 23000K

[c] experimental dispersions were calculated from the surface average diameters reported in Table 7 of ref. 8. For the initial and 600°C case the dispersion was calculated from the hydrogen adsorption and SAXS results, and for the 700 and 800°C cases the dispersion was calculated from the hydrogen adsorption, SAXS and TEM results.

DISCUSSION

The results of the three cases analyzed in the previous section showed that the simple atomic migration model can describe experimental results quite adequately. A few comments on the numerical values of the parameters are in order. From the values of α, the average interfacial energies can be computed since $\bar{\sigma} = \alpha R/V$. The values of α used for the three cases are 1.46 x 10^4 and 1.95 x 10^3 (K-nm) for Cases I and II and 2.146 x 10^3 to 5.880 x 10^3 (K-nm) for Case III. The corresponding values of $\bar{\sigma}$ are 1.33 x 10^4, 1.78 x 10^3 and 1.96 x 10^3 to 5.37 x 10^3 dynes/cm using V = 9.10 cm^3/g-atom for Pt. These values of interfacial energies, with the exception of the high value for Case I, are of the same order of magnitude as those reported by McLean and Hondros (10) for the Pt/Al$_2$O$_3$/air system. Some variation in α in the model is possible by varying E without greatly affecting $\bar{r}$ or D, but this results in changes in predicted PSD. The magnitude of E obtained for Case III is of the same order of magnitude as has been reported in the literature for sintering of Pt in nonoxidizing atmospheres (5).

In conclusion, the simple atomic migration model appears to describe the three cases examined quite adequately. More flexibility in fitting the data is possible if one considers the situation where the concentration of migrating species does not approach zero and varies with time (4). In order to conclude whether the atomic migration model describes the experimental data better than the crystallite migration model similar fitting of experimental data should be carried out for the later model.

REFERENCES

1. Ruckenstein, E. and Pulvermacher, B., AIChE J. 19, 356 (1973).
2. Ruckenstein, E. and Pulvermacher, B., J. Catal. 29, 224 (1973).
3. Flynn, P.C. and Wanke, S.E., J. Catal. 34, 390 (1974).
4. Flynn, P.C. and Wanke, S.E., J. Catal. 34, 400 (1974).
5. Wanke, S.E. and Flynn, P.C., Catal. Rev. (in press).
6. Wynblatt, P. and Gjostein, N.A., Scr. Met. 7, 969 (1973).
7. Wynblatt, P. and Gjostein, N.A., Progr. Solid State Chem. 9 (in press).
8. Renouprez, A., Hoang-Van, C., and Compagnon, P.A., J. Catal. 34, 411 (1974).
9. Wynblatt, P. and Gjostein, N.A., Scr. Met. 8, 169 (1974).
10. McLean, N. and Hondros, E.D., J. Mat. Sci. 6, 19 (1971).

EFFECTS OF GAS- AND LIQUID-PHASE ENVIRONMENTS ON THE SINTERING

BEHAVIOR OF PLATINUM CATALYSTS

K. Kinoshita, J. A. S. Bett and P. Stonehart

Advanced Fuel Cell Research Laboratories
Power Utility Division
United Aircraft Corporation
Middletown, Connecticut 06457

INTRODUCTION

Platinum has been used extensively as a catalyst in the petroleum industry for hydrocarbon cracking. More recently, Pt has been
considered for an electrocatalyst in fuel cells and for catalytic
mufflers in the exhaust systems of automobiles. The performance of
Pt as a catalyst in these heterogeneous catalysed systems is dependent upon the Pt activity and the real Pt surface area accessible
to the reactant (in the absence of catalyst deactivation by poisoning and mass transfer limitations). The high temperature application of supported Pt (Pt/Al_2O_3) as a heterogeneous catalyst in a gas-
phase environment causes sintering and loss of Pt surface area (1-7).
Unsupported Pt black sinters rapidly in a gas-phase environment (8,9)
but at a much lower temperature ($< 200°C$) than supported Pt. In contrast, unsupported Pt black (10-13) and Pt supported on carbon (14,
15) sinter in liquid-phase environments in the same temperature range
($100-200°C$). Since a loss of Pt surface area due to sintering results in a decrease of the total catalyst activity (constant Pt specific activity assumed) there is significant profit in maintaining
high surface area Pt catalysts in diverse operational environments.
Yet, despite the numerous publications presenting experimental data
for the sintering of supported and unsupported metal catalysts, the
mechanisms responsible for this phenomenon remain a subject of controversy.

Due to the application of Pt as an electrocatalyst, there has
been considerable effort in this laboratory to characterize and understand the sintering behavior of Pt in the liquid-phase environment. Based on the theoretical models for metal crystallite
growth, and their application to experimental results obtained on

117

supported and unsupported Pt catalysts in both gas- and liquid-phase environments, we have endeavored to shed some light on the Pt sintering mechanism in these media.

THEORETICAL MODELS FOR SINTERING

The sintering of Pt black electrocatalysts has been considered (10,11,13) in terms of classical sintering theories developed by Kuczynski (16) for powder metallurgy. From studies of gas-phase sintering of metal powders, it is recognized that three stages can be distinguished during the sintering process - (a) neck growth between the contacting particles, (b) densification brought about by diffusion and plastic deformation, and (c) the elimination of residual isolated voids.

Theories of sintering have identified four principle mechanisms controlling the process, which can be distinguished experimentally by the relationships (16) between the radii, x, of the neck, a, of the spheres, time, t, and temperature, T:

$$\frac{x^n}{a^m} = F(T)t \qquad\qquad (1)$$

where $F(T)$ is a function of temperature only and $n = 2$, $m = 1$ for viscous or plastic flow; $n = 3$, $m = 1$ for evaporation and condensation; $n = 5$, $m = 2$ for volume diffusion and $n = 7$, $m = 3$ for surface diffusion. Unfortunately, the present theories are only applicable to the initial stage of sintering involving the growth of necks between particles. It is also recognized that the irregular particle shapes for real catalyst particles are too complex to be represented by Equation 1 but nevertheless, a qualitative interpretation on the sintering of Pt black electrocatalysts has been obtained (10,11) in terms of surface diffusion as the rate limiting step.

The growth of metal crystallites on supports at high temperatures has been interpreted by either of two mechanisms. In both models the decrease in surface energy provides the driving force for crystallite growth, the difference being the mode of transport of metal as crystallites or atomic species. The theoretical models developed by Ruckenstein and Pulvermacher (17,18) and Phillip et al. (19) treated the growth of supported metal crystallites as the result of the coalescence of discrete metal crystallites migrating on the support. This process is analogous to two-dimensional Brownian motion of metal crystallites which undergo random collisions and subsequent coalescence. Rate expressions for the surface migration mechanisms have been derived as extensions of the Smoluchowski (20) coagulation theory. The general form of the integrated rate expression relating the metal surface area and the sintering time is given as

$$\frac{1}{S^n} = \frac{1}{S_o^n} + kt \tag{2}$$

where S_o and S represent the metal surface areas per unit volume before sintering and time, t, respectively. The constant, k, is directly proportional to ϕ, defined as the total volume of metal per unit area of support. Phillips et al. (19) considered the surface diffusion of metal crystallites as the rate determining step in crystallite growth. Ruckenstein and Pulvermacher (17,18) considered either surface diffusion or coalescence of crystallites as rate limiting, depending on the sintering conditions. The exponential, n, in Equation 2 varied between 2-5, depending on the rate limiting step for crystallite growth (i.e., surface diffusion of crystallites or coalescence) and the physical model. These results are summarized in Table I.

The second mechanism used to describe metal crystallite growth involves evaporation of atoms from small crystallites and condensation on larger ones in a process akin to the Ostwald ripening of dispersed precipitates in solution. The atoms may diffuse either on the surface of the support as adatoms or in the gas phase.

TABLE I. EQUATIONS FOR SINTERING RATES OF SUPPORTED METAL CATALYSTS

Reference	Mechanism	Rate determining process	n	k dependence on ϕ
Ruckenstein and Pulvermacher (17, 18)	Crystallite migration	Coalescence	2-3	Proportional
		Surface diffusion	3-5	Proportional
Phillips, Desloge and Skofronick (19)	Crystallite migration	Surface diffusion	7/2	Proportional
Chakraverty (21)	Ostwald ripening	Surface diffusion	4	Independent
	Ostwald ripening	Interface transfer	2	Independent
Wynblatt and Gjostein (23)	Ostwald ripening	Surface diffusion	4	Independent
	Ostwald ripening	Interface transfer	3	Independent
Flynn and Wanke (31)	Ostwald ripening	Interface transfer	3	Independent
Dunning (22)[a]	Ostwald ripening	Surface diffusion	3	Independent

[a] Solution for three-dimensional process.
ϕ = Total volume of metal per unit area of support.

The Ostwald ripening mechanism supposes that evaporation of atomic species from small crystallites and condensation on larger ones occurs by virtue of the difference in solubility as predicted by the Gibbs-Thomson equation. Chakraverty (21) has treated the cases where the rate determining step in Ostwald ripening is either surface diffusion of adatoms or the interface transfer of metal atoms from the support to the metal crystallite or vice-versa. Expressions similar to Equation 2 are obtained except the rate constant is independent of ϕ. Wynblatt and Gjostein (23) have amended Chakraverty's treatment, arriving at a similar conclusion, that the rate constant, k, is independent of ϕ. Flynn and Wanke (31) have also considered the sintering of supported metal catalysts, using the Ostwald ripening concept in their model. From an analysis of their model for the case where the rate limiting step is interface transfer, it is apparent that the rate constant, k, is independent of ϕ. Dunning (22) has solved the simpler case for Ostwald ripening of dilute dispersed precipitates and assuming that the solution for the three-dimensional problem is similar to the two-dimensional problem of catalyst sintering, it may readily be shown that Dunning's equation can also be expressed in the form of Equation 2 where n = 3 and k is independent of ϕ.

Thus, the major difference between the two sintering models describing the growth of supported metal crystallites is in the rate constant, k, of Equation 2; the Ostwald ripening model being independent of the metal loading, whereas the Smoluchowski model is directly proportional to catalyst metal loading. The various forms of Equation 2 for the different mechanisms have been summarized in Table I.

The theoretical models summarized in Table I refer to crystallite growth by the "non-inhibited" growth mechanism, a designation proposed by Wynblatt and Gjostein (23). To account for the experimentally observed values of n greater than predicted by the classical models, Wynblatt and Gjostein (23) proposed a different mode of crystallite growth referred to as "inhibited" growth. The same general processes described for non-inhibited growth apply, but for inhibited growth the metal crystallites tend to facet and thus require periodic nucleation events on their surfaces for continued growth. By invoking a requirement for nucleation of steps on crystal facets, for crystal growth either by crystallite migration or by atom diffusion, Wynblatt and Gjostein (7) have introduced an exponential dependence of the rate constant for sintering on r, the crystallite radius. Such an inhibition to interface transfer would yield the following rate law:

$$\frac{dr}{dt} = \frac{k}{r^m} e^{-Ar/RT} \tag{3}$$

The resulting solution of Equation 2 would have an effective exponent, n, which increased with time and could reach high values.

Based on the theoretical models discussed in this section, it was concluded that both the Smoluchowski collision and the Ostwald ripening models for sintering of supported metal catalysts result in rate laws (Equation 2) which have the same form. Since the accuracy with which the exposed metal area can be determined is not great, it is consequently difficult to identify the sintering mechanism from a determination of reaction order (exponent n in Equation 2). On the other hand, as pointed out by Dunning (22), a significant difference exists between the two mechanisms in the dependence of the rate constant, k, on ϕ.

EXPERIMENTAL SINTERING RESULTS

Unsupported Pt blacks have been shown by McKee (8) and Khassan et.al. (9) to suffer significant surface area losses, when heated in a reducing atmosphere at 100 to 200°C. These sintering rates are of comparable magnitude to sintering rates for Pt black in the liquid phase (96 wt% H_3PO_4) over the same temperature range (10,11), as indicated in Figure 1. McKee (8) reported that the sintering of Pt black in hydrogen followed the relationship

$$S = kt^{-0.11} \qquad (4)$$

where k is a function of temperature. It appears, therefore, that when the data in Figure 1 is projected in the form of Equation 2, the exponent, n, is considerably higher for the sintering of Pt blacks than proposed by the theories summarized in Table I.

Liquid metals such as mercury and amalgams experience a change in surface tension with potential (i.e., Lipmann Equation) and there

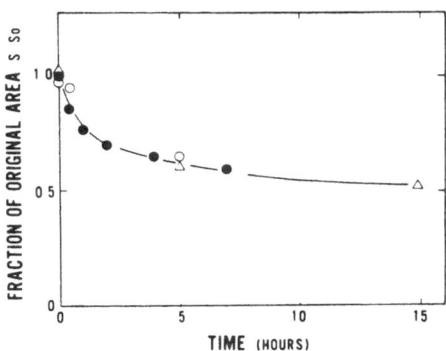

Fig. 1.   Sintering of Pt black at 150°C in hydrogen and in 96 wt% H_3PO_4. (△) Kinoshita et al. (10), 96 wt% H_3PO_4, 0.5V; (○) Stonehart and Zucks (11), 96 wt% H_3PO_4, 0.5V; (●) McKee (8), hydrogen.

is evidence to suggest that the surface energy of solids are also modified by the electrode potential (24,25). The potential dependence for the sintering of Pt black in 96 wt% H_3PO_4 (see Figure 2) is another manifestation of the interfacial tension (11). Oxides of platinum are formed on the Pt black electrode surface at 1.0V and large surface concentrations of adsorbed hydrogen are present at 0.1V, so changes in the rate of sintering due to the changing surface energy of the metal by varying the electrode potential can be modified by the surface concentration of another surface phase. A further example of adsorption effects controlling the rate of sintering for Pt blacks in 96 wt% H_3PO_4 is illustrated in Figure 3. It is evident that adsorbed carbon monoxide retards the Pt sintering rate and may be an example of the Rebinder effect operating on metals. Analogous to the effect of potential and chemisorbed species on the sintering rate of Pt black in a liquid-phase environment, heat treatment of Pt black in various gas compositions can markedly influence the gas-phase sintering rate. Stolyarenko and Vasev (26) observed that Pt black heat treated at 350°C for 30 minutes sintered less rapidly in an oxygen containing environment than in hydrogen. The authors' suggested that the sintering rate of Pt black in air is decreased by the presence of chemisorbed oxygen on the Pt surface.

Accompanying the surface area loss of Pt black during sintering is an increase in the average Pt crystallite size and a change in the morphology of the Pt agglomerate. Electron micrographs in Figure 4 show that before sintering in 96 wt% H_3PO_4 at 150°C, the Pt black crystallites appear to be loosely held together in large agglomerates with rather high porosity. After sintering at 0.8V for 256 hours, however, the Pt agglomerate had undergone considerable densification

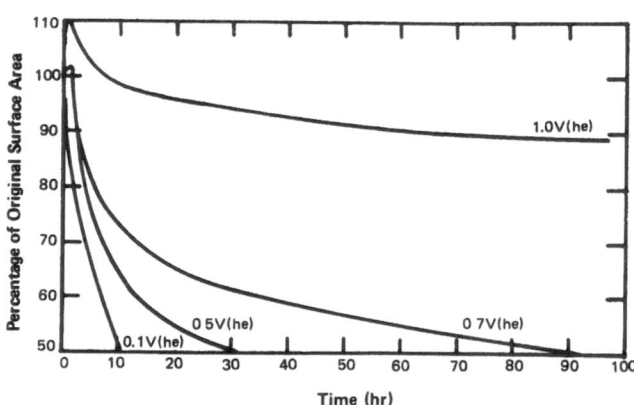

Fig. 2.   Effect of electrode potential on the loss of surface area of Pt black.   96 wt% H_3PO_4, 135°C, 24.5 m²/gm Pt.

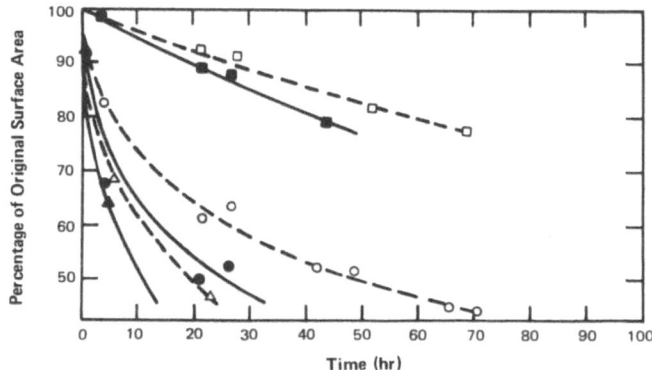

Fig. 3. Effect of temperature and CO on the loss of surface area
of Pt black. 96 wt% H_3PO_4, 0.5V, 24.5 m^2/gm Pt. □ ■,
105°C; O ●, 135°C; △▲, 150°C. Open points and dashed
lines for solutions saturated with CO.

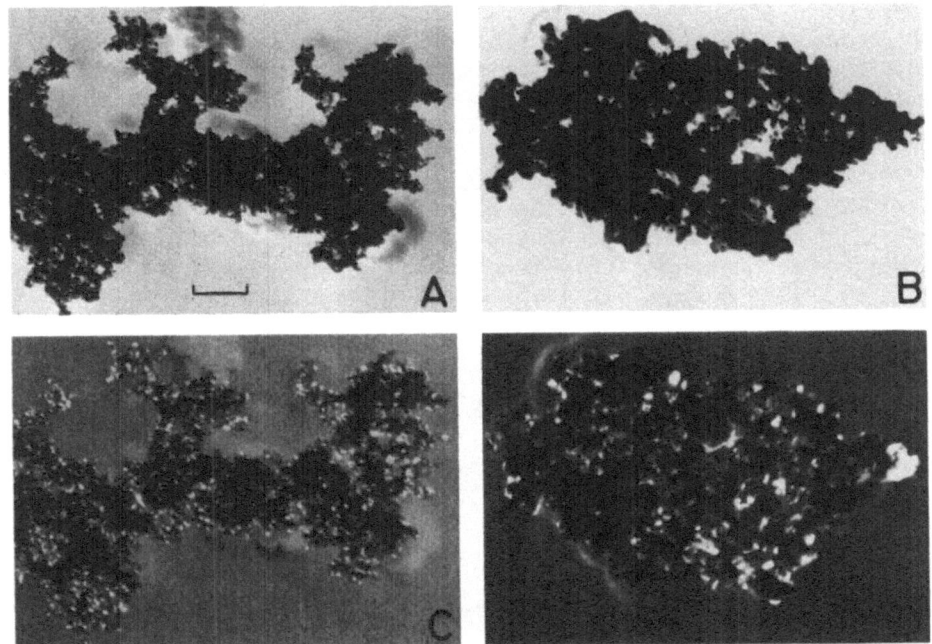

Fig. 4. Bright field (A,B) and dark field (C,D) electron micrographs of Pt black. Before (A,C) and after (B,D) sintering at 0.8V for 256 hrs in 96 wt% H_3PO_4, 150°C. Marker
equals 0.1 microns.

and smoothening of the previously "fluffy" Pt surface morphology has occurred. The application of dark field electron microscopy (27) clearly shows the growth of the Pt crystallite size when Pt black is sintered in 96 wt% H_3PO_4. Electron diffraction from Pt (111) crystallite planes which satisfy the Bragg condition for the particular orientation of the specimen appear as the bright spots in the dark field electron micrographs and the size of the diffraction spot is then indicative of the Pt crystallite size.

The increase in densification and the decrease in microporosity of Pt black sintered in 96 wt% H_3PO_4 (150°C) for 128 hours at 0.8V was measured by nitrogen adsorption-desorption isotherms at 78°K. The hysteresis between the adsorption and desorption branches of the isotherm allowed the determination of the capillary condensation from which it was possible to calculate the interparticle porosity of Pt black before and after sintering in H_3PO_4. The results, presented in Table II, for various partial pressures (P/P_o) of nitrogen at 78°K supported the visual observations from electron micrographs. Upon sintering in 96 wt% H_3PO_4, small voids in the Pt black agglomerate disappeared and not only had the total pore volume decreased but also the pore size distribution had shifted to larger diameters. This observation indicated that the small pores, < 10 Å, were eliminated before the large ones during sintering of Pt black in 96 wt% H_3PO_4. Gas-phase studies on the sintering of copper powders (28,29) showed a similar change in the porosity as reported for the liquid-phase sintering of Pt black.

The surface area loss of Pt supported on metal oxides (1-7,17, 23,31) due to sintering has been extensively investigated in both oxidizing and reducing gas-phase environments. In the present paper, Pt supported on graphitized carbon black (70 m^2/gm BET surface area) catalysts with widely different surface concentrations of Pt were studied (32) in an attempt to use this variable to distinguish be-

TABLE II.   INTERPARTICLE POROSITY OF Pt BLACK SAMPLES

Pt Black Sample	P/P_o	Uncorrected Kelvin Radius[a] (Å)	Interparticle Porosity (cm^3/gm Pt)
Unsintered	0.26	7	0.04
	0.75	35	0.17
	0.96	233	0.40
Sintered 0.8v in 96 wt% H_3PO_4 (150°C) for 128 hours	0.26	7	0.01
	0.75	35	0.06
	0.96	233	0.12

[a] calculated from Kelvin Equation (30). $r = -4.14/\log (P/P_o)$

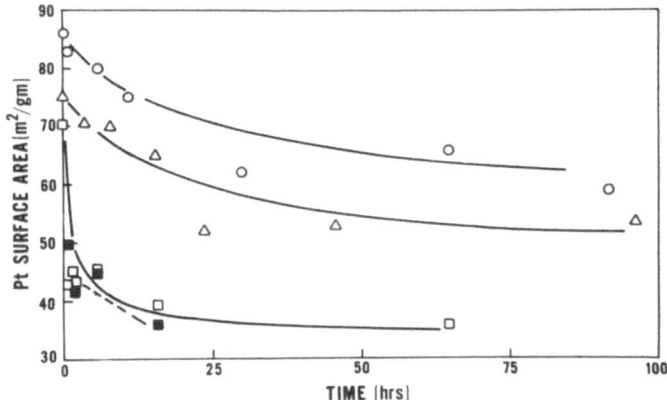

Fig. 5. Decrease in Pt crystallite surface area for Pt supported on graphitized carbon black at 600°C. (○) 5; (△) 12; (□) 20 wt% Pt in N_2. (■) 20 wt% Pt in H_2. Dashed line corresponds to decay from an initial 45 m^2/gm Pt datum after trap site saturation.

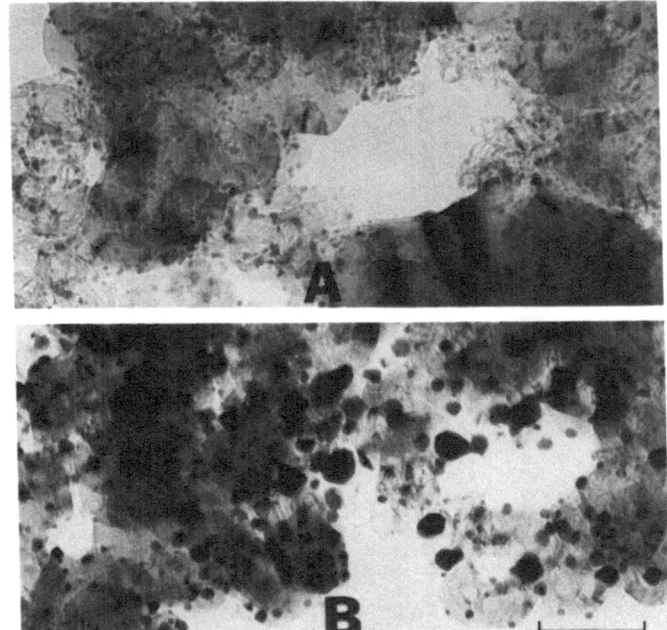

Fig. 6. Electron micrographs of 20 wt% Pt on graphitized carbon black. A, initial; B, after 16 hrs at 600°C in N_2. Marker equals 0.05 microns.

ween the Ostwald ripening and the Smoluchowski collision models for
sintering which were discussed in the previous section.

Figure 5 shows the decrease in Pt surface area for supported
catalysts containing three different surface concentrations of Pt.
At 600°C, sintering in both H_2 and N_2 atmospheres gave similar Pt
surface area changes with time. In contrast to the gradual decrease
in surface area with time for the 5 and 12 wt% Pt catalysts, the
20 wt% Pt catalysts suffered an initial rapid surface area loss.
Analysis of the data in Figure 5 indicated that high values of n were
required to fit the curves and when the n values were varied between
2 and 8 to fit the data according to Equation 2, the rate constants
derived from the experimental slopes of the functions showed a greater
dependence on ϕ than could be predicted by either sintering mechanism.

Confirmation that the decrease in Pt crystallite surface area
resulted from crystallite growth was obtained from a series of electron micrographs, with a typical result shown in Figure 6. Measurement of the Pt crystallite size distribution before and after heat
treatment for 16 hours at 600°C indicated the Pt crystallite distribution curve became broader with sintering. The surface area (27)
calculated from the Pt crystallite size distribution agreed with the
Pt surface area determined by electrochemical measurements (33).

The influence of temperature on the gas-phase sintering of 5 wt%
Pt on graphitized carbon is shown in Figure 7. At 500°C there was no
appreciable surface area decrease for the Pt crystallites. The Pt
surface area loss with sintering time at 600°C was smoothly monotonic,
whereas, at 700 and 800°C a rapid initial loss was observed, similar
to that present for 20 wt% Pt at 600°C.

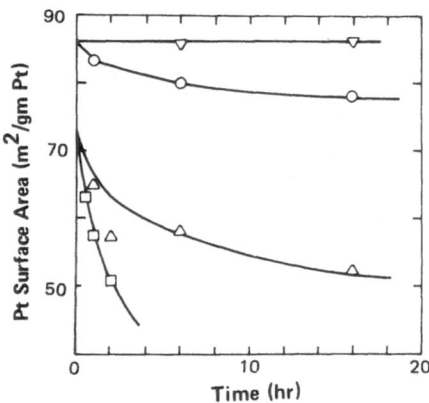

Fig. 7. Decrease in Pt surface area as a function of time and
temperature for 5 wt% Pt on graphitized carbon black. (∇)
500; ($\bigcirc$) 600; ($\triangle$) 700; ($\square$) 800°C.

Figure 8 shows the rapid initial decrease of the Pt crystallite surface area for supported catalysts containing 5 and 20 wt% Pt sintered in 96 wt% H_3PO_4 at 100 and 160°C. After a short time, the rate of surface area loss decreased and appeared to be dependent on temperature but not on platinum concentration of the catalyst. Similar to the results obtained for the sintering of supported Pt in the gas-phase, the sintering curves in Figure 8 indicated high values of n are required to fit the data by the simple power law,(Equation 2.) Bett et.al. (14) determined the values of n to vary from 11-13 for the 5 and 20 wt% Pt supported catalysts. In contrast to the gas-phase sintering of Pt supported on graphitized carbon, the rate constant, k, in Equation 2 is independent of the Pt concentration for liquid-phase sintering in H_3PO_4 (14). This conclusion was supported by additional experiments which showed that essentially the same rates of particle growth occurred in a number of different solvents (ethylene glycol, bromobenzene, water and toluene) of widely varying dielectric constants. Electron micrographs taken before and after sintering in hot concentrated H_3PO_4 confirmed that the decrease in Pt surface area was indeed due to Pt crystallite growth. Furthermore, the Pt surface areas calculated from measurements of the Pt crystallite size distributions (27) agreed with the electro-chemical surface area measurements (33).

The effect of potential on the sintering rates of supported Pt is illustrated in Figure 9. Over a wide potential range the data appear to fall within the same error limits. Unsupported Pt black in 96 wt% H_3PO_4, on the other hand, showed a large effect of potential (see Figure 2) on the sintering rate.

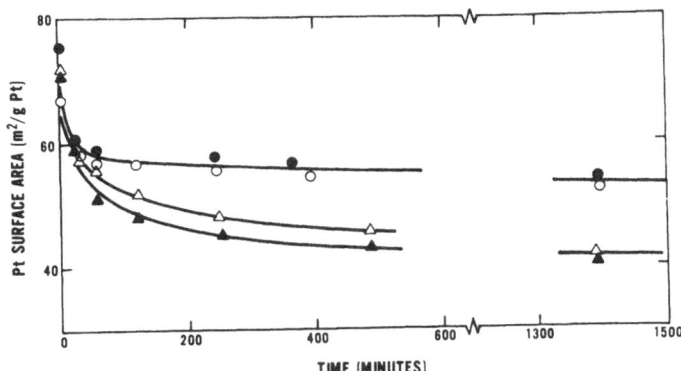

Fig. 8. Effect of temperature and Pt metal content on the surface area loss of Pt on graphitized carbon black. 96 wt% H_3PO_4, 0.8V. (●) 5 wt% Pt, 100°C; (○) 20 wt% Pt, 100°C; (△) 5 wt% Pt, 160°C; (▲) 20 wt% Pt, 160°C.

IMPLICATIONS OF EXPERIMENTAL RESULTS TO SINTERING MODEL

It is evident from attempts to fit the sintering data for Pt
blacks and Pt supported on graphitized carbons to equations of the
general form of 2 that the value of n required to fit the curves
(Figures 1,5,8) was high. The confidence with which the fit could
be established was not great, due to the high value of n coupled
with the generally low accuracy of surface area measurements. Bett
et al. (33) showed that from gas-phase sintering studies of supported catalysts with different Pt surface concentrations, it was possible to study the dependence of k on ϕ and hence, to distinguish
the sintering mechanism. This approach had obvious advantages over
elucidating the sintering mechanism from determinations of the exponent n in Equation 2, as suggested by Ruckenstein and Pulvermacher
(17,18), since a given value of n does not unambiguously define the
sintering mechanism (see Table I). Furthermore, the exponent n derived experimentally (7,14,32) is generally larger than the theoretically derived values and can not be explained by the derivation of
the rate law given by Equation 2.

To establish the sintering mechanism for supported Pt crystallites in gas-phase environments, Bett et al. (33) adopted the concept of "trap sites" on the carbon surface. This concept was suggested by Phillips et al. (19) who developed the kinetics for gold
particle growth on carbon in terms of a surface populated largely by
immobile particles trapped in potential wells requiring an activation energy to escape. Only the small number of particles with
energy in excess of this value are mobile and able to promote particle growth. One may, therefore, conjecture that with the 20 wt%
Pt on carbon catalyst containing 4.5×10^{11} crystallite/cm^2 carbon

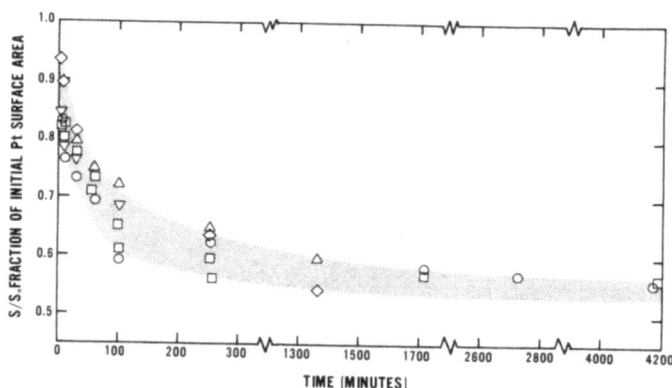

Fig. 9. Effect of potential on the loss of surface area for 5 wt%
Pt on graphitized carbon black in 96 wt% H_3PO_4, 160°C.
($\bigcirc$) 0.1V; ($\square$) 0.3V; ($\triangle$) 0.5V; ($\diamondsuit$) 0.8V; ($\triangledown$) 1.0V.

surface, the trap site density was exceeded. If this is so, then an initial population of unstable crystallites is produced, which would sinter rapidly at higher temperatures until the number of crystal- lites matched the trap site density. At this point, the regular surface diffusion controlled sintering process of the type described by Phillips et al. (19) could ensue with a higher activation energy required to move crystallites out of the trap sites. This point would appear to be reached after 2 hours of sintering for the 20 wt% Pt on carbon ,catalyst and the dashed lines in Figure 5 corresponds to decay from an initial 45 m^2/gm Pt datum after trap site saturation. With this in mind, the rate constant k for 20 wt% Pt was calculated using the differential form of Equation 2

$$\frac{-d\,S_o}{dt} = k\,S_o^5 \tag{5}$$

The value of k/ϕ so derived was in agreement with the corresponding values at lower Pt surface concentrations (Table III) indicating that surface diffusion of the Pt crystallites on the carbon surface was rate controlling. Since Pt black sintered rapidly at less than 200°C, this would imply that, for the sintering of supported Pt cata- lysts between 500 and 600°C, coalescence between crystallites under- going collision should be rapid and only crystallite migration need be considered as the rate determining step for gas-phase sintering. Wynblatt and Gjostein (23) have presented quantitative arguments to show that particle coalescence is rapid relative to particle migra- tion and therefore, the latter step should be the rate-controlling step in the migration-coalescence sequence. Also, large aggregates of coalescing Pt crystallites might be expected to appear in the electron micrographs of sintered catalysts (Figure 6) if coales- cence was the rate determining step, but such aggregates were not observed.

TABLE III.   RATE CONSTANTS FOR GAS-PHASE SINTERING OF Pt CRYSTAL- LITES ON CARBON AT 600°C - USING EQUATION 2, n = 4

Wt% Pt	ϕ $\left(\dfrac{cm^3\ Pt}{cm^2\ surface}\right)$ x 10^8	S_o (m^2/gm Pt)	k (m^2/gm)$^{-4}$hr^{-1} x 10^9	$\dfrac{k}{\phi}$
5	0.3	85	0.72	0.23
12	0.8	76	1.45	0.18
20	1.4	45	3.1	0.21

In contrast with the gas-phase sintering data, results for the liquid-phase sintering of Pt supported on graphitized carbon did not show any dependence of k on ϕ. This fact would suggest that the sintering mechanisms for supported Pt in a liquid- and gas-phase environment are different, based on conclusions reached in the previous section.

The surface area decrease by unsupported Pt blacks occurs in the same temperature range, 100-200°C, and with similar fractional surface area losses in both liquid- and gas-phase environments (Figure 1). Furthermore, the apparent activation energy for sintering of Pt black measured in a liquid environment, 25 kcal/mole (11), was close to that obtained by Khassan (9), 18 kcal/mole, in the gas phase. The liquid-phase sintering of supported Pt occurred in the same temperature range (100-200°C) as Pt black and moreover, with comparable apparent activation energy, 21 kcal/mole (14). In contrast, Pt supported on carbon required temperatures in excess of 600°C for significant growth of Pt crystallites in a hydrogen atmosphere and the apparent activation energy, 41 kcal/mole (32), was considerably higher. Since the apparent activation energies for the liquid-phase sintering of supported Pt on carbon and unsupported Pt black are comparable it is tempting to look for similarities in the sintering mechanism. For unsupported Pt black, it was concluded that mass transport in both gas- and liquid-phase environments occurred by the diffusion of Pt atoms on the Pt crystallite surfaces (9,10,11). The effect of liquids in reducing the temperature range for sintering of the supported Pt to the same range as that where the unsupported Pt black sinters suggests the possibility that the role of the liquid may lie in its ability to facilitate the transfer of the Pt adatom from the Pt surface to the carbon surface. In this case, surface diffusion of Pt atoms on the carbon surface (Ostwald ripening mechanism) would be the mode of mass transfer of Pt from smaller to larger Pt crystallites. Furthermore, consistent with the Ostwald ripening model for particle growth, the experimentally observed sintering rates for Pt supported on carbon in a liquid-phase environment were independent of the metal content.

In the presence of an electrolytic solution, the possibility exists for the diffusing species to be neutral adatoms or adatoms possessing an ionic change. The absence of any effect of potential on the sintering rate of supported Pt in 96 wt% H_3PO_4 (Figure 9) eliminates dissolution and re-deposition of Pt ions for the mechanism of Pt crystallite growth. This conclusion would follow from a consideration of the Pt/Pt^{++} equilibrium potential at 0.1V which would require a Pt ion concentration of the order 10^{-20} moles/l; a concentration too low to contribute to significant crystallite growth. Even if allowance was made for increases in solubility of small Pt crystallites as demanded by the Kelvin Equation, the Pt ion concentration would still be negligible. In addition, no measurable Pt loss from the supported catalyst was detected by chemical analysis before and after sintering in hot concentrated H_3PO_4, suggesting that mass transfer through the liquid-phase was not significant.

Flynn and Wanke (31) have pointed out that the high activation energy required for the removal of individual atoms from a crystallite is an argument against the Ostwald ripening model for sintering. For Pt, the heat of sublimation is 132 kcal/mole and the energy difference between an adatom on the substrate (alumina) and an adatom on a Pt crystallite is reported to be 100 kcal/mole in a reducing atmosphere (23). Nevertheless, there are some arguments which can be advanced to support the Ostwald ripening model for the sintering mechanism of Pt on carbon in a liquid environment. Geus (34) has pointed out that the presence of defect sites on the support surface can significantly increase the bonding strength of the adatom to the support and thereby reduce the activation energy required for an adatom transfer process. It is possible that the highly disordered surface of graphitized carbon black provides sites which enhance the removal of metal atoms from the crystallite. It is interesting to speculate on the influence of the liquid in facilitating the transfer of metal atoms from the crystallite to the support surface. Neikam and Vannice (35) investigated the phenomenon called hydrogen "spillover" in the Pt black/Zeolite/Perylene system and they postulated the "spillover" of adsorbed gases was due to the presence of perylene which acted as a bridge between the Pt and the support. The role of water in accelerating the reduction of tungsten trioxide containing Pt crystallites has been documented by Benson et al. (36). Water enhances the diffusion of the reducing species from the Pt crystallite to the tungsten trioxide surface. In a similar fashion, perhaps the liquid acts as a bridge for the transfer of Pt atoms from the crystallite to the carbon surface during the liquid-phase sintering of Pt on carbon.

The interpretations presented on the sintering of supported Pt are by no means free of conceptual difficulties. For example, the growth of Pt crystallites by crystallite migration and Pt adatom diffusion both occurring together in a supported catalyst is an interesting possibility which has not been considered here. Crystallite migration may be dominant for small crystallites and adatom diffusion playing a larger role with increasing radii (23). Unfortunately, the present theoretical models do not allow a prediction for the growth of crystallites by a mixture of both modes of sintering.

REFERENCES

1. H. Gruber, Anal. Chem. 34, 1828 (1962).
2. H. Gruber, J. Phys. Chem. 66, 48 (1962).
3. H. Maat & L. Moscou, Proc. 3rd Intl. Cong. on Catal.(W. Sachtler, G. Schuit & P. Zwietering, Eds.) p. 1277, North-Holland, 1965.

4. R. Herrmann, S. Adler, M. Goldstein & R. DeBaun, J. Phys. Chem. 65, 2189 (1961).
5. T. Hughes, R. Huston & R. Sieg, Ind. Eng. Chem. Proc. Res. Dev. 1, 96 (1962).
6. F. Huang & C. Li, Scripta Met. 7, 1239 (1973).
7. P. Wynblatt & N. Gjostein, Scripta Met. 7, 969 (1973).
8. D. McKee, J. Phys. Chem. 67, 841 (1963).
9. S. Khassan, S. Fedorkina, G. Emel'yanova & V. Lebedev, Zh. Fiz. Khim. SSSR 42, 2507 (1968).
10. K. Kinoshita, K. Routsis, J. Bett & C. Brooks, Electrochim. Acta 18, 953 (1973).
11. P. Stonehart & P. Zucks, Electrochim. Acta 17, 2333 (1972).
12. R. Thacker, Nature 212, 182 (1966).
13. P. Malachesky, C. Leung & H. Feng, U.S. Nat. Tech. Inform. Serv., AD Rep. 1972, No. 757714.
14. J. Bett, K. Kinoshita & P. Stonehart, to be published.
15. J. Connolly, R. Flannery & B. Meyers, J. Electrochem. Soc. 114, 241 (1967).
16. G. Kuczynski, Powder Metallurgy (W. Leszynski, Ed.) p. 11, Interscience, N.Y., 1961; Adv. Colloid Interface Sci. 3, 275 (1972).
17. E. Ruckenstein & B. Pulvermacher, J. Catal. 29, 224 (1973).
18. E. Ruckenstein & B. Pulvermacher, AIChE J. 19, 356 (1973).
19. W. Phillips, E. Desloge & J. Skofronick, J. Appl. Phys. 39, 3210 (1968).
20. M. Von Smoluchowski, Z. Phys. Chem. 92, 129 (1918).
21. B. Chakraverty, J. Phys. Chem. Solids 28, 2401 (1967).
22. W. Dunning, Particle Growth in Suspensions (A. Smith, Ed.) p. 1, Academic Press, London, 1973.
23. P. Wynblatt & N. Gjostein, Prog. Solid State Chem. 9 (1974).
24. P. Rebinder & E. Wenstrom, Acta Physicochim. 19, 36 (1944).
25. F. Bowden & J. Young, Research 3, 235 (1950).
26. L. Stolyarenko & A. Vasev, Elektrokhimiya 7, 1380 (1971).
27. D. Fornwalt & K. Kinoshita, Micron 4, 99 (1973).
28. G. Geach, Prog. Metal Phys. 4, 174 (1965).
29. A. Shaler, Trans. Amer. Inst. Min. Engrs. 185, 796 (1949).
30. E. Barrett, L. Joyner & P. Halenda, J. A.C.S. 73, 373 (1951).
31. P. Flynn & S. Wanke, J. Catal. 34, 390, 400 (1974).
32. J. Bett, K. Kinoshita & P. Stonehart, J. Catal. 35, 307 (1974).
33. J. Bett, K. Kinoshita, K. Routsis & P. Stonehart, J. Catal. 29, 160 (1973).
34. J. Geus, Chemisorption and Reactions on Metallic Films (J. Anderson, Ed.) Academic Press, New York, Vol. 1, p. 129, 1971.
35. W. Neikam & M. Vannice, J. Catal. 27, 207 (1972).
36. J. Benson, H. Kohn & M. Boudart, J. Catal. 5, 307 (1966).

SINTERING RETARDATION IN CATALYSTS

D. Lynn Johnson

Northwestern University

Evanston, Illinois 60201

INTRODUCTION

Dispersed particles have been observed to cause a reduced sintering rate in a number of materials (1-6). In particular, alumina promoted iron exhibits a retention of very high surface area in temperature and time regimes which would result in severe sintering of pure iron (1). This has been interpreted alternatively as an equilibrium fine particle size due to strain energy (1) and maintenance of particle size through a reduction in sintering kinetics (7). A third possibility, intermediate between these two, has been proposed recently (8) and will be discussed below.

The influence of dispersed oxide particles on sintering retardation of metals has been interpreted often as a dispersion hardening phenomenon with subsequent retardation of dislocation motion. Although there are undoubtedly circumstances in which this is true, there is also the possibility that dispersed particles pinned at the surface of the sintering particles could retard sintering by virtue of that pinning. Kuczynski and Lavendel (9) proposed a model for retardation of sintering by particles which are immobile as the surface moves normal to itself during sintering, and concluded that nonwetted particles would retard sintering more than wetted particles. In their model the particle had to be covered by the metal if the interface was to advance. A model has been proposed which describes sintering retardation by the action of mobile pinned particles (10), and will be described below.

STRAIN ENERGY MODEL

Schultz (1) proposed that tiny particles of $FeAl_2O_4$ are uni-

formly dispersed in the iron matrix in alumina-promoted iron catalyst, that each particle existing in an iron particle at a distance
greater than α from the free surface contributes an amount of
strain energy, E_d, to the system, and that the strain energy of
those particles within α of the surface is relaxed. Although his
treatment beyond this point is incorrect, this concept can give
rise to retarded sintering. If there are N_V dispersoid particles
per unit volume of solid catalyst, the fraction of those that
contribute strain energy is $1 - \alpha A/V$ where A/V is the ratio of the
surface area to the volume of the catalyst particles; this assumes
α is much less than the catalyst particle radius. The strain
energy per unit volume of catalyst is thus (8)

$$E_S = N_V E_d (1 - \alpha A/V) \tag{1}$$

The surface energy per unit volume of catalyst is $\gamma A/V$, giving, for
the total energy per unit volume,

$$E_T = N_V E_d + A/V (\gamma - \alpha N_V E_d) \tag{2}$$

According to Eq. (2), if $\alpha N_V E_d > \gamma$, the total energy would be
reduced by an increase in surface area of the catalyst particles.
While such an increase is not likely to occur during sintering,
since the area would be stabilized during catalyst preparation,
the area cannot spontaneously decrease as long as this inequality
exists. If the strain energy is reduced with time, either by migration of dispersed particles to the surface under the strain
energy gradients that exist, or by Ostwald ripening, the strain
energy term may become less in magnitude than γ, after which total
energy can be reduced by a reduction in the surface area, which will
occur spontaneously. Any further maintenance of high surface area
after this will depend upon kinetic factors.

PARTICLE DRAG MODEL

In this model (10) it is assumed that dispersed particles,
which are small in comparison to the catalyst particles, are uniformly dispersed upon the surface of the latter. Figure 1 shows
the details of the assumed geometry of the dispersed particle on a
substrate, the surface of which is moving in the upward direction.
It is assumed that all surfaces of the dispersed particle are segments of spheres, all surface and interfacial energies are isotropic, and all interfacial angles are always at their equilibrium
value. The force exerted by the surface on the particle is given
by

$$f = 2\pi a \gamma_1 \sin \alpha \tag{3}$$

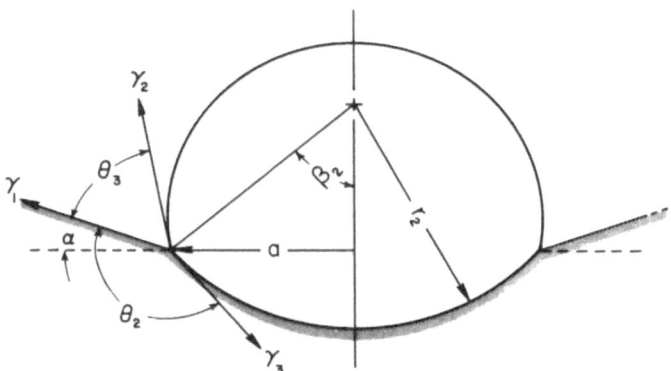

Fig. 1. Inert particle pinned to an advancing interface. The surface energies of the substrate and particle are γ_1 and γ_2, respectively; the interfacial energy between these is γ_3.

where a and α are defined in Figure 1, and γ_1 is the surface tension of the substrate material. A dimensionless force can be
defined as follows:

$$F \equiv \frac{f}{\gamma_1 a_o} = \frac{2\pi r_2}{a_o} \sin(\theta_2 - \alpha) \sin \alpha \qquad (4)$$

where r_2 and θ_2 are defined in Figure 1, a_o is the value of a for
$\alpha = 0$, and the particle volume is assumed to remain constant.

The velocity of the particle, and therefore the velocity of
the surface pinned by the particle, can be related to the rate of
diffusion of catalyst material in the interface between the particle and the catalyst. It is assumed that the driving force for
this diffusion is proportional to the force exerted by the surface
on the particle and inversely proportional to the average diffusion
distance, i.e., the distance from the line of contact at the par-
ticle-catalyst-vapor intersection and the bottom of the particle
depicted in Fig. 1. Under these assumptions, the dimensionless
velocity can be defined as follows:

$$U \equiv \frac{kTa_o^3}{\gamma\Omega\delta D_i} u = \frac{8\pi a_o^3 \sin(\theta_2 - \alpha) \sin \alpha}{r_2^3 \beta_2 \sin^3 \beta_2} \qquad (5)$$

where u is the velocity, δ is the thickness of the region of enhanced diffusion at the particle-catalyst interface, D_i is the

Table I

Particle Drag Model Mobility and Force Parameters

θ_2	θ_3	M at F = +1	M at F = -1	+ F_{max}	M at + F_{max}
60°	150°	0.331	0.347	1.738	0.305
150°	60°	0.822	2.028	6.952	0.353
110°	110°	0.936	1.074	3.628	1.004
150°	150°	1.204	1.265	4.055	1.633

interfacial diffusion coefficient for catalyst material, Ω is the atomic volume, and $\beta < \pi/2$. For $\beta_2 \geq \pi/2$, $\sin \beta_2$ is replaced by unity in this equation.

A dimensionless mobility can be defined as

$$M \equiv \frac{U}{F} = \frac{kTa_o^4}{\Omega \delta D_i} \cdot \frac{u}{f} \qquad (6)$$

Table I lists mobility and force parameters for this model for a nearly buried particle ($\theta_2 = 60°$, $\theta_3 = 150°$), a particle which is more nearly resting on the surface ($\theta_2 = 150°$, $\theta_3 = 60°$), a thick symmetrical lens ($\theta_2 = \theta_3 = 110°$), and a thin symmetrical lens ($\theta_2 = \theta_3 = 150°$). A number of conclusions may be drawn from the data of this table. Non-wetted particles can restrain greater forces than wetted particles for advancing surfaces, in agreement with Kuczynski and Lavendel's model. However, the mobility of non-wetted particles is very high for receding surfaces; the nearly wetted particle is far superior in restraining receding surfaces. Since, during sintering, some interfaces will be advancing and others will be receding, it is anticipated that more nearly wetted particles will be more effective inhibitors of sintering.

APPLICATIONS OF THE PARTICLE DRAG MODEL

Smoothing of a Sinusoidal Surface

The amplitude of a sinusoidal wave imposed upon a single crystal surface will spontaneously decrease at elevated temperatures under the action of surface tension forces. The particle drag model can be applied in a straightforward manner to the

surface smoothing models of Mullins (11). If it is assumed that
surface diffusion is the only transport mechanism which causes a
decrease in the amplitude of the sinusoidal surface, the amplitude
decay rate is given by

$$\dot{A} = \frac{\dot{A}_o}{1 + \frac{Na_o^4\omega^2}{R_D M}} \tag{7}$$

where $\dot{A}_o = -A\gamma\Omega^2\nu D_s\omega^4/kt$, the amplitude decay rate without particles, ν is the surface concentration of mobile atoms, D_s is the
surface diffusion coefficient of the catalyst, N is the surface
concentration of dispersed particles, ω is the angular frequency
of the sinusoidal surface, and $R_D \equiv \delta D_i/\Omega\nu D_s$. Besides the obvious
relationships shown in this equation, note that the volume of dis-
persed phase per unit area is proportional to Na_o^3, so that the
second term in the denominator is proportional to the dispersed
particle size for any given system at constant volume of dispersed
phase. Thus, spontaneous coarsening of the dispersed particles
would produce a decrease in $\dot{A}/\dot{A}_o$ with time.

Sintering of Spheres

The application of the particle drag model to the sintering
of spheres is less straightforward than the foregoing, but an
order of magnitude estimation of its expected importance can be
obtained. On the assumption that the dispersed particles are
small compared with the minimum radius of curvature of the neck
surface between a pair of spherical particles, and the further
assumption that the rate of approach of the particle centers, $\dot{y}$,
is governed by a combination of volume and grain boundary diffusion, the following equation can be obtained:

$$\dot{y} = \frac{\dot{y}_o}{1 + \frac{4Na_o^4 (D_v + 2\pi XbD_b/aA_n)}{XMa\delta D_i}} \tag{8}$$

where A_n and X are the neck surface area and neck radius normalized to the sphere radius, b is the width of the region of enhanced diffusion at the grain boundary, D_b is the grain boundary
diffusion coefficient, a is the sphere radius, and $\dot{y}_o$ is the
shrinkage rate without particles, given by (12)

$$\dot{y}_o = \frac{2\gamma\Omega(A_n D_v + 2\pi X b D_b/a)(X + R)}{\pi k T X^4 R a^3} \tag{9}$$

where R is the minimum radius of curvature of the neck surface normalized to the sphere radius. Putting realistic numbers into Eq. (8) indicates that particle drag would have negligible influence under the conditions assumed for this equation. The observed shrinkage rate retardation by dispersed particles, reported in the literature, must be caused by other factors.

If sintering of spheres is caused by surface diffusion, the neck growth rate with dispersed particles is given by

$$\dot{X} = \frac{\dot{X}_o}{1 + \dfrac{2\pi N a_o^4 X}{M R_D A_n R a^2}} \tag{10}$$

where $\dot{X}_o$ is the normalized neck growth rate without particles, given by (12)

$$\dot{X}_o = \frac{2\pi\gamma\Omega^2 \nu D_s (X - R + 2XR)}{k T a^4 R^2 A_n} \tag{11}$$

Here, reasonable values for the various parameters indicate that dispersed particles could significantly reduce neck growth due to surface diffusion.

CONCLUSIONS

Dispersed inert particles tend to impede the motion of free surfaces under capillarity induced morphological changes, such as sintering and surface smoothing. A number of predictions can be made on the basis of the models presented above. For a given dispersoid particle size, a particle which is more buried within the sintering material will have a lower mobility but also a lower maximum restraining force for an advancing surface than one which is more nearly resting on the surface. The latter particles offer little restraint to a receding surface. Since sintering involves both advancing and receding surfaces, it is likely that the more buried particles will be more effective in overall sintering retardation. For a given volume of dispersoid particles per unit area, larger, more widely separated, particles will be more effective in restraining surface motion than smaller, more closely spaced, particles. Surface dispersoids are potentially important in inhibiting surface diffusion sintering, but will be ineffective in inhibiting grain boundary and volume diffusion sintering by surface drag.

ACKNOWLEDGEMENTS

Supported by the Advanced Research Projects Agency through the Materials Research Center, Northwestern University.

REFERENCES

1. J. M. Schultz, J. Cat. 27, 64 (1972).
2. L. L. Seigle and A. L. Pranatis, Metal Progress 68, 86 (1955).
3. M. H. Tikkanen, B. O. Rosell, and O. Wiberg, Powder Met. No. 10, 49 (1962).
4. J. Brett and L. L. Seigle, Acta Met. 14, 575 (1966).
5. A. R. Hingorany, F. V. Lenel, and G. S. Ansell, Kinetics of Reactions in Ionic Systems, T. J. Gray and V. D. Frechette, eds. (Plenum Press, N. Y. 1969).
6. V. N. Antsiferov and L. A. Demidova, Nauch, Tr., Perm. Politekh. Inst. 80, 76 (1970).
7. N. Louat and J. Galligan, Scripta Met. 8, 197 (1974).
8. D. Lynn Johnson, Scripta Met. 8, 905 (1974).
9. G. C. Kuczynski and H. W. Lavendel, Internat. J. of Powder Met. 5, 19 (1969).
10. D. Lynn Johnson, to be published.
11. W. W. Mullins, J. Appl. Phys. 30, 77 (1959).
12. D. Lynn Johnson, J. Appl. Phys. 40, 192 (1969).

SINTERING OF SUPPORTED METALS

James C. Schlatter

General Motors Research Laboratories

Warren, Michigan 48090

ABSTRACT

Sintering can alter both the physical nature and the catalytic behavior of supported metals, and such changes have important practical implications. This paper includes a general discussion of experimental techniques, laboratory observations, and mathematical models used in examining catalyst sintering. Examples are taken primarily from literature reports on supported platinum. The performance of sintered catalysts will also be discussed, and recent results from our laboratories will indicate the effect of sintering on automotive catalytic converter performance.

INTRODUCTION

Small metal particles in contact with one another will tend to fuse at elevated temperatures. This phenomenon has important practical benefits in many metal fabrication processes, and ceramists and metallurgists use the word "sintering" to describe this coalescence of powder particles (1). One result of the coalescence is a decrease in the metal surface area. Since heterogeneous catalysis is a surface process, sintering is usually detrimental to the performance of metal catalysts.

To maximize the effectiveness of solid catalysts, then, there is a great incentive for preparing and maintaining the metal in the form of very small particles. The most desirable particle size is generally on the order of 1-10 nm; crystallites of such dimensions have a significant fraction of their atoms at the surface and thus available for interaction with reactant molecules. The relationship

141

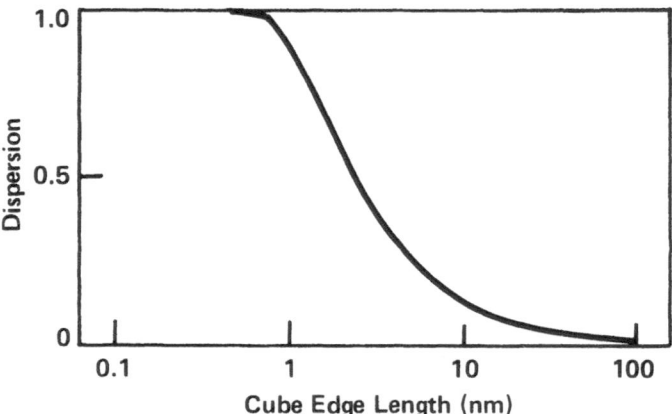

Fig. 1. Dispersion vs. crystallite size for a hypothetical simple crystal (from ref. 2).

between dispersion (ratio of surface atoms to total metal atoms in the sample) and crystallite size is shown in Fig. 1 for a cube-shaped crystal with a simple cubic lattice of 0.25 nm spacing (2).

Metal particles of less than about 20 nm diameter have very low thermal stability when touching one another. Platinum black, for example, with an initial average particle diameter of ∿20 nm, sinters at room temperature if exposed to hydrogen following an oxygen exposure (3,4). Palladium black (7 nm diameter) has also been observed to sinter in hydrogen at temperatures below 100°C (5).

The need for highly dispersed catalytic metals, combined with the impracticality of using metal powders, has led to the widespread use of porous, high surface area ceramic materials to support and retain the minute particles. Alumina, silica, and silica-alumina with areas on the order of 100 m^2/g are common catalyst supports, and spheres or cylinders of 3 mm diameter are typical geometries. By providing stability for highly dispersed metallic systems, the support markedly decreases the total amount of metal required to obtain a given level of catalytic activity. For example, fresh platinum black has approximately 10 m^2 of platinum surface per gram, whereas supported platinum can exceed 200 m^2 of exposed platinum per gram of metal. In addition, the supported metal is stable up to approximately 500°C. Consequently, metal loadings on the support can usually be less than 1 wt % and often less than 0.1 wt %. The support can affect catalyst properties other than metal dispersion [e.g., reaction selectivity (6), reactant transport (7), and poisoning (8)], but these are not subjects for discussion here.

The remainder of this review will describe the characterization, preparation, and thermal stability of dilute, highly dispersed, supported metal systems which are typical of industrial catalysts.

CHARACTERIZATION

The small size and low concentration of the active component (metal) on the support present a formidable challenge to the experimenter trying to describe a supported catalyst. This area has been the subject of considerable attention in the petroleum and petrochemical industries as well as in academic laboratories and, more recently, in the automobile industry; consequently, many experimental techniques have been brought to bear on the problem. Papers on the physical nature and chemical behavior of supported metals have been reviewed several times, most recently by Cinneide and Clarke (9), Sinfelt (2), and Whyte (10); but it is worthwhile to describe briefly here the experimental tools which are important in studies of catalyst durability.

Techniques for characterizing the state of the metal on the support are primarily concerned with determining the metal particle size or the metal surface area; either measurement describes the degree of dispersion of the metal. In general, physical techniques indicate particle size; chemisorption methods yield a number related to the surface area. If some simplifying assumptions are made (11), the particle size measurement can be converted to the corresponding metal area and vice versa. This calculation allows comparisons to be made between results from the various experimental tools.

Perhaps the most commonly used physical technique for determining the metal particle size on supported catalysts is x-ray line broadening (12). Basically, the method relies on the fact that x-ray diffraction lines are broadened when the particles are smaller than about 100 nm. The width of the lines can be converted into an average particle diameter for particles larger than about 4 nm; smaller particles give diffraction lines too broad for accurate analysis. Nevertheless, the range from 4 to 100 nm includes many of the supported metals of interest, and x-ray measurements are relatively simple to make on a routine basis. With somewhat more elaborate sample and instrument preparations, the metal crystallinity within a small area ($\sim$100 μm diam) of the catalyst surface can be examined for details not seen in the usual line broadening experiment (13).

Another physical technique in common usage for catalyst characterization is the electron microscope. Although the required sample preparation is sometimes rather involved (14), the image obtained in favorable cases gives detailed information not only about the average metal particle size but also about the distribution of sizes

in the catalyst. Unlike x-ray diffraction, electron microscopy is
not limited to particles of >4 nm diameter; in at least one case
single atoms have been imaged (15). However, the fact that discrete
particles are "seen" in the microscope makes data analysis time consuming and tedious because many particles must be counted and sized
to ensure that a statistically representative sample is used to characterize the catalyst. Uncertainties have been reported both for
high (16) and low (17) dispersions of platinum on alumina.

A novel use of a scanning electron microscope for catalyst studies has been developed by Baker and co-workers in England (18, 19).
The technique involves a special cell that admits a controlled atmosphere to the sample during viewing; the dynamics of the surface interactions are then recorded on videotape for later analysis. The
resolution is on the order of ten nm (20).

The use of x-rays for studying supported metals is not limited
to line broadening measurements. Small-angle x-ray scattering
that results from heterogeneities in electron density within the
catalyst (i.e., differences between metal and support electronic
structures) can be analyzed to yield values for not only the mean
values but also the variance of the metal particle size distribution.
The marked interferences from the pores in typical supports have been
minimized either by filling the pores with a liquid of electron density similar to that of the solid support (17,21) or eliminating
the pore structure by pressing the catalyst at 10 GPa (10^5 atm) (22).

X-ray scattering has also been used to determine nearest neighbor distances in supported platinum crystallites (23). A recent review covers in detail the use of this method for obtaining radial
distributions (24).

Another type of x-ray analysis for determining radial distributions involves a high resolution measurement of x-ray absorption.
Fourier analysis of the fine structure of the absorption curve near
an element's characteristic absorption edge is used to determine the
atomic environment of the absorbing species. Such a technique has
been used by Lytle and Sayers to study a base metal automotive exhaust catalyst (25).

All of these physical methods have been discussed in terms of
their "ideal" behavior, disregarding the experimental difficulties
and inherent limitations to their use. Perhaps the most significant
limitation is sensitivity; this is a problem with any of the afore-
mentioned techniques. At the lowest levels of interest for automotive catalysts (0.1 wt % metal or less), most of the instruments cannot distinguish between the catalyst and the bare support. With the
electron microscope, it becomes difficult to find the metal particles
on the support, and obtaining statistically meaningful results is

laborious at best. The shallow depth of field at high resolutions
can result in uncertainties in the interpretation of electron photomicrographs of particles smaller than about 2.5 nm (16). Nevertheless, the potential value of the particle characterization dictates
that each technique be evaluated for the particular case of interest.

A feature common to all of these physical techniques is that
they provide only an indirect measure of the catalytic surface area.
They respond to metal particle volume, cross section, or electron
density, but not to area. Of course, a knowledge (or assumption) of
particle shape and size distribution allows one to calculate the
area from physical measurements, but gas adsorption is often a much
simpler and more direct route for area determination. In situations
where both physical and adsorption measurements have been compared
(11,14,15,17,23,26,27), the results showed good agreement in spite
of the necessarily simplistic assumptions that are commonly made.

The total surface area (mostly support area) of a catalyst is
most often measured by the method of Brunauer, Emmett, and Teller
(28). The so-called BET method results from the successful modeling
of the condensation process when a gas (typically N_2, Ar, or Kr) is
physically adsorbed on a solid surface, usually at liquid nitrogen
temperature (77K). Such measurements are routinely done for samples
with areas from 500 m^2/g down to a few hundred cm^2/g.

At the metal loadings of practical interest, however, the total
surface area of the catalyst bears little relationship to the actual
catalytic area, i.e., the active metal surface. Fortunately, this
important but relatively small area (generally less than 1% of the
total sample area) is accessible to adsorption techniques. In order
to minimize interference from the support, an adsorbate is chosen
which will interact specifically with the component of interest. For
supported noble metals the common choice is H_2, although CO and O_2
are used also. At room temperature and above, there is little physical adsorption; thus the uptake of gas is almost solely the result
of chemisorption on the metal. From this adsorption one can calculate the exposed metal area and the average particle size (11). Such
measurements, and their uncertainties, have been discussed elsewhere
(10,29-31).

Although adsorption measurements reveal little information about
the range of particle sizes, there is general agreement that chemisorption of the proper adsorbate does reflect the amount of accessible catalytic surface. In addition, the technique is not limited to
particles of a certain size, nor are sensitivity limitations as severe as with physical methods. Because adsorption of the reactants
is an important step in heterogeneous catalysis, chemisorption is
particularly desirable as a tool for characterizing catalytic materials.

Other physical techniques are useful in describing the nature of supported catalysts, although they are not intended to give information about metal particle size per se. Electron probe microanalysis uses an electron beam to excite x-radiation characteristic of the elements in the beam; thus an elemental analysis can be obtained for a surface region as small as 1 μm in diameter (32). This is vital information in characterizing the distribution of the active component [and possible contaminants (33)] within a supported catalyst, and analysis by microprobe is a relatively easy way to get the required data. Just as for other physical methods, sensitivity becomes a limiting factor when the metal loading is low.

The actual chemical state of the catalytic component is an unresolved question in many instances, and spectroscopic tools are currently emerging as potential probes for determining the chemical environment at the catalyst surface. Auger electron spectroscopy (34, 35,36), Mössbauer spectroscopy (37,38,39) and x-ray photoelectron spectroscopy (36,40) are examples of modern techniques which have recently been applied to catalyst studies.

PREPARATION

Having considered the experimental tools for describing the state of an active metal on a support, we turn our attention to the preparation of the catalyst itself. There are many patents and papers on catalyst preparation and formulation, and they could justify an extensive review in their own right. This section will not be an exhaustive description of preparative techniques; rather it will describe in general terms some of the procedures and important variables involved.

For automotive (and most other) catalysts, there are several reasons for seeking to understand and to optimize the preparative procedures. An obvious one is the fact that expensive and relatively rare metals (platinum and palladium) are being used in the catalyst formulations. It thus becomes especially worthwhile to disperse these metals as completely as possible on the support in order to minimize the quantities required. Even disregarding economics, it is important for converter performance that the catalytic activity (and thus metal area) for a given volume of catalyst be high in order to minimize converter size requirements. Finally, preparations are sought that will enhance the catalyst's ability to withstand the rigors of the exhaust environment for extended periods of time. Among these stability requirements is resistance to sintering. The more stable the catalyst, the less the converter needs to be over-sized to compensate for deterioration in efficiency.

Three basic procedures have been used for depositing metals and metal oxides on oxide supports -- impregnation, ion exchange, and precipitation. The first is the most common method, and the basic technique is quite simple (41,42). The support material, in the form of either powder, granules, or pellets, is exposed to a solution (usually aqueous) containing the desired active component in a soluble form. The solution penetrates the porous structure of the support and leaves the dissolved salts behind when the catalyst is subsequently dried. Multiple impregnations are sometimes used when a heavy metal loading or a combination of metals is desired on the support. Further treatments, to be discussed shortly, convert the deposited compound(s) to the catalytically active species. As an example, platinum supported on alumina is often made by wetting the alumina with an aqueous solution of chloroplatinic acid (H_2PtCl_6), drying, and reducing the platinum to its metallic state by heating the sample in hydrogen. The platinum content of the catalyst is the same as the original amount in the solution (unless excess liquid is discarded).

Many variables in the impregnation procedure can affect the resulting catalyst. In particular, the concentration, exposure time, and adsorption properties of the solute are critical factors governing the distribution and amount of metal on the support (43,44). This is because the support acts as a chromatographic column when the impregnating solution first enters the pores. The activating salt is quickly adsorbed at the pore openings, and some time is required to enable the salt to diffuse into the interior of the pellet and eventually distribute itself evenly throughout the support. If the support is removed and dried after only a short exposure to the solution, the active component is likely to reside only at the exterior of the support particles.

The adsorption characteristics of the impregnating solution can be adjusted by including a second solute that competes with the activating solute for adsorption sites on the support (44,45). In this way the active species is forced to penetrate further into the pores to locate empty surface sites. By exploiting the adsorption characteristics of certain organic acids in competition with dissolved metallic salts, Michalko was able to prepare spherical alumina beads (3mm diam) with a variety of metal distributions, including a thin shell of metal completely imbedded below the surface of the support beads (46). Theoretically, for positive-order reactions influenced by mass transfer resistances, the active component is used most efficiently when deposited selectively at the exterior of the pellet (47). This is not true for negative-order reactions such as carbon monoxide oxidation, however; in those cases the optimal metal distribution may be deeper into the support pellet (49,50).

Ion exchange provides a different approach to the deposition of catalytic materials on hydrous oxide supports. Whereas impregnation often involves the metal as an anion (e.g., platinum in H_2PtCl_6 is complexed as $[PtCl_6]^{-2}$), the ion exchange process requires that the metal be in the cation ($[Pt(NH_3)_4]Cl_2$, for example) (48). In such a form, the metal-containing ion can exchange with hydrogen ions present on the surface of the support. The exchange is enhanced by raising the pH of the solution; so the metal must generally be tied up as a complex ion to avoid precipitation of the metal oxide or hydroxide. Ammine complexes are attractive because the non-metallic residue of the catalyst activation procedure is volatile. The final metal content of the catalyst is determined either by direct analysis or by analysis of the solution before and after ion exchange.

Platinum on silica catalysts made via ion exchange have better initial dispersions and resistance to sintering than similar catalysts made by impregnation (27,48). Alumina has a much lower capacity than silica for ion exchange (48), but the improved properties of ion exchanged metals on silica are claimed for alumina-supported systems as well (51). Because of its simplicity, however, the impregnation technique is the most widely used procedure for making supported metal catalysts.

A third method, precipitation, is commonly used to form hydrous oxide catalysts such as titania, silica-alumina, molybdena-alumina, and chromia-alumina (41). It has been used on occasion to make platinum on alumina, too (26,52). The general procedure is to add a precipitating agent to a solution which contains salts of the desired component(s). For supported platinum catalysts, chloroplatinic acid mixed with alumina sol can be gelled by adding ammonia. As in ion exchange preparation, there is an advantage to the fact that ammonia forms volatile products during heat treatments. There is some indication (26) that gelation and impregnation lead to different types of platinum on the alumina surface. Unless specified otherwise, statements in this paper regarding catalyst characteristics will refer to alumina-supported metals made by impregnation. Most catalysts are made by this technique.

For any type of support, the final form of the metal particles is influenced by more than just the deposition procedure. Both the activation procedure and the subsequent use of the catalyst can have considerable influence on its characteristics. Following application of the metal salt to the support, the wet solid must be processed further before being used as a catalyst. Although the pretreatment can affect the metal distribution as a result of different solvent evaporation rates in different-sized pores (43), the primary effect of initial heat treatments is manifested in the average crystallite size of the metal on the support.

Several pretreatment parameters have been shown to influence the resultant catalyst. The purpose of the pretreatment is two-fold -- to activate and stabilize the catalyst. The basic procedure for supported noble metals usually involves drying at a relatively low temperature (100°C - 200°C), calcining in air at an elevated temperature (400°C - 700°C), and reducing in hydrogen at the high temperature, although the calcination step is sometimes omitted. Drying removes much (or most) of the solvent from the catalyst and thus decreases the mobility of the metallic species. Calcining like-wise aids in fixing the active component, often by decomposing the deposited species. The high temperature firing also stabilizes the support; so it is sometimes done prior to impregnation (42). For maximum catalyst stability in use, the temperature is chosen to ex-ceed the normal operating temperature of the process. The noble metals are generally most effective in their metallic state; so the reduction step completes the catalyst activation.

All phases of the pretreatment affect the final catalyst, and the presence of water vapor in the early stages seems to have an especially strong influence. Dorling et al. (27) made an extensive study of the effects of pretreatment conditions on Pt/SiO_2 catalysts and recommended that samples be dried very thoroughly prior to the reduction step. Calcination above 120°C before reduction led to an increase in metal particle size in the final catalyst; so the firing in air was usually omitted. The authors noted, however, that Pt/Al_2O_3 samples were not as susceptible to such damage during calcina-tion, perhaps because the platinum salt (H_2PtCl_6) is more strongly adsorbed on Al_2O_3 than on SiO_2. Mills et al. (53) have shown that prolonged (>100 hr) heating of an unreduced Pt/Al_2O_3 catalyst in air at 450°C decreases its capacity for H_2 chemisorption (after reduc-tion) by about 60%. They also showed that water vapor (or oxygen) in the hydrogen stream used for reduction is detrimental to the per-formance of the final catalyst. Thus the susceptibility of supported platinum catalysts to damage from water vapor is related to the state of the platinum; once reduced, the metal is much more stable toward a wet gas. This is of obvious importance for automotive applications, since the incoming exhaust typically contains at least 10 mol % water vapor.

There is, then, a balance to be sought in the catalyst prepara-tion procedure; the drying must be of sufficient duration and tem-perature to eliminate water vapor from the system before reduction, yet not so severe as to decrease the ultimate metal area of the cata-lyst. The subtlety of this balance can be noted in reports of labor-atory preparation procedures. Dorling et al. (27) saw effects of H_2 flow rate on the rate of reduction. Corolleur et al. (54) found that samples reduced at 200°C in 200 cc/min H_2 had a mean metal par-ticle diameter of 6 nm compared to 20 nm for samples reduced at 50°C in 1 cc/min H_2. These authors also noted an improvement in metal

dispersion when a N_2 stream was used to dry the catalyst before reduction. Freel (55) obtained improved reproducibility when he heated
his catalysts slowly (4°C/min) in H_2 to the reduction temperature of
500°C. A dependence of catalyst characteristics upon the batch size
being treated (56,57) has even been noticed. All of these observations are most likely related to the presence of trace amounts of
water evolving from the support which can affect the mobility of the
metallic species during the pretreatment.

In summary, there is a great variety of procedures for making
supported catalysts, and the activating pretreatment that follows
deposition of the catalytic component on the support can be a critical step. Slight changes in technique may cause important altera-
tions in the resulting catalyst; so the procedure must be carefully
optimized for each formulation and application. For emission control
catalysts, this optimization process is an area of continuing research, and it remains to be seen whether the catalyst that is initially (i.e., when freshly installed) optimal remains so after the
prescribed 50,000 miles of driving. Thus research on catalyst preparation leads naturally into research on catalyst durability; the
latter presents some of the most persistent problems in catalysis today. One such problem, high temperature durability, is the subject
of the remainder of this review.

SINTERING -- EXPERIMENTAL OBSERVATIONS

The industrial importance of supported metal catalysts has resulted in several studies of sintering in these systems. It should
be emphasized that, to students of catalysis, "sintering" denotes
the overall process whereby catalytic surface area is lost as a result of high temperature exposure. Thus, sintering may include the
coalescence of small metal particles, but it also must include a
mechanism for transporting the dilute metal throughout the support.
In practice, both the support and the active component are subject
to thermal damage by sintering, but the former is generally less
susceptible than the latter (58,59).

Catalyst composition is an obvious factor in thermal stability.
Whereas unsupported platinum black has a delicate surface that sinters rapidly and severely even under relatively mild conditions (4,
60,61), noble metals dispersed on refractory oxide supports have
been observed to retain a significant fraction of their initial exposed surface even after two hours at 780°C (59). Most supported
catalysts cannot withstand extended exposure to such a temperature
without damage, though.

Once the catalyst is made, sintering in use is governed predominantly by temperature, time, and atmosphere. Herrmann et al. (62)

reported that, not surprisingly, the metal surface area in a platinum/γ-alumina catalyst decreases with increasing temperature and time in nitrogen. For samples of 0.375 and 0.774 wt % platinum, the rate of decrease in area was second-order in the area remaining, that is,

$$\frac{dS}{dt} = -k \, S^2. \qquad (1)$$

S is the exposed metal area (determined by hydrogen chemisorption), t is time, and k is the second-order rate constant for sintering. Experiments at 564°C, 594°C, and 625°C yielded an activation energy of about 70 kcal/mol for the sintering process.

Maat and Moscou (59) treated 0.6% platinnum/γ-alumina in air at 780°C and observed the same second-order sintering kinetics; they did not report an activation energy. Kirklin and Whyte (21) sintered 0.6% Pt/η-Al$_2$O$_3$ in a hydrogen atmosphere; and although these authors did not determine the sintering order, the fact that their data (taken from Table III of reference 21) do not fit a second-order equation may mean that a completely different process determines the sintering rate in a reducing atmosphere. The reported activation energy (8.3 kcal/mol) is much lower than that measured for sintering in N$_2$.

Somorjai et al. (22) had earlier reached the same conclusion after sintering 5% Pt/η-Al$_2$O$_3$ in reducing or oxidizing atmospheres. The growth of platinum particles is faster under oxidizing than under reducing conditions (22,33,53), and Somorjai found that the apparent activation energy for sintering was considerably lower in a reducing atmosphere (14-27 kcal/mol) than in an oxidizing atmosphere (44-55 kcal/mol). Area losses estimated from particle sizes given in (22) do not follow second-order kinetics in either test atmosphere.

Other sintering data for Pt/Al$_2$O$_3$ heated in hydrogen have shown a linear relationship between the logarithm of the amount of CO chemisorbed (i.e., platinum area), and the logarithm of the heating time (63,64). The reducing atmosphere again was observed to give only a slight temperature dependence (63). Among the papers cited, none shows a significant effect of metal loading on sintering characteristics (22,62,64); however, other authors have noted an increased rate of area loss when the metal loading was increased (65,66).

The wide range of reported rates and area dependencies indicates that sintering is a complex process that is subject to influences of subtle changes in the environment. One such change is the rate of cooling from the sintering temperature (61). Rapid cooling was observed to preserve a high metal surface area that apparently existed

as a result of dissociation and mobility of metal particles at elevated temperatures. Slow cooling permitted the metal particles to
coalesce and lose exposed area.

Other authors have reported structural changes in the metal
crystallite surface upon heating, but not a decrease in area. Workers in Japan have noted distinct effects of the atmosphere on activity loss (for hydrogenation) of 6.62 wt % palladium supported on
silica-alumina (67). High temperature in oxygen caused the metal
particle size to increase, whereas in hydrogen the only change was
an annealing of lattice defects with a resultant loss in activity
but not surface area. Certain multivalent cations (Al^{3+}, Th^{4+}, La^{3+})
added to the supported palladium appeared to prevent sintering in
oxygen (68).

Comparisons of sintering in hydrogen and oxygen have been made
in our laboratories using a 0.44 wt % palladium on alumina catalyst
(69). In those experiments, the chemisorption of hydrogen on an
oxygen-covered surface was used as a measure of the metal area (70).
After 4 hr at 650°C in oxygen, the hydrogen uptake was virtually
unchanged from that of the fresh catalyst, but a similar treatment
in hydrogen decreased the uptake by 75%. The most intriguing aspect
of this work was that a high temperature exposure to oxygen could re-
store the chemisorption capacity of the hydrogen-sintered sample;
thus the apparent sintering of palladium in hydrogen was completely
reversible. These results still await a complete explanation.

The atmosphere in an automotive converter has also been observed
to cause structural changes in the surface of platinum supported on
alumina (13). Perhaps more significantly, alternating oxidizing and
reducing conditions and the presence of water vapor promote sintering
at relatively low (<550°C) temperatures (53,71).

SINTERING -- EFFECTS ON CATALYST PERFORMANCE

It stands to reason that the catalytic activity per unit weight
of metal will decrease with sintering because the fraction of atoms
at the surface decreases. In most instances this is the primary result of metal particle growth, and it is the primary reason that
sintering is of concern to catalyst users. Its specific effect on
the efficiency of an automotive converter will be discussed below.

Changes in metal particle size can also have a more subtle influence on certain reactions; in those, the catalytic activity per
unit surface area of metal changes with metal dispersion. As Boudart and co-workers have demonstrated, such situations offer especially intriguing opportunities for research because they provide
systems in which catalytic activity might be correlated with specific

surface structures (72). Manogue and Katzer (73) have pointed out
that decreasing the crystallite size can affect the specific activity
of a catalyst by 1) changing the abundance of particular surface sites
especially favorable for a certain reaction, 2) changing the sensitivity of the catalyst to its environment, or 3) changing the inherent properties of the metal itself as a result of having so few
"bulk" atoms. The surface structure (e.g., relative number of edge
and surface atoms) of metal crystallites is expected to change most
dramatically as the particles grow from one to five nm in diameter
(74); so supported metals are uniquely suited for studying the effects of such changes. Published reports showing the response of
various reactions to changes in metal dispersion have been reviewed
by several authors (20,21,75-77), and activity measurements will
continue to be sensitive probes for geometric and electronic features of catalytic surfaces.

In contrast, a catalytic converter for oxidizing residual pollutants in auto exhaust is rather insensitive to changes in the catalyst. One reason for this lack of sensitivity is that fresh noble
metal automotive catalysts typically have an excess of metal surface
available. This excess, perhaps necessary to speed warm-up and increase lifetime, is superfluous under steady-state conditions when
the conversion is limited by mass transfer rather than reaction
rates. This was demonstrated by Barnes and Klimisch (78), who found
that fresh samples of platinum or palladium varying from 0.05 to 0.3
wt % on 3 mm alumina spheres gave nearly equivalent emission control.

Nevertheless, severe sintering has been observed to have an effect on converter performance. Work done in automotive (79), petroleum (80-82), and catalyst industry laboratories (83) has shown
the damaging effects of high temperature exposure on monolithic oxidation catalysts. The mechanical properties of the support may deteriorate at elevated temperatures (84), but metal area loss is more
likely than support damage under normal operating conditions (79,81,
82).

The following results (85) show how sintering (in the absence
of poisoning) of a pelleted platinum catalyst affects its performance in a converter operated according to the 1975 Federal Test Procedure (FTP). For these experiments a single large batch of 0.15
wt % Pt/Al_2O_3 was divided into four parts. Each part was given a
different treatment in flowing air at high temperature to sinter the
platinum (but not the support) without exposing it to exhaust contaminants and without changing its concentration or distribution on
the support. Specific metal areas were measured by hydrogen adsorption (86), and the results are given in Table 1. Each catalyst in
turn was then used to fill a small 1.3 ℓ experimental converter
placed at the "Y" in the exhaust system of a 1974 Chevrolet. The
reported results for the FTP performance are based upon the average
of at least three tests with each sample.

Table 1

Catalyst Characterization

Treatment	Pt area (cm^2/g catalyst)	Relative Pt area	Dispersion	Average particle diameter (nm)*
4 hr, 500°C	2800	1.00	0.73	1.5
4 hr, 750°C	1600	0.58	0.42	2.6
5 hr, 900°C	100	0.04	0.03	36.
22 hr, 900°C	40	0.01	0.01	110.

* estimated from dispersion

It was noted that the metal area would be unlikely to affect the efficiency of an exhaust catalyst significantly once the converter had reached operating temperature (450-650°C). This is because the transport of reactants from the gas stream to the active catalyst surface is considerably slower than the surface reaction itself; so it is conceivable that sintering could occur without markedly changing the observed steady-state efficiency of a converter. However, the converter takes a certain amount of time to reach its steady operating temperature after a cold start. During this time the inherent activity of the catalyst is the controlling factor, and sintering could very well play an important role in determining the converter's efficiency.

The FTP test is a schedule of various operating modes designed to simulate driving in urban traffic (87). An "FTP cycle" begins at zero velocity, proceeds through a well-defined combination of accelerations, decelerations, and steady speeds, and returns to zero velocity. The entire FTP test consists of 23 such cycles requiring a total of 31 minutes of driving; the first 18 cycles (starting with a cold engine) are all different, while the last five are a repetition of the first five.

Figs. 2 and 3 show how the conversion efficiencies for carbon monoxide (CO) and hydrocarbons (HC) vary with time for the four catalysts (identified by relative platinum area) in the early portion of the FTP test. As anticipated, the conversions in the first cycle, when the converter is relatively cool, are strongly affected by the metal area. As the converter heats up, the differences become much smaller and, in the case of CO, virtually disappear. The differences which persist for hydrocarbons may seem slight, but it should

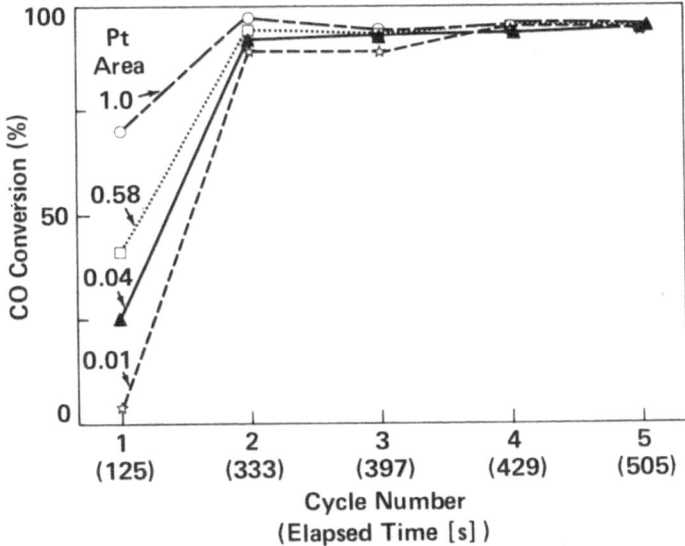

Fig. 2. CO conversions (averaged over each FTP cycle) in the early part of an emissions test for catalysts of varying metal area.

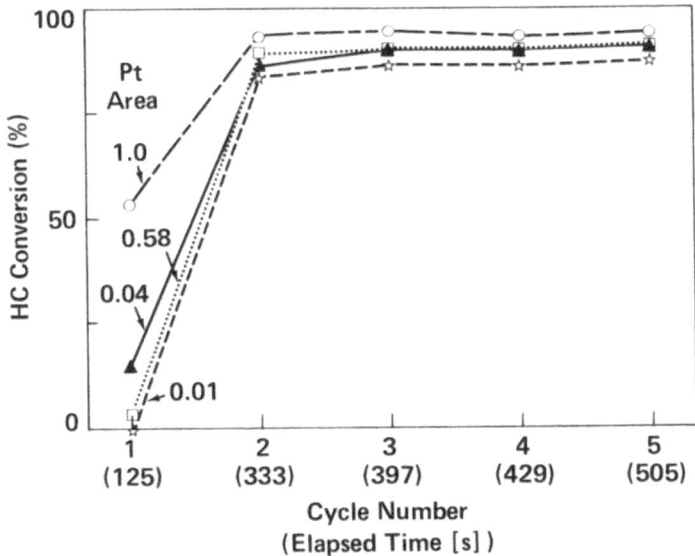

Fig. 3. HC conversions (averaged over each FTP cycle) in the early part of an emissions test for catalysts of varying metal area.

be remembered that a change from 90% to 80% conversion translates
into a doubling of the unreacted emissions. The apparent negative
HC conversion shown in Fig. 3 is the result of hydrocarbons adsorb-
ing on the catalyst as it cools after a test and then desorbing in
the first cycle of the succeeding test. (The low HC activity of the
"0.58" catalyst in the first cycle is unexplained.)

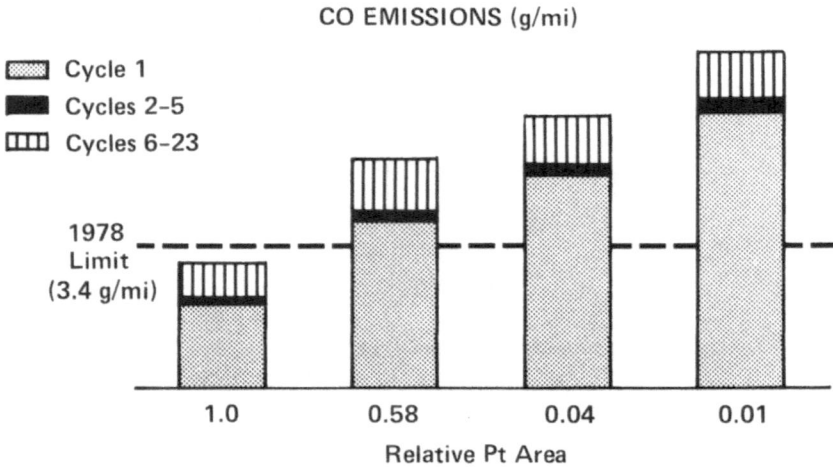

Fig. 4. Breakdown of CO emissions during FTP test for catalysts of
varying metal area.

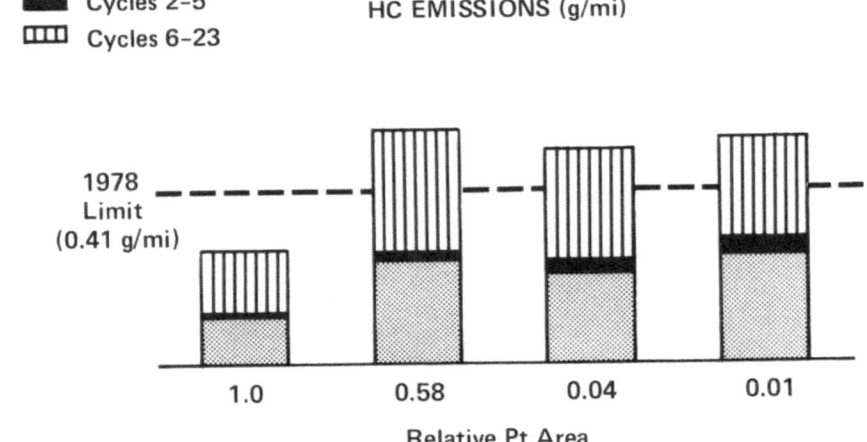

Fig. 5. Breakdown of HC emissions during FTP test for catalysts of
varying metal area.

Seeing that the conversions over the sintered catalysts quickly approach those for the high area sample, one might ask whether the differences observed in the first cycle make a significant contribution to the total FTP results. Figs. 4 and 5 show the breakdown of the overall emissions (in g/mi) into cold (#1), transitional (#2-5), and hot (#6-23) cycles; and it is clear that the first cycle accounts for a large fraction of the FTP emissions. For CO, in fact, the present 1978 limit is exceeded in the very first cycle on the highly sintered samples, while performance in the hot cycles is similar for all four catalysts. The HC conversion in both hot and cold cycles seems to deteriorate with sintering, although the correlation with metal area is not exact. Some exhaust hydrocarbons, the paraffins in particular, are relatively slow to oxidize; so they may remain sensitive to metal area even while other HC reactions are mass transfer limited over the hot catalyst.

We thus conclude that, for a supported platinum catalyst in a small experimental converter, sintering degrades the converter performance. The primary effect is seen in the conversion during warm-up. Despite losing more than 95% of the metal surface, though, the catalyst retains remarkable steady-state efficiency. It remains to be seen how these observations carry over to the production-size (4.3 ℓ) converter and to palladium or platinum-palladium combination catalysts (85).

SINTERING -- THEORETICAL TREATMENTS

For supported catalysts, the sintering process itself has not been the subject of detailed fundamental investigation until relatively recently. In 1973, Ruckenstein and Pulvermacher published an elaborate mathematical treatment of sintering of supported metal crystallites (88,89), Wynblatt and Gjostein (90-92) and Flynn and Wanke (93,94) have since used somewhat different approaches to the same problem. The requisite of a sintering model is that it must account both for transport of metal across relatively large distances on the support and for incorporation of migrating metallic species into growing particles. Each of the three theoretical treatments cited above deals in its own fashion with these two basic components of sintering, and the experimental evidence to date does not overwhelmingly favor one model over the others.

The Ruckenstein-Pulvermacher analysis supposes that sintering occurs via migration, collision, and fusion of metal crystallites on the support surface. Either the migration or the merging step can be rate-determining. The authors observe that, according to their model, the rate of change of the metal surface area "S" can be described by an equation of the form

$$\frac{dS}{dt} = -KS^n. \qquad\qquad (2)$$

K is a constant, and the exponent n can vary from 2 to 8 depending upon the way in which the rate determining step (diffusion or merging) is assumed to depend upon the metal particle size. Ruckenstein and Pulvermacher cite examples of both diffusion limited (n = 2 or 3) and merging limited ($4 \le n \le 8$) sintering experiments from the literature.

Sintering is treated as a two-dimensional evaporation-condensation process by Flynn and Wanke. The metal-support system is pictured as seeking to establish an equilibrium between the metal atoms on the metal particles and those which have escaped to the support surface. The rate at which atoms dissociate from metal particles is taken as being independent of the particle size, while the rate of atom recapture is proportional to the particle diameter; so the larger particles will grow at the expense of the smaller ones. The Flynn-Wanke model indicates that the size of the metal particles and the width of the distribution are critical parameters in determining the exponent n for fitting sintering data to equation (2). The value of n is expected to vary during the course of a sintering experiment.

The Wynblatt-Gjostein approach is to consider the events which occur on an atomic scale during sintering. The concepts of crystal nucleation and growth, particle migration and coalescence, and surface or vapor transport of metal are all discussed in terms of their relative importance in the overall rate of metal area loss. Order of magnitude rates of the various steps in a proposed sintering sequence are calculated using data obtained primarily on pure metals or supported metal films rather than on porous supported catalysts. As such, the parameters are close to being independently determined, rather than being the result of a fitting process applied to data from catalyst sintering.

Judging from their calculations and the available experimental literature, Wynblatt and Gjostein conclude that particle growth can occur in either a "non-inhibited" or an "inhibited" mode. The Ruckenstein-Pulvermacher model (particle migration, collision, and coalescence) and the Flynn-Wanke model (atom escape, diffusion, and recapture) are examples of non-inhibited particle growth. The idea of inhibited growth arose from Wynblatt and Gjostein's observation (by electron microscopy) that platinum particles on a flat alumina surface appear to approach a limiting size during a long term (∿70 hr) exposure at 700°C. They attribute this behavior to faceting of the metal crystallites; thus crystallite growth would be inhibited by the requirement of a nucleation process for each new layer of atoms added to the particle.

The three theories just described represent three different approaches, but they all agree on at least a few points. The major one is that large supported metal crystallites will grow at the expense of small ones at elevated temperatures. The pore structure of typical catalyst supports will tend to stabilize the metal dispersion, as will strong metal-support interactions. All the authors can point out the parameters in their models which account for platinum being more stable in reducing than in oxidizing atmospheres, but this appears to be an ex post facto correlation with experiments.

The proposed models, of course, do not agree in all aspects, and some of the differences should be amenable to experimental resolution. The major difference between the Ruckenstein-Pulvermacher and Flynn-Wanke models is the assumed transport mechanism -- particle migration vs. atom diffusion, respectively. Ruckenstein and Pulvermacher argue (89) that the interactions betwen metal atoms and the support surface would be too weak to make dissociation energetically favorable, and metal evaporation rates would be far too slow to account for observed sintering rates. Wynblatt and Gjostein agree that sintering by either surface or gas phase platinum atom transport is ruled out under reducing conditions (92); however, they determine from their calculations that dissociation and surface diffusion (of PtO_2) would be a very effective transport mechanism in an oxidizng atmósphere.

The particle migration model of Ruckenstein and Pulvermacher is difficult to reconcile with certain experimental observations (93), namely: 1) metal dispersions have occasionally been observed to increase at high temperatures in oxygen, 2) platinum particles have grown to dimensions larger than the support particles (30 nm) in some experiments, and 3) there is little direct evidence from microscopy work that particle motions occur to the extent required to explain catalyst sintering. The Wynnblatt-Gjostein analysis shows that, for metal crystallites smaller than ∿5 nm, particle mobility is likely to account for the sintering observed under reducing conditions; however, decreasing diffusivities for increasing particle sizes make particle migration unable to account for sintering of particles larger than ∿7.5 nm (91). Wynblatt and Gjostein also reason that, contrary to the Ruckenstein-Pulvermacher hypothesis, the coalescence of two colliding particles will always be faster than their migration; so coalescence cannot be rate determining in catalyst sintering.

In view of the complexity of supported metal sintering and the fact that the mechanism is likely to depend strongly upon the particular catalyst and experimental conditions chosen, it is not surprising that sintering data fitted to a power law [eq. (2)] give exponents ranging from 2 (59,62) to 13 (90). Flynn and Wanke calculate (94) that the sintering order in one of Wynblatt and Gjostein's experiments (90) changes from 13 to 4 as sintering progresses. In contrast to this, the experimental data (Fig. 3 of ref. 90) show a

a constant order of ∿13 for the entire 70 hr experiment. Thus the
Flynn-Wanke model (fitted to the data at zero time and 15 hr) predicts a somewhat more rapid loss of area at times >15 hr than is
observed experimentally. Wynblatt and Gjostein agree, though, that
in any given instance the sintering order might change as the catalyst and conditions change.

The available data on sintering kinetics are generally not adequate to provide conclusive confirmation of any particular mechanism;
all three pairs of authors concur on this point. The atom transport
models seem best able to account for the high area that resulted when
a supported platinum catalyst was quickly cooled from 800°C (61), but
crystallite diffusion appears necessary to account for sintering in
reducing atmospheres. Probably the best test for a model is its
ability to predict the particle size distribution as a function of
sintering time, but such data are difficult to obtain on realistic
systems. The breadth of the distribution, as reflected in the different average radii obtained from different measurement techniques,
can in principle be used as an indicator of the sintering mechanism
(95); but the interpretation of such data depends upon the model
used.

More extensive experimental work must be done before the relative merits of the various theories can be assessed. In particular,
improved experiments using electron microscopy have great potential
for answering questions about how the particle size distribution re-
sponds to high temperature exposure. Platinum dispersed within a
microporous support presents experimental (and theoretical) difficulties beyond those encountered in using a flat alumina plate for a
support, but even simple systems can yield valuable information a-
bout the influence and relative importance of variables in a sintering experiment.

Thus, despite the ubiquity of thermal damage to supported metals,
there remains much to be learned about the outwardly simple process
of metal transport and agglomeration on a support. The roles of the
metal, the support, the atmosphere, the stabilizers (if any), and
the temperature all provide fertile grounds for further research.

REFERENCES

1. Kuczynski, G. C., Advan. Colloid Interface Sci. 3, 275 (1972).
2. Sinfelt, J. H., Annu. Rev. Mater. Sci. 2, 641 (1972).
3. Benson, J. E., and Boudart, M., J. Catal. 4, 704 (1965).
4. Vannice, M. A., Benson, J. E., and Boudart, M., J. Catal. 16, 348 (1970).
5. Sermon, P. A., J. Catal. 24, 460 (1972).
6. Sinfelt, J. H., Advan. Chem. Eng. 5, 37 (1964).

7. Satterfield, C. N., Mass Transfer in Heterogeneous Catalysis (MIT Press, Cambridge, 1970).
8. Hegedus, L. L., and Summers, J. C., 4th N. Amer. Mtg. Catal. Soc., Toronto, 1975, paper #22.
9. Cinneide, A. D. O., and Clarke, J. K. A., Catal. Rev. 7, 213 (1973).
10. Whyte, T. E., Catal. Rev. 8, 117 (1973).
11. Spenadel, L., and Boudart, M., J. Phys. Chem. 64, 204 (1960).
12. Klug, H. P., and Alexander, L. E., X-ray Diffraction Procedures (Wiley, New York, 1954), p. 491.
13. Miyazaki, K., J. Catal. 28, 245 (1973).
14. Adams, C. R., Benesi, H. A., Curtis, R. M., and Meisenheimer, R. G., J. Catal. 1, 336 (1962).
15. Prestridge, E. B., and Yates, D. J. C., Nature (London) 234, 345 (1971).
16. Flynn, P. C., Wanke, S. E., and Turner, P. S., J. Catal. 33, 233 (1974).
17. Renouprez, A., Hoang-Van, C., and Compagnon, P. A., J. Catal. 34, 411 (1974).
18. Baker, R. T. K., Harris, P. S., and Thomas, R. B., Res/Develop. 24, 22 (1973).
19. Baker, R. T. K., Thomas, R. B., and Notton, J. H. F., Platinum Metals Rev. 18, 130 (1974).
20. Baker, R. T. K., Barber, M. A., Harris, P. S., Feates, F. S., and Waite, R. J., J. Catal. 26, 51 (1972).
21. Kirklin, P. W., and Whyte, T. E., Jr., preprint, Div. Petrol. Chem., 163rd Nat. ACS Mtg., Boston, 1972, p. C32.
22. Somorjai, G. A., Powell, R. E., Montgomery, P. W., and Jura, G., Small Angle X-ray Scattering, H. Brumberger, Ed. (Gordon and Breach Sci. Publ., New York, 1967), p. 449.
23. Ratnasamy, P., Leonard, A. J., Rodrique, L., and Fripiat, J. J., J. Catal. 29, 374 (1973).
24. Ratnasamy, P., and Leonard, A. J., Catal. Rev. 6, 293 (1972).
25. Lytle, F. W., and Sayers, D. E., Calif. Catal. Soc., Pasadena, 1973.
26. Adler, S. F., and Keavney, J. J., J. Phys. Chem. 64, 208 (1960).
27. Dorling, T. A., Lynch, B. W. J., and Moss, R. L., J. Catal. 20, 190 (1971).
28. Brunauer, S., Emmett, P. H., and Teller, E., J. Amer. Chem. Soc. 60, 309 (1938).
29. Freel, J., J. Catal. 25, 149 (1972).
30. Fruma, A., Int. Chem. Eng. 12, 503 (1972).
31. Wilson, G. R., and Hall, W. K., J. Catal. 17, 190 (1970).
32. Roth, J. F., and Reichard, T. E., J. Res. Inst. Catal., Hokkaido Univ., 20, 85 (1972).
33. Klimisch, R. L., Summers, J. C., and Schlatter, J. C., 3rd N. Amer. Mtg. Catal. Soc., San Francisco, 1974; Advan. Chem. Ser., in press.
34. Bhasin, M. M., J. Catal. 34, 356 (1974).

35. Williams, F. L., and Baron, K., 4th N. Amer. Mtg. Catal. Soc.,
    Toronto, 1975, to appear in J. Catal.
36. Ross, P. N., and Stonehart, P., 4th N. Amer. Mtg. Catal. Soc.,
    Toronto, 1975, paper #50.
37. Delgass, W. N., and Boudart, M., Catal. Rev. $\underline{2}$, 129 (1968).
38. Ross, P. N., Jr., and Delgass, W. N., J. Catal. $\underline{33}$, 219 (1974).
39. Garten, R. L., and Ollis, D. F., J. Catal. $\underline{35}$, 232 (1974).
40. Kuczynski, G. C., Carberry, J. J., and Martinez, E., J. Catal.
    $\underline{28}$, 39 (1973).
41. Folkins, H. O., and Miller, E., Ind. Eng. Chem. $\underline{49}$, 241 (1957).
42. Stillwell, W. D., ibid., 245.
43. Maatman, R. W., and Prater, C. D., ibid., 253.
44. Vincent, R. C., and Merrill, R. P., J. Catal. $\underline{35}$, 206 (1974).
45. Maatman, R. W., Ind. Eng. Chem. $\underline{51}$, 913 (1959).
46. Michalko, E., U. S. Patents No. 3 259 454 and 3 259 589.
47. Minhas, S., and Carberry, J. J., J. Catal. $\underline{14}$, 270 (1969).
48. Benesi, H. A., Curtis, R. M., and Studer, H. P., J. Catal. $\underline{10}$,
    328 (1968).
49. Hegedus, L. L., and Petersen, E. E., Cat. Rev. Sci. Eng. $\underline{9}$, 245
    (1974).
50. Wei, J., and Becker, E. R., 167th Nat. ACS Mtg., Los Angeles,
    1974; Adv. Chem. Ser., in press.
51. Gil'debrand, E. I., Int. Chem. Eng. $\underline{6}$, 449 (1966).
52. McHenry, K. W., Bertolacini, R. J., Brennan, H. M., Wilson,
    J. L., and Seelig, H. S., Proc. 2nd Int. Cong. Catal., Paris,
    2295 (1960).
53. Mills, G. A., Weller, S., and Cornelius, ibid., 2221.
54. Corolleur, C., Gault, F. G., Juttard, D., Maire, G., and Muller,
    J. M., J. Catal. $\underline{27}$, 446 (1972).
55. Freel, J., J. Catal. $\underline{25}$, 139 (1972).
56. Williams, A., Butler, G. A., and Hammonds, J., J. Catal. $\underline{24}$,
    352 (1972).
57. Dalla Betta, R. A., and Boudart, M., Proc. 5th Int. Cong. Catal.,
    Palm Beach, 1329 (1972).
58. Scholten, J. J., and von Montfoort, J. Catal. $\underline{1}$, 85 (1962).
59. Maat, H. J., and Moscou, Proc. 3rd Int. Cong. Catal., Amsterdam,
    1277 (1964).
60. Hassan, S. A., J. Appl. Chem. Biotechnol. $\underline{24}$, 497 (1974).
61. Emelianova, G. I., and Hassan, S. A., Proc. 4th Int. Cong.
    Catal., Moscow, 1329 (1968).
62. Herrmann, R. A., Adler, S. F., Goldstein, M. S., and DeBaun,
    R. M., J. Phys. Chem. $\underline{65}$, 2189 (1961).
63. Hughes, T. R., Houston, R. J., and Sieg, R. P., Ind. Eng. Chem.
    Proc. Design Develop. $\underline{1}$, 96 (1962).
64. Gruber, H. S., Anal. Chem. $\underline{34}$, 1828 (1962).
65. Graham, J. R., and Lard, E. W., 3rd N. Amer. Mtg. Catal. Soc.,
    San Francisco, 1974, paper #10.
66. Flynn, P. C., and Wanke, S. E., 4th N. Amer. Mtg. Catal. Soc.,
    Toronto, 1975, to appear in J. Catal.

67. Furuoya, I., and Shirasaki, T., Kogyo Kagaku Zasshi 72, 1223 (1969).
68. Furuoya, I., Shirasaki, T., Echigoya, I., and Morikawa, K., ibid., 1431.
69. Schlatter, J. C., "Reversible Changes in the Apparent Area of Supported Palladium," Report GMR-1754.
70. Benson, J. E., Hwang, H. S., and Boudart, M., J. Catal. 30, 146 (1973).
71. Pitkethly, R. C., discussion following ref. 59.
72. Boudart, M., Aldag, A. W., Ptak, L. D., and Benson, J. E., J. Catal. 11, 35 (1968).
73. Manogue, W. H., and Katzer, J. R., J. Catal. 32, 166 (1974).
74. Poltorak, O. M., and Boronin, V. S., Zh. Fiz. Khim. 40, 2671 (1966).
75. Boudart, M., Amer. Sci. 57, 97 (1969).
76. _____, Advan. Catal. Relat. Subj. 20, 153 (1969).
77. Ostermaier, J. J., Katzer, J. R., and Manogue, W. H., J. Catal. 33, 457 (1974).
78. Barnes, G. J., and Klimisch, R. L., SAE Auto. Eng. Mtg., Detroit, 1973, paper #730570.
79. Shelef, M., Dalla Betta, R. A., Larson, J. A., Otto, K., and Yao, H. C., 74th AIChE Nat. Mtg., New Orleans, 1973.
80. Voltz, S. E., Morgan, C. R., Liederman, D., and Jacob, S. M., Ind. Eng. Chem. Prod. Res. Develop. 12, 294 (1973).
81. Liederman, D., Voltz, S. E., and Snyder, P. W., Ind. Eng. Chem. Prod. Res. Develop. 13, 166 (1974).
82. Johnson, M. F. L., Mooi, J., Erickson, H., Kreger, W. E., and Breder, E. W., SAE Auto. Eng. Mtg., Toronto, 1974, paper #741079.
83. Briggs, W. S., and Graham, J. R., SAE Auto. Eng. Cong., Detroit, 1973, paper #730275.
84. Roth, J. F., and Gambell, J. W., ibid., paper #730277.
85. Schlatter, J. C., and Barnes, G. J., to be published.
86. Benesi, H. A., Atkins, L. T., and Mosely, R. B., J. Catal. 23, 211 (1971).
87. Federal Register, Vol. 37, No. 221 (November 15, 1972), Part II, p. 24215.
88. Ruckenstein, E., and Pulvermacher, B., AIChE J. 19, 356 (1973).
89. _____, J. Catal. 29, 224 (1973).
90. Wynblatt, P., and Gjostein, N. A., Scr. Met. 7, 969 (1973).
91. _____, Progr. Solid State Chem. 9, (1974).
92. Wynblatt, P., Dalla Betta, R. A., and Gjostein, N. A., to be published in the Proceedings of the Battelle Colloquium on the Physical Basis of Heterogeneous Catalysis.
93. Flynn, P. C., and Wanke, S. E., J. Catal. 34, 390 (1974).
94. _____, ibid., 400.
95. Pulvermacher, B., and Ruckenstein, E., J. Catal. 35, 115 (1974).

QUASI-EQUILIBRIA AND SINTERING IN THE COBALT OXIDE - ZIRCONIA CATALYST - SUPPORT SYSTEM

M. Bettman and H. C. Yao

Scientific Research Staff, Ford Motor Company

P. O. Box 2053, Dearborn, Michigan 48121

ABSTRACT

Catalysts of varying concentrations of cobalt oxide on ZrO_2 supports were prepared by impregnation and calcination to 600°C. In very dilute catalysts the cobalt oxide consists of a two-dimensional, dispersed "δ-phase" of as yet undetermined stoichiometry. At higher concentrations, a saturated δ-phase is in equilibrium with bulk, or "β-phase" particles. The latter appear to be Co_3O_4 spinel particles. The saturated δ-phase has about 4.5×10^{18} Co atoms per m^2, i.e., about 60% of the estimated concentration of surface cobalt atoms on polycrystalline Co_3O_4.

Further heating to 800°C causes Ostwald ripening of the β-phase particles, presumably by surface diffusion via the δ-phase. This suggests very poor wetting of the support by the β-phase.

Diffusion couples of Co_3O_4 - porous ZrO_2, held at 850°C for many days, also illustrate the establishment of the δ-phase. By-products of this experiment are the estimated values of 0.22 wt. % for the solubility, and 10^{-15} - 10^{-14} cm^2/sec for the volume diffusion coefficient of cobalt (as an undetermined oxide) in monoclinic zirconia, at 850°C.

INTRODUCTION

The most pertinent prior work seems to be that of O'Reilly and MacIver (1) (OM) and of Tomlinson, et. al., (2) (T). (OM) investigated a series of γ-alumina supported chromic oxide

165

catalysts, as a function of chromia concentration, by epr. The catalysts were prepared by impregnation with aqueous chromic nitrate, followed by calcination to 500°C. Their alumina substrates had BET areas of 160 m^2/g. The spectrum was found to be a superposition of two epr lines. One of these predominate at low concentrations and becomes weaker with increasing chromic oxide concentrations. (OM) associated this line with a magnetically dilute, i.e., highly dispersed phase which they named the "δ-phase." The other line, similar in behavior to that of bulk chromia, was associated with a clustered phase, designated as a "β-phase." They found that with increasing chromium concentration, the amount of δ-phase reached a maximum and then decayed again.

(T) made magnetic susceptibility measurements on a series of γ-alumina supported cobalt oxide catalysts of increasing concentrations. Their preparation method was very similar to that of (OM). (T) noted that very dilute catalysts had high magnetic moments per cobalt atom, reminiscent of divalent cobalt. Moderately concentrated catalysts had magnetic properties very similar to those of bulk Co$_3$O$_4$. Their analysis of their data led to curves of relative amounts of δ- and β-phases, as a function of cobalt concentration. Up to about 1.2 wt. % Co, most of it is in a δ-phase. Beyond that, the δ-phase concentration decays, and reaches zero at a total cobalt concentration of 7 wt. %.

In some earlier work, one of us (3) investigated cobalt oxide catalysts supported by monoclinic zirconia, with a view to possible application as automobile exhaust emission catalysts. The methods of investigation were catalytic activity in the combustion of ethylene, nitric oxide chemisorption, and the line width of the Co$_3$O$_4$ (311) X-ray diffraction line, the only one not obliterated by zirconia lines. From these line widths, a mean particle size, hence a likely surface area of the three dimensional Co$_3$O$_4$ spinel particles could be deduced. This area correlated reasonably well with the measured catalytic activity, compared to that of pure Co$_3$O$_4$. On the other hand, the NO chemisorption measurements suggested cobalt oxide areas which were many times larger, by as much as a factor of ten, or more. This observation applied not only to samples prepared by impregnation methods, but also to mechanical mixtures of Co$_3$O$_4$ and ZrO$_2$ that were subsequently heated to 700 C for sixteen hours. This suggested a large-area δ-phase, accounting for most of the NO chemisorption, but of low catalytic activity, in equilibrium with a β-phase of more modest area, accounting for most of the catalytic activity. The latter is assumed to be essentially pure Co$_3$O$_4$ spinel, on the basis of the single observable diffraction line, and on the basis of its high catalytic activity. It was then established that in a very dilute catalyst, presumed to consist of δ-phase only, the catalytic activity per cobalt ion, in the oxidation of ethylene, was about 100 times lower than that of a surface cobalt ion on polycrystalline Co_3O_4.

Our interpretation of the cobalt oxide-zirconia system differs from the concepts of (OM) and (T) in that we consider the δ-phase to be in local equilibrium with β-phase particles. Very dilute, equilibrated catalysts are believed to contain the δ-phase only. With increasing cobalt concentration, the δ-phase saturates and the β-phase nucleates. A very small β-phase particle, possessing excess surface free energy, is expected to be in equilibrium with a super-saturated δ-phase in its vicinity. This would allow for a mechanism of Ostwald ripening of β-phase particles via diffusion through the δ-phase.

Some assumptions and qualifications underlie this picture: 1) The samples must have been at a high enough temperature for appreciable surface diffusion to take place. 2) The temperatures must be kept below the level where copious interdiffusion of catalyst and support take place, in cases where the two are miscible. 3) The unsaturated δ-phase may be patchy and domain-like, even on a given (hkl) face of the substrate. The saturated δ-phase can have concentrations which vary from one (hkl) face of the substrate to another. Our determination of the saturation value, on a polycrystalline substrate, is therefore a mean value which is not rigorously defined, and could change somewhat from one sample to another.

The present work was undertaken for the purpose of defining the phase boundary between the δ- and β-phases, i.e., to determine the concentration of the saturated δ-phase.

EXPERIMENTAL

Zirconia was made by decomposition of zirconium nitrate (Johnson Matthey, specpure grade) at 300°C, followed by calcination at 800°C for six hours. This resulted in a BET area of about 11 m^2/g. For the impregnation with cobalt oxide, a measured amount of $Co(NO_3)_2$ solution was added to the dried and weighed ZrO_2. The solids were then dried slowly in a desiccator under vacuum and then calcined at 600°C for six hours. The nitric oxide was purified by several freezing evaporation cycles, retaining 2/3 of the middle fraction for the chemisorption experiments (4). Research grade CO and ultrapure grade Ar (Matheson Co.) were used without further purification. The volumetric measurements for BET area and NO and CO chemisorption measurements were performed in a conventional constant volume adsorption apparatus, previously described (5).

For a macroscopic diffusion experiment, thin disks of monoclinic zirconia were prepared by adding 0.3% polyvinylbutyral binder to small particle size (about 50 Å) monoclinic ZrO$_2$, pressing in a steel die, and calcining at 905°C for 68 hrs. The small-particle ZrO$_2$ was derived from two sources. 1) By boiling the Johnson Matthey specpure "zirconium nitrate" overnight and 2) by precipitating zirconyl nitrate, of indifferent purity, with ammonia, filtering the resulting amorphous hydrated oxide, and converting it to the monoclinic form by hydrothermal digestion at 250°C in an autoclave for two hours. Material 2 led to greenish tinted disks with a higher BET area than the pure white ones (6.11 vs. 2.2 m^2/g). All disks had dimensions of approximately 0.6 cm diameter x 0.018 cm thickness. They were vacuum coated on one flat side with about 5 x 10^{-5} cm of cobalt metal. They were then placed into an air furnace, at 850°C, for various periods up to two weeks. It is assumed that the cobalt quickly oxidized to Co$_3$O$_4$, some of which then diffused into the porous zirconia to form a saturated δ-phase. After the 850°C treatment the disks were broken, and cobalt concentration profiles were determined by microprobing across the broken cross section, using a fully focussed spot of a 20 KV, 100 nA beam in an ARL EMX probe. Standards for the probe were made by aqueous impregnation methods.

The X-ray diffraction work was carried out on a computer controlled Norelco powder diffractometer with a diffracted-beam, graphite monochromator, using a fine-focus copper tube. Step scanning over the Co$_3$O$_4$ (311) line was carried out at intervals of 0.05° in 2θ, using counting times of 400 seconds for the weakest (most dilute) samples, and 100 seconds for the others. The low angle tail of the Co$_3$O$_4$ (311) line suffers from interference from the nearby, weak, ZrO$_2$ (10$\bar{2}$) line. A correction was carried out by a difference method, utilizing a sample which contained no β-phase (6).

RESULTS AND DISCUSSION

A. Chemisorption

The results of the NO and CO chemisorption are plotted in Fig. 1. The logarithm of the number of micromoles of chemisorbed NO or CO per m^2 of substrate are plotted versus the logarithm of the number of micromoles of cobalt per m^2 in each sample. The chemisorption values chosen are those of the "irreversibly adsorbed" gases (7), i.e. roughly, the gases which cannot be pumped off at room temperature. The log-log curves are seen to break up into two approximately linear portions. We interpret the break as the point where the δ-phase has saturated, and the β-phase nucleates. Beyond this point, the slope is much decreased, because only a portion of the additional cobalt atoms, in the β-phase particles, are surface atoms that can be "seen" by the chemisorbed gas.

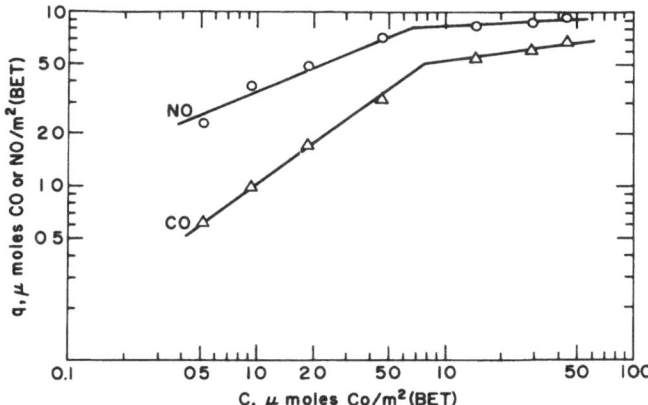

Fig. 1.   Chemisorption Results

     If the values on the ordinate are divided by the values on the
abscissa of Fig. 1, we see that the multiplicity for NO chemisorption on the most dilute sample is greater than four moles NO for
each mole of Co. This interesting phenomenon has been noted before
(5), but not studied in detail. With increasing cobalt concentration this multiplicity decreases to about unity (actually, 1.2)
at the intersection point. For CO chemisorption, the multiplicity
starts at 1.19 for the sample most dilute in Co, and decreases to
0.66 at the intersection of the two curves.

                        B.   Ostwald Ripening

     The experiment on the particle size coarsening of the β-phase
is summarized in Table 1. The two samples listed were not part of
the series used for the chemisorption experiments. They were prepared in a slightly different manner, but from the same starting
materials. The ZrO_2 was pretreated to 850°C, but maintained a
similar surface area of 12 m^2/g. After cobalt impregnation and
16 hr calcination to 600°C, the mean β-phase particle diameter was
determined from the width of the Co_3O_4 (311) diffraction line.
The samples were then calcined for 16 hrs at 800°C, and the line
width was measured again. The narrowing was very appreciable (see
Table 1).

     In the past, some transmission electron microscopy has been
carried out on some cobalt oxide-zirconia samples. The zirconia
generally looks like a rock pile, consisting of polycrystals.
Many of the pores consist of the interstices between "rocks," and
are of similar dimensions but somewhat smaller than the rocks. In
other words, we would expect an abundance of pores in the range of
500 Å or less in the two samples of Table 1. The fact that most

Table 1

Ostwald Ripening of the β-Phase Upon Heat Treatment

Sample	Wt. % Cobalt[a]	Particle Size,[b] Å, After: 600°C[c]	800°C[c]
4-149A	1.5	270	2190
4-149B	4.0	327	1590

(a) $\dfrac{g\ Co}{g\ Co + g\ ZrO_2}$ x 100; ignoring oxidized state of the Co.

(b) from width at half height of Co_3O_4 (311) X-ray diffraction line; see text.

(c) highest temperature reached, in air, for 16 hrs.

of the Co_3O_4 particles coarsen to much larger dimensions suggests poor wetting. After forming a surprisingly concentrated δ-phase the remaining cobalt oxide agglomerates into a β-phase which does not seem to wet the δ-phase at all.

C. Diffusion Experiment

The results of the microprobe analysis of two macroscopic diffusion couples are shown in Figures 2 and 3. The two samples were treated at 850°C for 308 hours and 218 hours, respectively. The substrates had specific surface areas of 2.2 and 6.11 m^2/g, respectively. The ordinates are in wt. % of cobalt (as metal), the abscissas in μm of distance from the cobalt oxide-zirconia interface. A curve of the type $C = C_o [1-\theta(KX)]$ has been fitted to some of the points, those which are relatively close to the boundary. C_o is the concentration of cobalt on the zirconia side of the boundary, and is also the expected concentration throughout the zirconia at infinite time. $\theta(KX)$ is the error function. This formula would hold in the case of true volume diffusion, assuming a concentration-independent diffusion coefficient. Actually, we are dealing here with surface diffusion along an interior, porous surface. Thus, K is related to a pseudo diffusion length, which has no simple physical interpretation in this case. At larger X values, the experimental points fall sharply below the curve, suggesting that the surface diffusion rate decreases with decreasing cobalt concentration. A possible explanation is that the vanguard of cobalt atoms, at low concentrations, are trapped at some surface sites, or on high index planes.

The C_o values, divided by the specific surface area of the

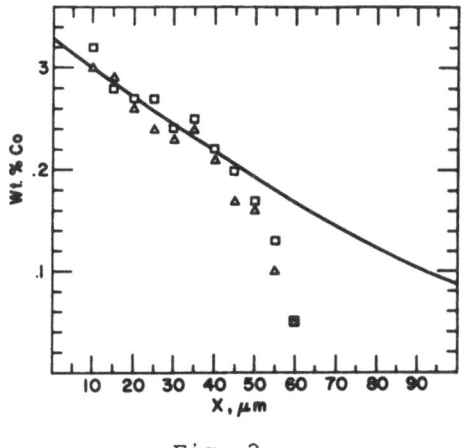
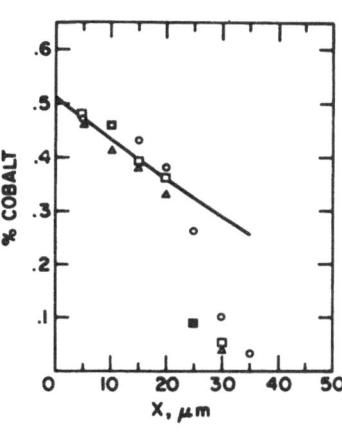

Fig. 2   Fig. 3

Fig. 2.   Cobalt concentration profile in a disk of 2.2 m^2/g porous
          monoclinic ZrO$_2$, vs. distance from Co$_3$O$_4$-ZrO$_2$ interface;
          after 308 hrs at 850°C; different point markings repre-
          sent microprobe traverses along different paths.

Fig. 3.   Same, for a 6.11 m^2/g disk, after 218 hrs at 850°C.

substrates, were expected to yield the desired saturation concentrations of the δ-phase. This calculation leads to 25 and 14.3
μmol/m^2, respectively, for the samples with the 2.2 and 6.11 m^2/g
substrates. Besides being unequal, these values are also appreciably too high compared to the about 7 μmol/m^2 for the saturated δ-
phase, deduced from the chemisorption experiments. The pellet
diffusion results suggest that the implied assumption of zero
volume solubility of cobalt oxide in zirconia is not quite accurate.
If, on the contrary, we assume that the zirconia grains near the
cobalt oxide interface have had enough time, at 850°C, to reach
saturation volume solubility as well as a saturated δ-phase, two
simultaneous linear equations will give the values of both the
volume solubility and the δ-phase saturation concentration, as
follows.

$$2.2x + y = 0.00325$$

$$6.11x + y = 0.00514$$

where x is the saturation concentration of the δ-phase in g Co/m^2-
zirconia, and y is the volume solubility, g Co/g ZrO$_2$. The coefficients of x are the measured specific surface areas. The values
on the right side of the equations are taken from the ordinate inter-
cepts of Figures 2 and 3. The solution gives x = 4.83 x 10^{-4} g
Co/m^2 ZrO$_2$, or 8.2 μmol/m^2 for the saturated δ-phase, and y =

0.0022 g Co/g ZrO_2 for the volume solubility of cobalt (calculated as metallic cobalt, but, no doubt, in the form of an oxide) at 850°C. The value for the saturation concentration of the δ-phase, obtained from this analysis, is reasonably close to the value obtained from the chemisorption experiments. The assumption that volume saturation had been reached near the Co_3O_4-ZrO_2 interface requires diffusion coefficients of cobalt oxide in zirconia, at 850°C, of the order of 10^{-15} to 10^{-14} cm^2/sec. This estimate is based on ZrO_2 particle dimensions of 0.4 μm, and a diffusion time of 308 hrs. Some additional microprobe data, not shown here, for a 6.11 m^2/g disk held at 850°C for only 5 days, showed a C_O value of only 0.46 wt %, as opposed to the 0.514 wt % (see above) for a similar disk held at 850°C for 9 days. This suggests that 5 days were insufficient to produce full volume saturation of cobalt oxide in the zirconia grains nearest the Co_3O_4-ZrO_2 interface. It follows that the estimated volume diffusion coefficient, at 850°C, of about 10^{-15}-10^{-14} cm^2/sec, is approximately correct, i.e., it is not just a minimum value.

ACKNOWLEDGEMENTS

The authors are indebted to J. Tabock for the microprobe work, E. B. Schermer for the cobalt evaporation, and to N. A. Gjostein and S. M. Kaufman for valuable discussions.

REFERENCES

1. O'Reilly, D. E., and MacIver, D. S., J. Phys. Chem. 66, 276 (1962); Poole, C. P., and MacIver, D. S., Adv. in Catal. 17, 223 (1967).

2. Tomlinson, J. R., Keeling, Jr., R. O., Rymer, G. T., and Bridges, J. M., Actes Du Deuxieme Cong. Intern. De Catal., Paris, 1960, p. 1831.

3. Bettman, M., unpublished.

4. Otto, K., and Shelef, M., J. Catal. 14, 226 (1969).

5. Yao, H. C., and Shelef, M., Proc. of the Symposium "The Catalytic Chemistry of Nitrogen Oxides," Oct. 7-8, 1974, General Motors Corp., Warren, Mich.

6. Klug, H. P. and Alexander, L. E., "X-Ray Diffraction Procedures for Polycrystalline and Amorphous Materials, 2nd Ed., John Wiley, N. Y. 1974.

7. Yao, H.C. and Shelef, M.J. Phys. Chem. 78, 2490 (1974).

A STUDY OF CAPILLARITY AND MASS TRANSPORT

ON THE Al_2O_3 SURFACE

F. H. Huang, R. A. Henrichsen* and Che-Yu Li
Department of Materials Science and Engineering
Cornell University, Ithaca, New York 14853
*General Refractories Research Center,
P.O. Box 1673, Baltimore, Maryland 21203

ABSTRACT

Periodic gratings were produced on single crystal Al_2O_3 surfaces. Upon annealing the tendency of the gratings to facet was found to depend on the surface orientation. The rate of the decay of the amplitude of sinusoidal gratings and that of the width of the grain boundary grooving were measured as a function of temperature and orientation to give the rate constants of capillarity induced mass transport.

INTRODUCTION

This paper reports the results of mass transport studies on Al_2O_3 surfaces using both the decay of sinusoidal gratings and the thermal grooving of grain boundaries. It also reports the anisotropic nature of the Al_2O_3 surface as revealed by the morphology of periodic gratings as a function of surface orientation.

Recent work on the rate of ripening of platinum particles on a single crystal Al_2O_3 surface[1] has shown that it is controlled by mass transport on the Al_2O_3 surface and depends on the surface orientation. These results are for platinum particles of the size of the order of 1000Å. Since the ripening of metal particles can decrease the effectiveness of a ceramic supported catalyst, it is possible that the performance of this type of catalyst will be affected by the ceramic surface orientation and by processes which can alter the ceramic surface orientation. The latter processes are, for example, capillarity induced morphological changes, which can produce preferred surface orientations. The properties of the

173

Al$_2$O$_3$ surface are known to be anisotropic.[2,3] The results of the
present work therefore provide useful background information for
the understanding of the performance of a ceramic supported
catalyst. They can also be used for the determination of rate
controlling mechanisms for the initial stage sintering of Al$_2$O$_3$
particles.

Surface diffusion on Al$_2$O$_3$ has been investigated by measuring
the kinetics of grain boundary grooving[4,5] and other capillarity
induced processes.[6,7] These investigations give wide differences
in the value of surface diffusivity. It is possible that the
anisotropic nature of the Al$_2$O$_3$ surface may be partly responsible
for the disagreement because more than one surface orientation is
included in the reported work. In this study a systematic investigation of the orientation and temperature dependence is carried
out. The study of the orientation dependence is made possible
using the technique of the decay of sinusoidal gratings. Experiments on thermal grooving of grain boundaries are performed to
compare with the results in the literature.

EXPERIMENTS

The Decay of Sinusoidal Gratings:

Single crystal sapphire plaques were obtained from Union
Carbide, Materials Systems Division with various surface orientations. The orientations are specified by the manufacturer as
0°, 60°, and 90° from the c-axis. These orientations were within
a few degrees of the (0001), (11$\bar{2}$3) and (11$\bar{2}$0) planes, respectively.
Periodic gratings of wave lengths 7, 8, 9, 10, 11.3, 12.7 microns
were etched into the specimen surface to produce sinusoidal gratings upon annealing. In some of the specimens the gratings were
produced by chemical etching. THe experimental techniques are
described elsewhere.[8,9,10] In the specimens from which the data
in Figures 5 and 6 were taken the gratings were produced by ion
milling. The mask for the latter process consisted of a layer of
chromium film approximately 8000Å thick and a layer of Kodak Thin
Film Resist a few microns thick and was prepared in the same way
as that for chemical etching.[8,9,10] The ion million was carried
out in a Ion Micro-Milling Instrument, Model III, Commonwealth
Scientific Corporation. Argon ions were used at a voltage of 6 kV
and a current of 100 microamps. The usual milling time is two to
four hours.

Gratings of each wave length were produced in four different
directions at 45° intervals. The annealing was carried out in a
N.R.C. Model 2914 high temperature vacuum furnace. The temperature of the furnace was recorded and controlled using a tungsten-
3% rhenium vs tungsten-25% rhenium thermocouple. The vacuum in

the work chamber is maintained in the 10^{-6} torr range during annealing. A sapphire crucible made of the same material as the specimens was used to contain the specimens. The morphology and the amplitude of the gratings after each annealing were recorded on photographic film using a Zeiss Interference Microscope. Details of the experimental equipment and procedures are given in reference 8.

Grain Boundary Grooving:

The specimens for the grain boundary grooving measurements were also single crystal sapphire plaques containing a few low angle grain boundaries and were also obtained from the Union Carbide, Materials System Division. The surface orientation, specified by the manufacturer as 60° from the c-axis was within 3° of (11$\bar{2}$3) plane. The annealing is carried out in the same furnace as described previously. The specimens are placed in a high purity alumina crucible (Marganite, ΔRR) during annealing. The groove widths were recorded on photographic film after each annealing using a Zeiss Interference Microscope.

RESULTS AND DISCUSSION

The Decay of Sinusoidal Gratings:

A. Morphology

Typical morphology of the gratings after annealing is shown in Figure 1. The gratings shown are on (0001) surface. The morphology of the gratings on the two other surfaces, (11$\bar{2}$3) and (11$\bar{2}$0), can be found in reference 8.

It is seen from the figure that sinusoidal gratings (lower left photograph) are produced only in the direction where the gratings are parallel to the line joining the pole of the surface with the pole of the (0001) plane in the center stereogram. When the grating direction is normal to the line joining the two poles (lower right photograph) the grating will evolve upon annealing into facets to expose the nearest low index plane, (0001). The facets in Figure 1 are identified as the (0001) plane. These experimental results demonstrate the anisotropic nature of the Al_2O_3 surface and suggest the possibility of the existence of a cusp in the γ plot at (0001).

Faceting was not observed on the other two orientations, (11$\bar{2}$3) and (11$\bar{2}$0), in the present work. However, the anisotropic nature of the surface was demonstrated by the existence of unsymmetrical gratings when the direction of the gratings was not parallel to the line joining the pole of the surface and the pole of the nearest low index orientation.

Figure 1:   Interference micrographs showing morphologies of gratings on the (0001) surface.   Center stereogram represents mismatch
of surface and (0001) plane.

B.   Kinetics Data and Analysis

The amplitude of the sinusoidal gratings as a function of
time is,[11,12]

$$A(t) = A_o \exp[-(B\omega^4 + C\omega^3)t] \qquad (1)$$

where A_o = initial amplitude, t = time, ω = frequency of the grat-
ings, $B = \upsilon\gamma D_s\Omega^2/kT$ and $C = C_o\gamma D_v\Omega^2/kT$ and υ = surface concentration of diffusing species, D_s = surface diffusivity, Ω = volume of
the diffusing species, γ = surface tension, C_o = bulk concentration,
D_v = volume diffusivity and kT has the usual meaning.

The negative slope, S, of the ln A vs t plots is,

$$S = -\frac{d\ln A}{dt} = (B\omega^4 + C\omega^3) \qquad (2)$$

According to Equation (2) when the function S/ω^3 is plotted against
ω, one obtains a straight line; its slope is "B" and the y-intercept
is "C".[12]

The kinetics data of the decay of the amplitude of the sinusoidal grarings obtained at several temperatures on (0001), (11$\bar{2}$3)

and $(11\bar{2}0)$ surfaces are given in the form of S/ω^3 vs ω plots in
Figures 2-6. In the figures the y-intercept is chosen to be the
maximum allowable value of "C" based on the grain boundary grooving
data.[8] This choice of the values of "C" will facilitate the determination of the slope of the S/ω^3 vs ω plots to obtain the value of
"B" parameter.

The slope of the S/ω^3 vs ω plots contains the surface diffusivity. Therefore a higher value of D_S will correspond to a higher
slope in these figures. The 1600°C data are plotted with an expanded S/ω^3 scale in Figures 2-4 to show the reproducibility of
the data. It should be noted that in the figures the surface orientation is specified by the nearest low index plane and also by the
deviation of the surface orientation from the nearest low index
plane $\Delta\theta$.

The values of the "B" parameter evaluated are given in Table
I.

The values of the constants required to calculate the surface
diffusivity D_S from the "B" parameter are taken to be: $\upsilon = \Omega^{-2/3}$,
$\Omega = 2.11\times10^{-23}\text{cm}^3$ corresponding to the volume of $1/2$ (Al_2O_3) and
$\gamma = 905$ erg/cm[3].[4] The data in Figure 5 and Figure 6 give

$$D_S = 1.05\times10^6\exp[-(107\pm3\text{ kcal/mol})/RT]$$

$$D_S = 2.77\times10^6\exp[-(119\pm7\text{ kcal/mol})/RT)$$

for (0001) and $(11\bar{2}0)$ surfaces, respectively.

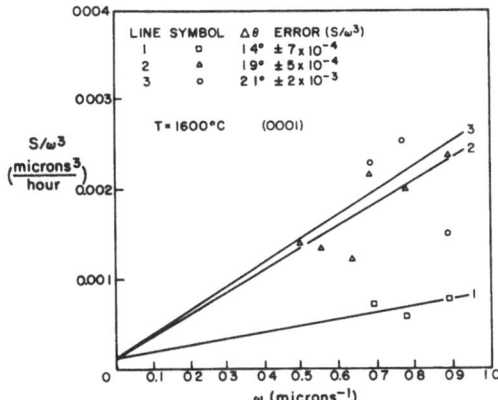

Figure 2: Plot of S/ω^3 vs ω for
(0001) surface annealed at 1600°C.
The deviations of the surface orientation from the nearest low index plane, $\Delta\theta$, are different for
each specimen. Data line 2 is
the same as the data line 2 in
Figure 5.

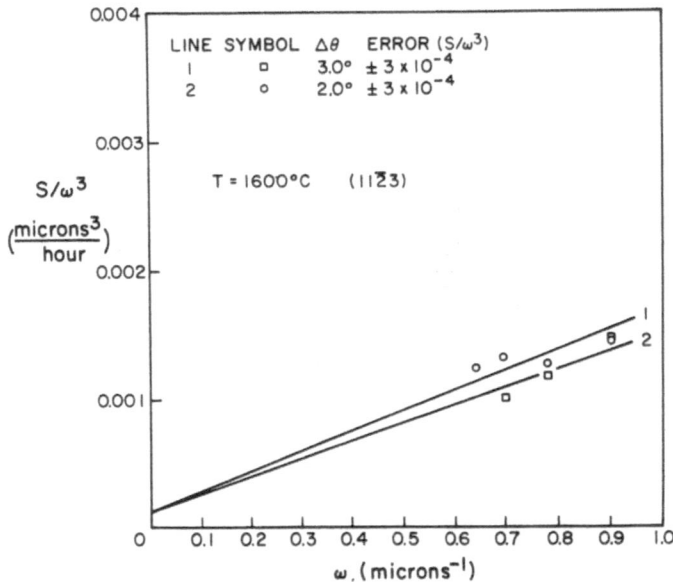

Figure 3:   Plot of S/ω^3 vs ω for $(11\bar{2}3)$ surface annealed at 1600°C.

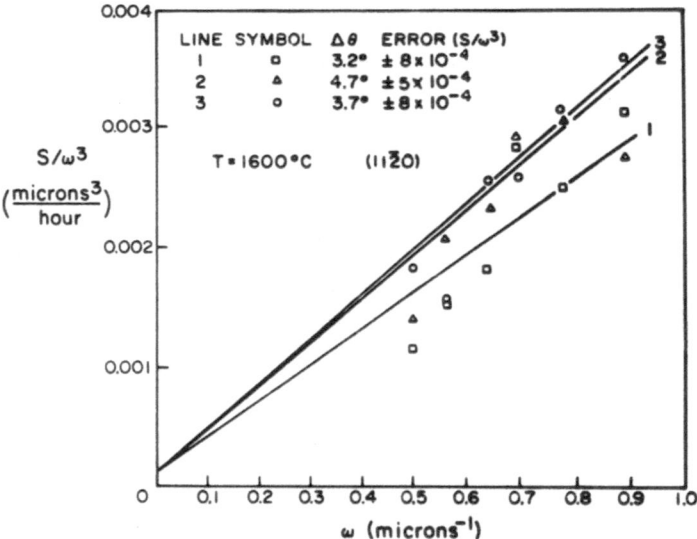

Figure 4:   Plot of S/ω^3 vs ω for $(11\bar{2}0)$ surface annealed at 1600°C. Data line 2 is the same as the data line 2 in Figure 6.

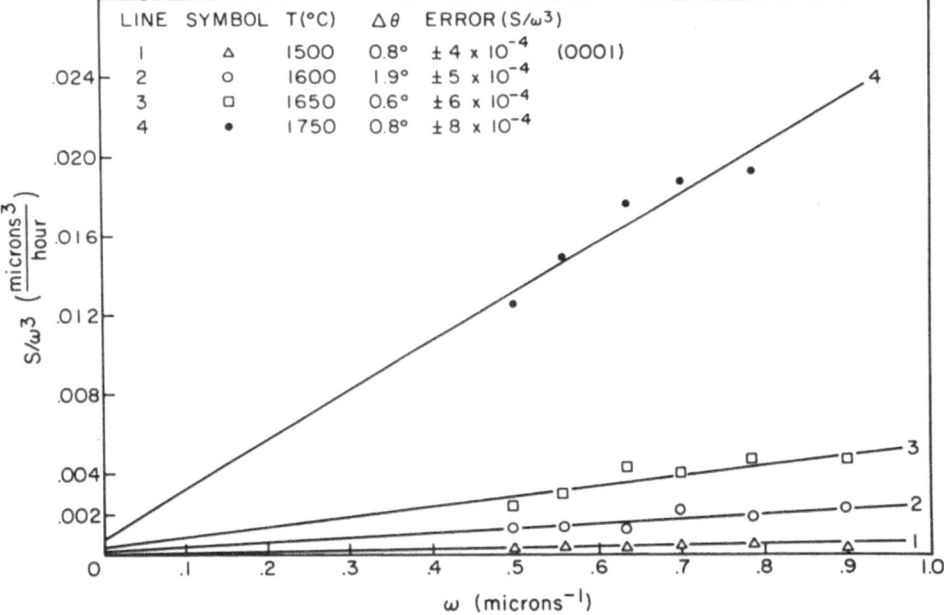

Figure 5:   Plot of S/ω^3 vs ω for (0001) surface annealed at 1500°C, 1600°C, 1650°C and 1750°C. Points on y-axis correspond to the maximum volume diffusion possible as explained in text.

Table I: "B" Parameters from Figures 2-6

Figure	Line	Temperature	Orientation	$B\mu m^4/hr$
2	1	1600°C	(0001)	0.00070
2	2	1600°C	(0001)	0.0024
2	3	1600°C	(0001)	0.0026
3	1	1600°C	(11$\bar{2}$3)	0.0015
3	2	1600°C	(11$\bar{2}$3)	0.0016
4	1	1600°C	(11$\bar{2}$0)	0.0030
4	2	1600°C	(11$\bar{2}$0)	0.0034
4	3	1600°C	(11$\bar{2}$0)	0.0035
5	1	1500°C	(0001)	0.00060
5	2	1600°C	(0001)	0.0024
5	3	1650°C	(0001)	0.0053
5	4	1750°C	(0001)	0.0250
6	1	1500°C	(11$\bar{2}$0)	0.00060
6	2	1600°C	(11$\bar{2}$0)	0.0034
6	3	1650°C	(11$\bar{2}$0)	0.0058
6	4	1750°C	(11$\bar{2}$0)	0.0337

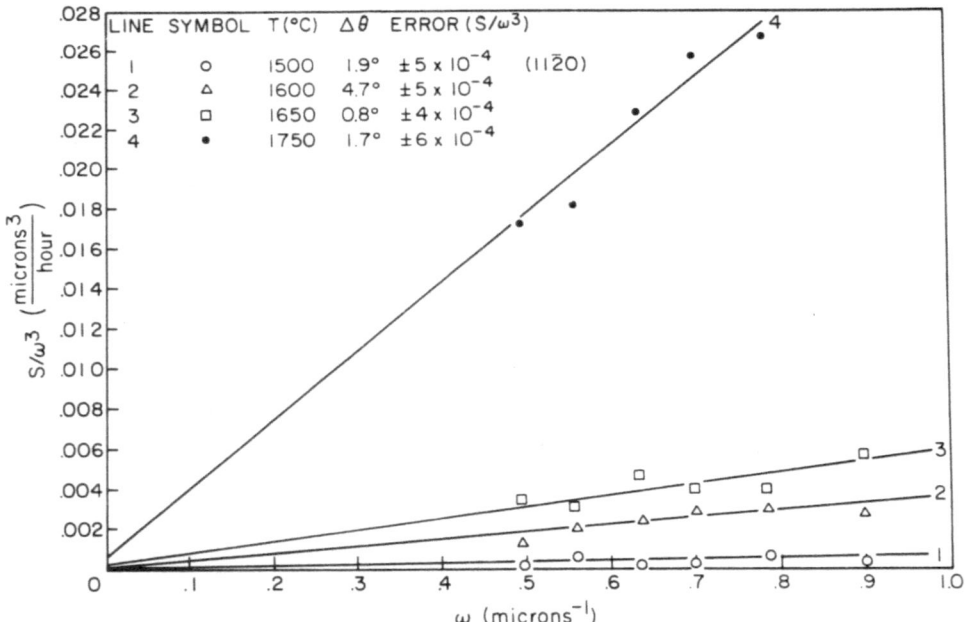

Figure 6:   Plot of S/ω^3 vs ω for (11$\bar{2}$0) surface annealed at 1500°C, 1600°C, 1650°C and 1750°C.   Points on the y-axis are the same as in Figure 2.

From the values of the "B" parameter and "C" parameter it can be concluded that in the range of the wave length of the gratings, surface orientation and temperature included in the present work, the decay of the amplitude of sinusoidal gratings on the Al_2O_3 surface is controlled by mass transport along the surface. Present data suggest that surface diffusion on the Al_2O_3 surface not only depends on the gross surface orientation but is also sensitive to the value of $\Delta\theta$, which measures the deviation of the surface orientation from the nearest low index plane as shown in Figure 1. In general the larger the value of $\Delta\theta$, the larger the value of the "B" parameter.

The kinetics data discussed above are taken from gratings whose direction is nearly parallel to the line joining the pole of the surface and that of the nearest low index plane. Since the direction of the facets (Figure 1) and that of mass transport are both normal to this direction, the direction of the mass transport which produces the amplitude decay is therefore parallel to that of the facets. It may be speculated that facets of the same direction but of atomic scale exist [8] and these atomic steps can serve as paths for fast diffusion. The density of these small facets will increase with increasing $\Delta\theta$. This possibility is consistent with the result that a larger "B" parameter is found on surface with higher $\Delta\theta$.

On the $(11\bar{2}3)$ surface the gratings in all four directions are of adequate quality to evaluate the values of the "B" parameter.[8] The "B" value obtained from the gratings which are normal to the direction joining the two poles discussed previously is 50% lower than that for the other three directions. This directional dependence is also consistent with the argument presented in the previous paragraph.

Grain Boundary Grooving:

The grain boundary grooving experiments yield grain boundary width, W, vs time, t, data. Depending on the mass transport mechanism, the grain boundary width will show different time dependence. There are controversies in the literature concerning the analysis and the interpretation of grain boundary grooving data.[8,13,14] In the present work the grain boundary width data will be analyzed in the form of $\log(W^n - W_0^n)$ vs log $(t - t_0)$ plots where n is a constant depending on the mass transport mechanism and the subscript zero refers to a particular pair of experimental data at early times. The justification for this method of data analysis can be found in reference 8.

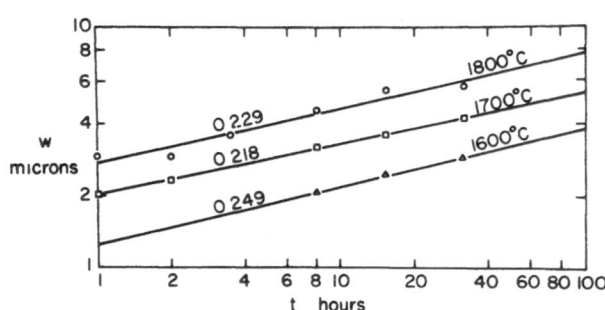

Figure 7: Plot of log W vs log t for experimental (Al_2O_3) grain boundary grooving data.

When volume diffusion is controlling a plot of $\log(W^n-W_o^n)$ vs $\log(t-t_o)$ gives a slope of one for n=3 and a slope greater than one for n=4. When surface diffusion is controlling the $\log(W^n-W_o^n)$ vs $\log(t-t_o)$ plots give a slope of one for n=4 and a slope less than one for n=3. For the case of mixed volume and surface diffusion mass transport this plot gives a slope greater than one for n=4 and a slope less than one for n=3.

The log W vs log t plots of grain boundary grooving data are presented in Figure 7. The graphs at 1700°C and 1800°C have slopes less than 0.25. A slope of 0.25 corresponds to surface diffusion controlled mass transport.[11] In Figures 8(a), (b), and (c) the same data are given in $\log(W^n-W_o^n)$ vs $\log(t-t_o)$

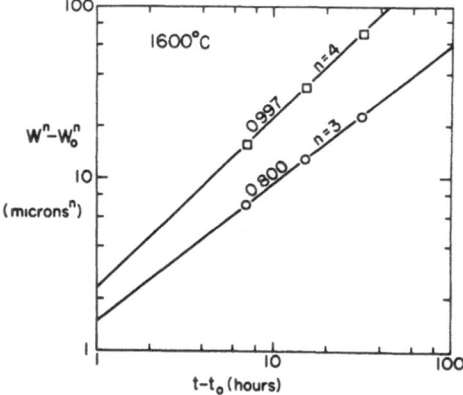

Figure 8a: Plot of $\log(W^n-W_o^n)$ vs $\log(t-t_o)$ for experimental data at 1600°C.

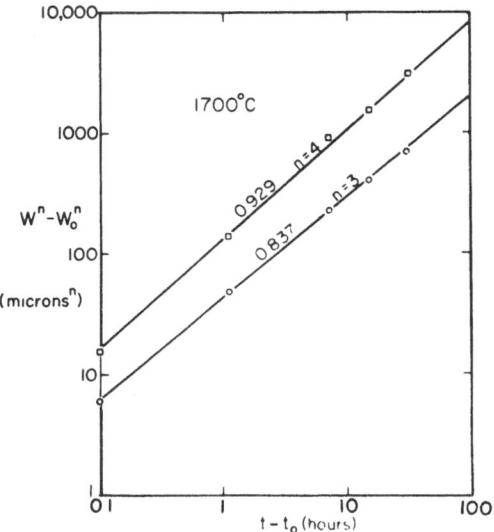

Fig. 8b: Plot of $\log(W^n-W_o^n)$ vs $\log(t-t_o)$ for experimental data at 1700°C.

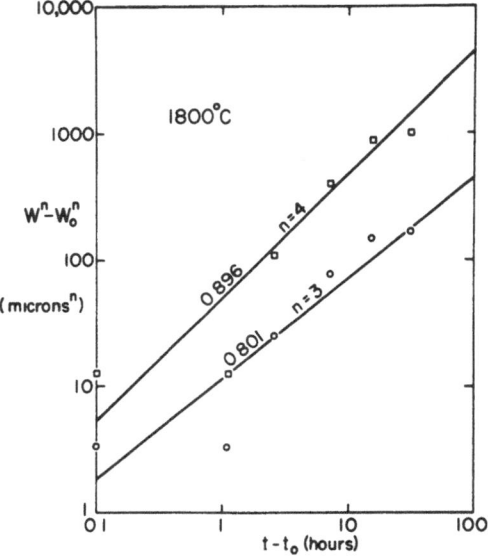

Figure 8c: Plot of $\log(W^n-W_o^n)$ vs $\log(t-t_o)$ for experimental data at 1800°C.

plots where W_o and t_o are obtained from the lines in Figure 7 at
$t = 0.9$ hr. It is seen for n=4 the slopes of the plots are closest to one and for n=3 the slopes are far below one. These results suggest that surface diffusion is the controlling mass
transport mechanism for grain boundary grooving at these temperatures for the (11$\bar{2}$3) surface. Together with the data for the decay of sinusoidal grating on surfaces with other orientations,
we conclude that it is doubtful that volume diffusion is important
as a mass transport mechanism for grain boundary grooving. The
conclusion given in reference 5 is different from the above. A
detailed discussion of both reference 5 and the implication of the
present results for determining the important mass transport
mechanisms in the initial stage of sintering of Al_2O_3 powder compacts are given in reference 8.

The small deviation from the slope of one for n=4 at 1700°C
and 1800°C cannot be attributed to the contribution of the volume
diffusion processes. If this were the case, the slope of the
plots in Figure 8(a), (b), and (c) for n=4 should be greater than
one as discussed previously.

The parameter "B" can be evaluated using the data in Figures
8(a), (b), and (c) and the following equation,[11]

$$W^4 - W_o^4 = (4.6)^4 \, B \, (t-t_o) \qquad (3)$$

The values of "B" parameter are given in Table II.

Table II: "B" Parameters From
Figures 8 (a), (b), and (c) and Equation 3

$T(°C)$	$B(\mu m^4/hr)$
1600	0.0050
1700	0.029
1800	0.108

Using the values of the constants given previously the surface diffusivity for 1/2 (Al_2O_3) can be calculated from the grain
boundary grooving data as

$$D_s = 1.3 \times 10^7 \, \exp[(122 \pm 4 \text{ kcal/mol})/RT]$$

The activation energy for surface diffusion for the (11$\bar{2}$3)
surface from grain boundary grooving experiments is close to that
for the (0001) and (11$\bar{2}$0) surfaces obtained in the experiments on
the decay of sinusoidal gratings. The differences in pre-exponential constants for these surface diffusivities may not be entirely

due to the effects of orientation. For example by comparing the
value of "B" in Table I to that in Table II for 1600°C and the
(11$\bar{2}$3) surface, it is seen that the "B" parameter from the grain
boundary grooving experiments is larger. This difference may
partly result from the environmental effects during annealing.
The sapphire plaques for the grain boundary grooving studies were
contained in a Morganite Δ RR Al$_2$O$_3$ crucible during annealing,
whereas the sapphire pla ues for the experiments of the decay of
the sinusoidal gratings were contained in a sapphire crucible. Deposits were seen on the grain boundary grooving specimen annealed
at 1600°C indicating the presence of an impurity atmosphere. The
deposits have been analyzed to contain Si, Ca and Ta. The sapphire
plaque which serves as a lid for the sapphire crucible for the
latter experiments contained in a Morganite crucible shows similar
deposits at 1600°C while the plaques for the experiment of sinusoidal gratings contained within the sapphire crucible are free
of deposits.

One motivation of performing the grain boundary grooving experiments is to obtain data to compare with results in the literature. The grain boundary grooving data by Robertson and Ekstrom[4]
and Shackelford and Scott[5] have been analyzed in reference 8 to
give values of D$_s$ of 4.3x10^9 exp (-139,000/RT) and 1.2x10^5 exp
(-101,000/RT) respectively. The annealing environments of these
experiments were close to that for the present grain boundary
grooving experiments. Since the present work has demonstrated
that the effects of orientation and annealing environments are not
very strong, it is suspected that the difference between the
present results and those in references 4 and 5 may be due to the
considerable scatter in the latter data.

SUMMARY

For the interest of the morphology of Al$_2$O$_3$ surface and
capillarity induced changes, the present work has found,

(1) The Al$_2$O$_3$ surface is anisotropic. The morphology of the
surface grating suggest the existence of a cusp in the γ plot at
the (0001) plane.

(2) The grain boundary grooving data on the (11$\bar{2}$3) surface
at 1600°C, 1700°C, and 1800°C demonstrate that surface duffusion
is the controlling mass transport mechanism.

(3) The data of the decay of sinusoidal gratings for the
three orientations investigated suggest that surface diffusion is
also controlling when the grating wave length is smaller than 30
μm.

ACKNOWLEDGEMENT

This work was supported by the Army Research Office - Durham. The use of the facilities at the Materials Science Center of Cornell University is acknowledged. The authors wish to express their appreciation for the assistance of Mr. B. G. Addis.

REFERENCES

1. Huang, F.H. and Li, Che-Yu, Scripta Met. $\underline{7}$, 1239 (1973).

2. a. Park, J.Y., M.S. Thesis, Cornell University (1971).
    b. Park, J.Y. and Li, C.Y., Met. Trans. $\underline{3}$, 1670 (1972).

3. DiMarcello, F.V., Key, P.L., Williams, J.C., J. Am. Ceramic Soc. $\underline{55}$, 509 (1972).

4. Robertson, W.M. and Ekstrom, F. E., <u>Materials Science Research</u>, Vol. 4, (T.J. Gray and V.D. Frechette, ed.) Plenum Press, New York, 1969, p. 273.

5. Shackelford, J.F. and Scott, W.D., J. Am. Ceramic Soc. $\underline{51}$, 688 (1969).

6. Prochazka, S. and Coble, R.L., Phys. Sintering $\underline{2}$, 15 (1970).

7. Yen, C.F. and Coble, R.L., J. Am. Ceramic Soc. $\underline{55}$, 507 (1972).

8. Henrichsen, R.A., Ph.D. Thesis, Cornell University (1973).

9. Henrichsen, R.A. and Li, C.Y., Rev. Sci. Inst. $\underline{39}$, 1770 (1968).

10. Henrichsen, R.A. and Li, Che-Yu, <u>Proceedings of Third International Conference on Sintering and Related Phenomena</u>, Notre Dame, Indiana, 1972, Plenum Press.

11. Mullins, W.W., <u>Metal Surfaces</u>, ASM, Metals Park, Ohio, 1963.

12. a. Maiya, P.S., Ph.D. Thesis, Cornell University (1966).
    b. Maiya, P.S. and Blakely, J.M., J. Appl. Phys. $\underline{38}$, 698 (1967).

13. a. McAllister, P.V., Ph.D. Thesis, Univ. of Utah, (1970).
    b. McAllister, P.V. and Cutler, I.B., Met. Trans. $\underline{1}$, 313 (1970).
    c. McAllister, P.V. and Cutler, I.R., Adv. in Chem. Phys. $\underline{21}$, 669 (1971).
    d. McAllister, P.V. and Cutler, I.B., J. Am. Ceramic Soc. $\underline{33}$, 351 (1972).

14. Gjostein, N.A., Met. Trans. $\underline{1}$, 315 (1970).

SINTERING OF PELLETED CATALYSTS FOR AUTOMOTIVE EMISSION CONTROL

I. AMATO, D. MARTORANA AND B. SILENGO

FIAT - D.C.R. - Laboratori Centrali

TORINO - ITALY

INTRODUCTION

The development of the catalytic converter system is one of the approaches followed in order to reach a substantial reduction of the pollutants from the exhaust gases of the automobile.

Particulate catalysts consist of noble metal active components dispersed on high surface area ceramic supports; such a support plays an essential part as regards the activity, stability, thermal, and other critical properties of the finished catalyst. Aside from its purely mechanical function (that is, to act as a base for the catalytic component), other effects of a support are:
- to give a large exposed surface for the active agent;
- to provide surface sites not provided by the active agent.
The support must have high surface area, hardness, attrition resistance, and a broad range of possible pore structures. It is well-known that varying conditions make the auto exhaust emission control unique among catalytic processes: the catalytic converter must operate in limited space and almost completely in an unsteady state and yet it must perform adequately in the temperature range of 500-1000°C.

Alumina, and particularly gamma alumina (1), has been the universal choice as a support for the auto exhaust catalysts, since this material represents the best compromise between the physical and chemical properties, required for these supports.

If one assumes the overall thermal stability problem from a fundamental view-point, the thermal transformation of active

transition aluminas into inactive low surface alpha alumina requires a substantial lattice rearrangement, typically from a
cubic close-packed structure to hexagonal close-packed structure
(11). Resistance to such thermal transformations, which is of
great importance in auto exhaust applications, can be achieved
by stabilizing the tetrahedral aluminum sites by addition of a
suitable additive (2 - 3). Foreign ions incorporated into gamma
alumina, can hinder the aluminum diffusion and thereby stabilize
the crystal lattice of gamma alumina. Of course, cations which
form a spinel, no longer show good stabilization at higher temperature and often may even accelerate recrystallization. Only
those cations which, although present in the lattice of gamma
alumina, are not spinel formers, show a stabilizing effect on long
heating at 1000°C and above.

Another way to stabilize gamma alumina is by addition of
silica that probably forms, by reaction with alumina, a glassy
phase which is extremely viscous. That viscous phase coats the
alumina grains surface and retards the migration of alumina and
thus retards the transformation of the phases (5).

The scope of this paper is an examination of the thermal
stability of different kinds of particulate catalysts with or
without stabilization. This analysis was performed by the evaluation of the chemical and physical properties during sintering at
high temperatures.

1. EXPERIMENTAL RESULTS

Thirty catalysts subdivided as follows have been examined:
4 extruded catalysts, without stabilization; 10 pelletized catalysts, without stabilization; 8 pelletized catalysts, stabilized
with silica; 4 gelified catalysts, stabilized with silica; 4 gelified catalysts, stabilized with BaO.

All these catalysts have, as active components, noble metals.
The catalysts have been annealed, in static air, at temperatures
in the range 800-1250°C. In this range, changes of the following
support properties have been evaluated:
1.1 Crystalline structure. 1.2. Density. 1.3. Surface area.
1.4 Pore size distribution.

The maximum duration of thermal treatment, at each temperature, has been fixed at thirty hours. For this time interval, it
was found that all the samples investigated had reached the limiting values of the four above-mentioned properties.

1.1. CRYSTALLINE STRUCTURE

The evolution of crystalline structure of catalysts after thermal treatments at 1100°C for times between four and thirty hours has been examined using the X-Ray diffraction analysis. The results of this analysis are reported in Table 1. On the basis of these results it is possible to verify that:
- in the unstabilized catalysts after four hours at 1100°C the prevalent crystalline phase is alpha alumina; after thirty hours the alumina is practically all in the alpha phase;
- in the catalysts stabilized with BaO, alpha alumina has not been detected; after thirty hours there is a presence of delta and theta alumina; moreover, two new compounds from the reaction of BaO with Al_2O_3 have been detected;
- in the catalysts stabilized with silica after thirty hours the prevalent phase is alphy alumina; in spite of the intimate contact between silica and alumina, the presence of mullite was not detected (6).

Table 1 - Crystalline structure evaluation of the catalysts annealed at 1100°C.

Initial composition	after 4 hours		after 20 hours		after 30 hours	
γ Al_2O_3	α Al_2O_3 γ Al_2O_3 δ Al_2O_3 θ Al_2O_3	prevalent phase	α Al_2O_3 γ Al_2O_3 δ Al_2O_3 θ Al_2O_3	prevalent phase	α Al_2O_3	
γ Al_2O_3 + SiO_2	γ Al_2O_3 δ Al_2O_3 θ Al_2O_3	prevalent phase	γ Al_2O_3 δ Al_2O_3 θ Al_2O_3 α Al_2O_3		α Al_2O_3 θ Al_2O_3 δ Al_2O_3 γ Al_2O_3	prevalent phase
γ Al_2O_3 + BaO	γ Al_2O_3 BaO·Al_2O_3	prevalent phase	γ Al_2O_3 BaO·Al_2O_3	prevalent phase	γ Al_2O_3 δ Al_2O_3 θ Al_2O_3 BaO·Al_2O_3 BaO·6Al_2O_3	prevalent phase

In the unstabilized catalysts the cubic close-packed lattice
of gamma alumina is converted, via a free transformation, to the
hexagonal packed lattice of alpha alumina. In the catalysts stabilized with BaO, the Barium ions diffusing partially in the de-
fective lattice of gamma alumina, can occupy the vacant positions
of lattice or they can replace the aluminum ions in the lattice.
It follows that Barium cations become incorporated into gamma
alumina with a very good distribution (Fig. 1), and they will hinder the aluminum diffusion at least over a small temperature
range thereby, they act as a stabilizer (7). Consequently the
rearrangement of the oxygen lattice becomes difficult, and the
transformation of gamma alumina to alpha alumina at higher temperatures is delayed. In the catalysts stabilized with silica the
silicon ions are not incorporated in the gamma alumina lattice, but
the silica forms a glassy phase on the surface of alumina grains
that inhibits the gamma → alpha alumina recrystallization.

Figure 1 - Barium distribution in gamma alumina support by X-ray
           emission (100 X).

1.2. Density

The density of the catalysts has been measured with a mercury
picnometer; the density variation, $\Delta D/D_0$ (D_0 is the initial density), of the catalysts is reported in Figure 2 versus the temperature.

Examination of these data suggests the following:
- no difference, in terms of $\Delta D/D_0$, is shown by the catalysts obtained with different methods of fabrication;
- three different curves of density variation are shown, one for

the unstabilized catalysts, one for the catalysts stabilized
with BaO and, finally, one for the catalysts stabilized with
silica;
- in the 800-1050°C temperature range the lowest density variation
is shown by the catalysts containing silica;
- in the case of temperatures >1050°C the lowest density variation
is by the catalysts stabilized with BaO.

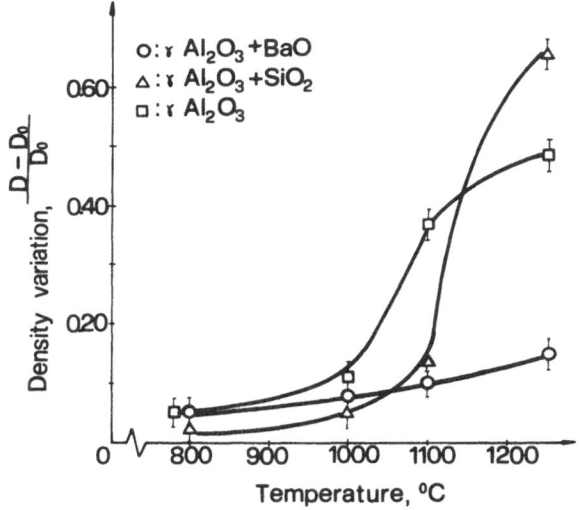

Figure 2 - Density variation of the catalysts after annealing at a
fixed time of 30 hours.

1.3. Surface Area

The use of high surface area supports provides the best possible dispersion and utilization of the noble active components
and the achievement of the catalysts stability as well. If the
noble metal is dispersed on a low area alumina, it is present in
the form of crystallites of size 50 Å (8). On the other hand when
the noble metal is dispersed on a high surface alumina it is in
the most resistant state with respect to sintering and activation
at high temperatures (9). The dispersion and stability imparted
to the active phase by high surface alumina supports seem to be
due both to the fine pore structure of the support and to reactive sites on the surface. These sites, the amount of which is

related to the surface area, can also affect the catalyst activity
because they can act as adsorption sites or reactants that will
then diffuse to catalytic sites. It is possible to verify that
the catalyst activity rises from zero to a maximum as surface
area (BET) increases from near zero to a value typical of a given
catalyst. Further increases in area have essentially no influence
on activity (10). It follows that any process that decreases
available area below this requirement will cause a decreased activity. Therefore, it is important that the supports have a high
surface area, but it is even more important that the catalysts
have not a large loss in surface area during the service.

The surface area of the catalysts examined has been measured
with a sorptometer following the BET method. The surface area
variations, $(S_0-S)/S_0$ (S_0 is the initial surface area), plotted
versus temperature are shown in Figure 3. There is an analogy between the curves reported in Figure 3 and those relative to density variation (Figure 2). Also in this case the difference in
the catalysts exists as a function of the presence and type of the
stabilizer. It is important to point out that in the 800-1100°C,
the lowest variations are shown by the catalysts stabilized with
BaO.

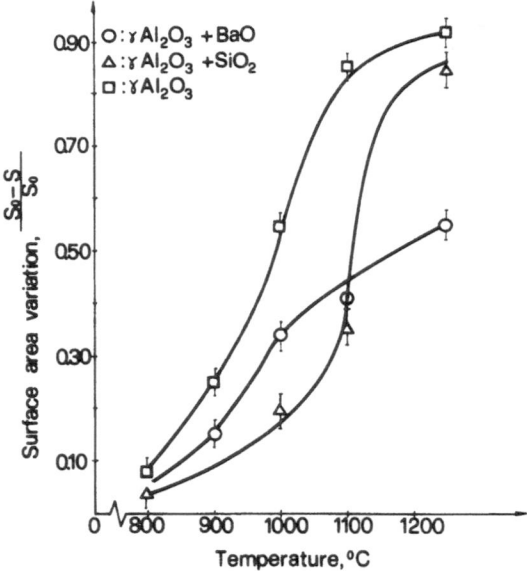

Figure 3 – Surface area variation of the catalysts after annealing
           at a fixed time of 30 hours.

Because the thermal treatment induces in the support examined phase transformations with a consequent decrease of surface area, the hypothesis can be advanced that surface area (BET) is related to the temperature in the following way:

$$S \; \alpha \; f \; (\frac{1}{T^*-T}) \qquad\qquad [1]$$

that is S can be related to $1/(T^*-T)$ by a specific nonlinear function f; in the equation [1] S is the surface area, T is the temperature and T^* has the physical significance of phase transformation temperature and represents the temperature at which the complete transformation of gamma alumina into alpha alumina is reached. To investigate the type of f, the experimental curves have been examined: it results that a possible f could be an exponential function therefore, the following equation cay be hypothesized:

$$S = S' \; \exp \; (- \; \frac{C}{T^*-T}) \qquad\qquad [2]$$

where S' and C are constants for each type of catalyst. Since to plot the experimental data (Figure 3) it has been chosen as dependent variable $\Delta S/S_0$ (ΔS is the difference between the initial surface area S_0 and the surface area at a given temperature), the equation [2] becomes:

$$\frac{S_0-S}{S_0} = 1 - k \; \exp \; (- \; \frac{C}{T^*-T}) \qquad\qquad [3]$$

where k is a constant characteristic for each type of catalyst. In logarithmic form, the equation [3] becomes:

$$\log(1 - \frac{\Delta S}{S_0}) = \log(\frac{S}{S_0}) = \log k - \frac{C}{2,303(T^*-T)} \qquad\qquad [4]$$

Plotting log S/S_0 versus $1/(T^*-T)$, the curves in Figure 4 were obtained giving to T^* particular values, characteristic for each type of catalyst, chosen in such a way as to obtain the experimental points on a straight line.

The T^* values obtained in this way are as follows:
- 1100°C for unstabilized catalysts;
- 1300°C for catalysts stabilized with SiO_2;
- 1600°C for catalysts stabilized with BaO.

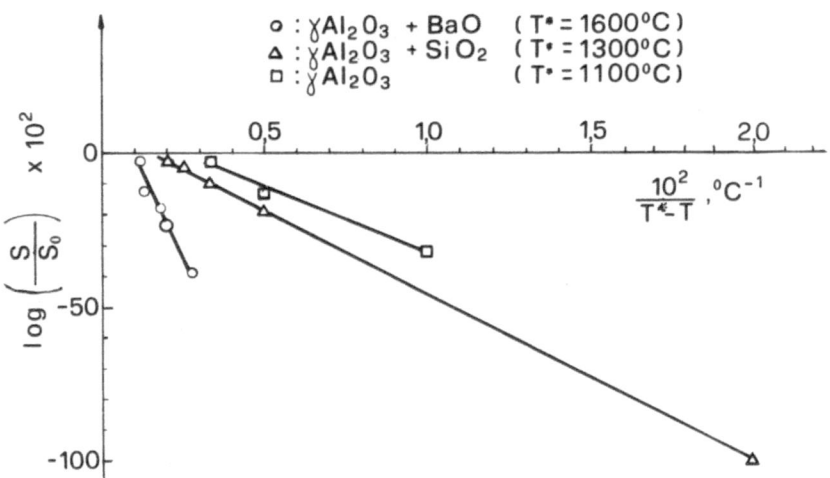

Figure 4 - Log (S/S₀) versus 1/T*-T for the catalysts.

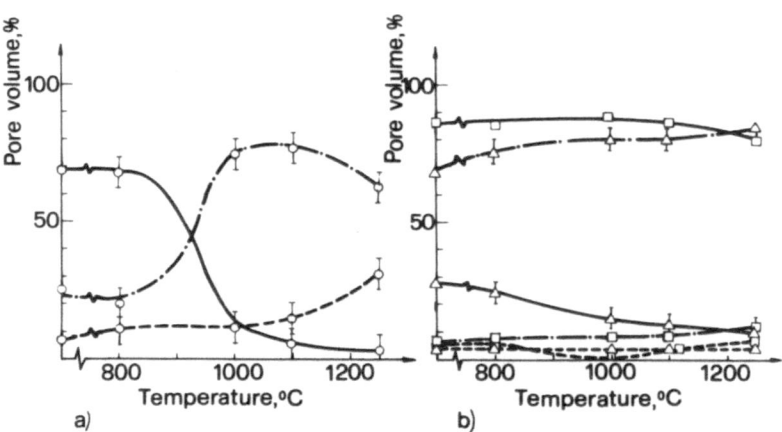

Figure 5 - Evolution of the pore size distribution after annealing
          at a fixed time of 30 hours; _____: micropores, - . - :
          intermediate pores, -----: macropores.
          a) Pelleted catalysts, not stabilized;
          b) ⁻: - gelified catalysts, with silica;
             ⊼: - gelified catalysts, with BaO.

The T* values so obtained are greatest for the catalysts stabilized and in particular, for those stabilized with BaO. These
values of T* have been compared with the temperatures of $\gamma \rightarrow \alpha$
transformation, experimentally determined by annealing the samples
at various temperatures and verifying, by X-ray analysis, the temperature at which the complete $\gamma \rightarrow \alpha$ transformation occurs. It
has been verified that for all catalysts examined the T* values
experimentally obtained are about the same as the values indirectly determined from equation [4]. With such approach it is
possible to know, by the plots of surface area (BET) versus T,
the temperature at which the complete $\gamma \rightarrow \alpha$ phase transformation
occurs. These temperature values show the different thermal stability of the examined catalyst families. Above T* the law does
not describe the phenomenom. Because the density variation shows
analogous behaviour as surface area, this approach can be extrapolated to the density.

                    1.4. Pore Size Distribution

     The total porosity, strictly connected with the surface area,
the distribution of pore sizes in catalysts and the spatial arrangement of their pore network, are very important for catalysis,
because it is through these pore systems that the reactants diffuse over the specific sites. Because of reasonably significant
differences in behaviour, the pores are generally divided into
three types: micropores (diameter smaller than 200 Å), intermediate pores (diameter from 200 to 800 Å) and macropores (diameter greater than 800 Å). Ideally all the potentially active sites
would be located within the micropores and they are readily accessible by means of pores sufficiently large in diameter; the
gas diffusion in them would be of the ordinary, bulk diffusion
type, rather than the Knudsen type (11-12).

     The evolution of pore size distribution with temperature has
been evaluated for the following catalysts:
- n. 2 pelletized unstabilized catalysts;
- n. 3 gelified catalysts, stabilized with BaO;
- n. 2 gelified catalysts, stabilized with silica.
The results of this evaluation are illustrated in Figure 5. The
examination of the curves leads us to the following conclusions:
- the pore size distribution in the stabilized catalysts remains
  almost constant as the temperature increases; there are small
  variationions in the values of micro and intermediate porosity;
- the pore size distribution in the non-stabilized catalysts
  varies greatly in the 850-950°C temperature range; at these
  temperatures, in fact, there is a strong reduction in the micropore fraction and a corresponding increase in the fraction of
  intermediate pores.

Conclusion

The crystalline structure investigation has shown that the stability of the examined catalysts, at 1100°C, decreases in the order: catalysts stabilized with BaO, catalysts stabilized with silica; unstabilized catalysts. The evaluation of the physical properties (density, surface area) has shown that in the 800-1000°C, temperature range, the catalysts containing silica present the lowest properties variation; for the temperatures $\leq$ 1100°C the catalysts containing BaO show the maximum stabilization. The variation of the pore size distribution with increasing temperature is negligible until the temperature of 1250°C is reached for the stabilized catalysts; the non-stabilized catalysts shown, in the 850-950°C temperature range, a complete variation of porosity configuration.

Moreover, it has been verified that the surface area decrease $(S_0-S)/S_0$ with temperature, follows an exponential law, and is a function of $1/(T^*-T)$; T^* can be taken as the temperature at which the complete $\gamma \rightarrow \alpha$ alumina transformation occurs, and can be indirectly calculated from the experimental curves: surface area (BET) versus temperature.

References

1. K. Wefers and G. M. Bell: "Oxides and Hydroxides of Aluminum"; Technical Paper Alcoa Research Laboratories, n. 19, p. 43, 1972.
2. A. E. Smith and A. O. Beech: "Termostable Catalysts for the Dehydrogenation of Hydrocarbons"; U.S. Patent, 2, 422, 1972.
3. H. Krischner, K. Torkar and P. Hronish: Monat. Chem., 99, p. 1733, 1968,
4. H. S. Blok and C. L. Thomas: U.S. Patent, 2, 216, 262.
5. R. K. Iler: Jounral of the American Ceramic Society. 47 (7), p. 339, 1964.
6. F. M. Wahl, R. E. Grim and R. B. Graf: "Phase Transformations in Silica - Alumina Mixtures as Examined by Continous X-ray Diffraction"; Am. Mineralogist, 46 (9-10), p. 1064, 1961.
7. R. M. Levy and D. J. Bauer: J. Catalysis, 9, p. 76, 1967.
8. J. F. Roth and J. W. Gambell: "Control of Automotive Emission by Particulate Catalysts"; SAE, 730277, Jan. 1973.
9. J. Roth: "Copper Based Auto Exhaust Catalysts: Mechanisms of Deactivation and Physical Attrition"; ACS Meeting Los Angeles, March 1971.

10. W. A. Cannon and C. E. Welling: "The Application of Va-
     nadia Alumina Catalysts for the Oxidation of Exhaust Hydro-
     carbons"; SAE Meeting, Jan. 12-16, 1959.

11. H. E. Osment: "Active Aluminas as Catalyst Support for
     Treatment of Automotive Exhaust Emissions"; SAE 730276, Jan.
     1973.

12. M. M. Dubinin: "Porous Structure of Adsorbent and Catalysts";
     Advances in Colloid and Interface Science, 2, p. 217, 1968.

EFFECT OF AMBIENT ATMOSPHERE ON SINTERING OF αAl_2O_3 SUPPORTED Pt CATALYSTS

J. Zahradnik,* E. F. McCarthy,** G. C. Kuczynski and
J. J. Carberry
University of Notre Dame, Notre Dame, Indiana

Abstract

Samples of 0.036 and 0.046% Pt supported on α-Al_2O_3 were sintered in air, argon and nitrogen, at temperatures between 650° and $800^\circ C$. Air sintered samples revealed a signal effect of average crystallite size upon CO oxidation (structure sensitivity) at low CO partial pressures, while structure insensitivity was manifested for the same reaction at higher CO partial pressures. Sintering kinetics in air followed approximately second order kinetics, third being revealed for sintering in argon. In a N_2 atmosphere, interesting anomalies were observed.

Introduction

Sintering or rather coalescence of metal particles dispersed upon refractory supports is of no small concern to the scientists interested in catalysis and reactor engineers. For the growth of supported crystallites is accompanied by:

a) a reduction in specific surface area of the catalytic metal.

b) possible changes in surface structure, coordination number or morphology.

*Present address: Institute of Chemical Process Fundamentals, Czechoslovak Acad. of Sciences, Rozojova 135 16502 Prague 6, Czechoslovakia.

**Present address: Amoco Oil Company, Amoco Chemical Center, Post Office Box 400, Naperville, Illinois 60540

Consequence (a) leads to a reduction in the global rate R of the
catalyzed reaction defined

$$R = \frac{\text{moles reacted}}{\text{time x total weight of catalyst}}$$

the "total" weight of catalyst includes support, promoters, etc.
In contrast the specific rate r

$$r = \frac{\text{moles reacted}}{\text{time x no. of surface sites}}$$

may or may not vary with the degree of sintering. The specific
rate or turnover number will depend on (b) and the sensitivity of
the reaction to that consequence.

The number of surface sites is expressed by dispersion D

$$D = \frac{\text{no. of surface atoms}}{\text{total no. of atoms}}$$

Dispersion D can be measured by:

(1) chemisorption[1] of gases such as H_2 or CO
(2) titration[2], e.g.

$$Pt \; O + \frac{3}{2} H_2 \rightarrow Pt \; H + H_2O$$

or

$$Pt \; O + 2 \; CO \rightarrow Pt \; CO + CO_2$$

Thus sintering of supported Pt group metals of sizes which escape
X-ray detection can be followed by measuring dispersion D. The
behavior of the turnover number or specific rate as a function of
sintering severity sheds light upon the fundamental catalytic
process. In accordance with Boudart's definition[3], if r is independent of D, the reaction is structure insensitive or facile,
otherwise it is called demanding or structure sensitive.

Sintering

In this study dispersion $D = A/A_0$ where A and A_0 are total
areas of metal at time t and o respectively, was measured as a
function of time, temperature and the nature of the ambient gas
phase, for catalysts containing 0, 036% and 0.046% Pt deposited on
α-Al_2O_3 (BET area ~ 4.2 m^2/g.)

Dispersion was measured at room temperature by H_2 titration.
Samples were sintered in air, argon and nitrogen at 600°, 700°,
750°, and $800^\circ C$. The results in the form of A/A_0 as a function
of time plots are given in Figs. 1,2,3 and 4. The sintering
kinetics conform to the phenomenological rate law

$$\frac{dA}{dt} = -KA^n \tag{1}$$

where K is a constant.

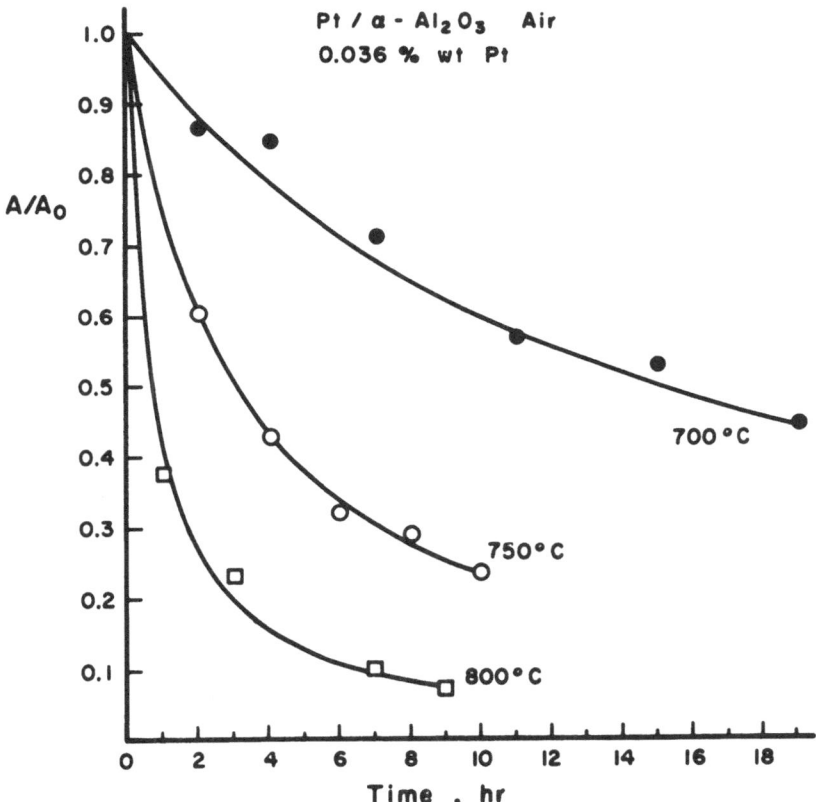

Fig. 1. A/Ao as a function of time t, for a catalyst containing 0.036% Pt sintered in air.

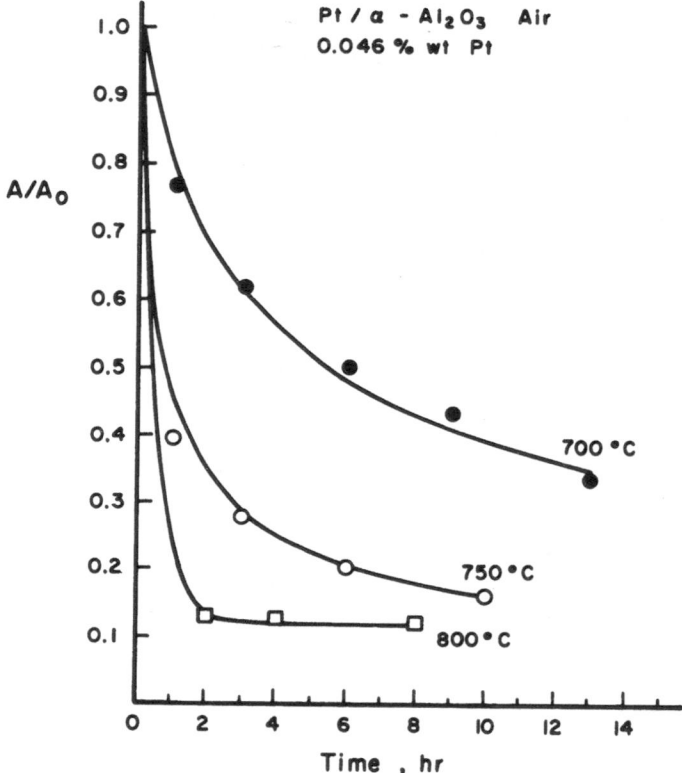

Fig. 2. A/Ao as a function of time t, for a catalyst containing
0.046% Pt sintered in air.

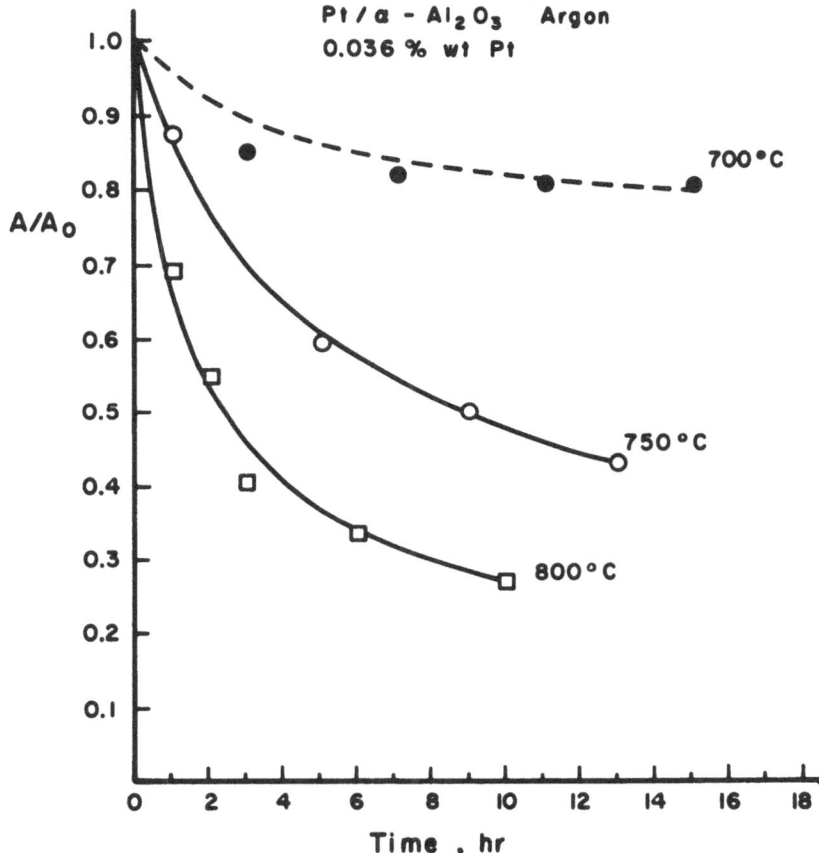

Fig. 3. A/Ao as a function of time t, for a catalyst containing 0.036% Pt sintered in argon.

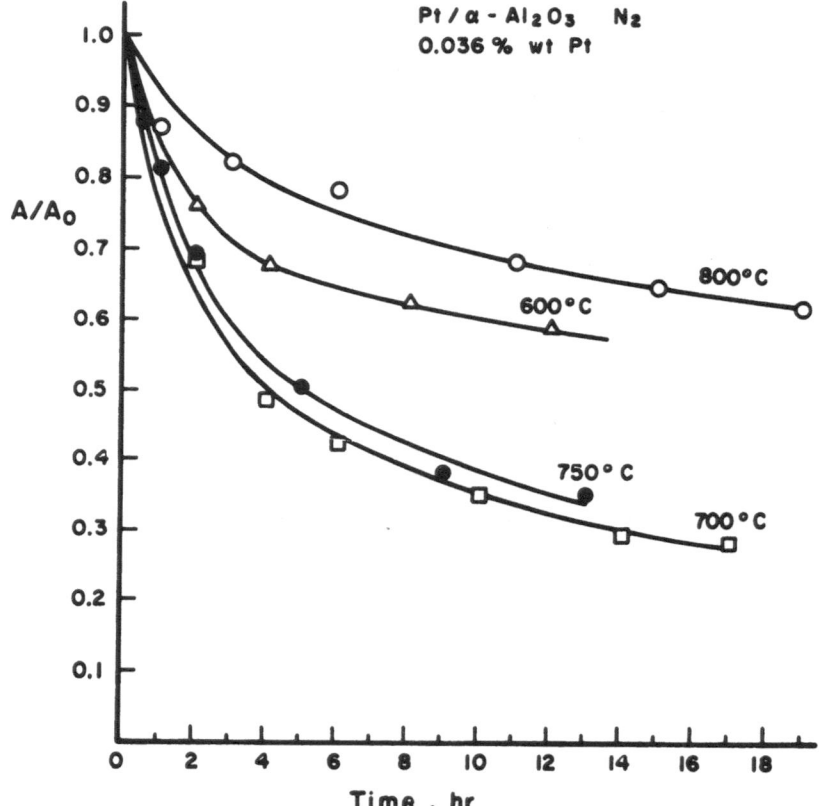

Fig. 4. A/Ao as a function of time t, for a catalyst containing 0.036% Pt sintered in nitrogen.

Table I

Kinetic Parameters used in Sintering Equation 1

Catalyst	Atmosphere	$T^\circ C$	n	$K \times 10^{-5}$	E kcal/mol	$A_o m^2/g$	$d_o Å$
Pt/αAl₂O₃	air	700	2	46.15		145	19
0.036% Pt		750	2	223.7	66.5	145	19
		800	2-3	846.3		162	17
Pt/α-Al₂O₃	air	700	3	5.50		71	39
0.046% Pt		750	3	48.50	70.0	62.5	45
		800	-			69	44
Pt/α-Al₂O₃	Ar	700	high			185	15
0.036% Pt		750	3	0.535	58	177	16
		800	3	2.005		179	16
Pt/α-Al₂O₃	Ar	750	3	8.2		69.6	40
0.046% Pt							
Pt/α-Al₂O₃	N₂	600					
0.036% Pt		700	3	2.71		116	24
		750 800	3	1.07		167	17
			6	1.02		161	17

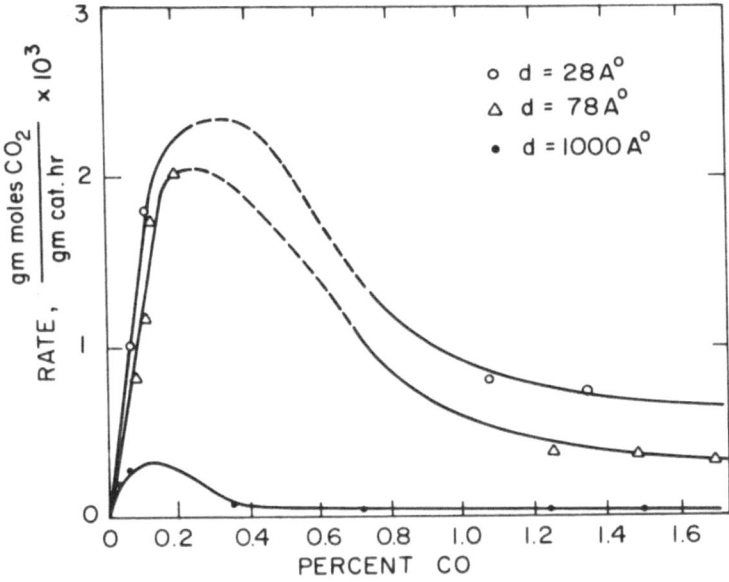

Fig. 5. Effect of particle size on globol rate of CO oxidation at 180°C.

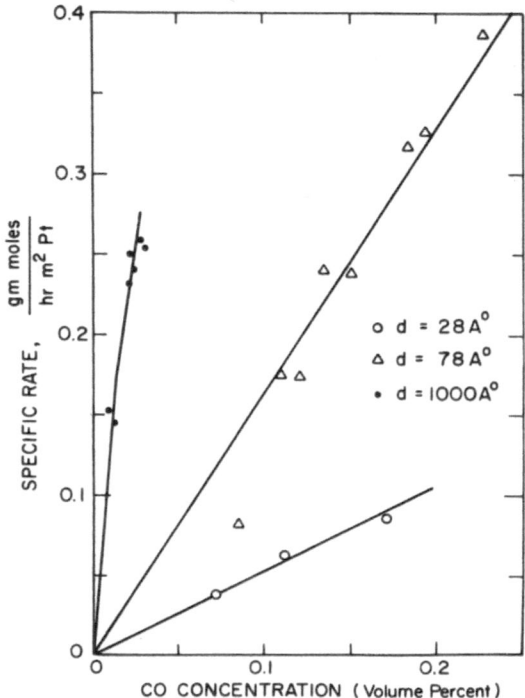

Fig. 6. Effect of particle size on specific rate of CO oxidation at 180°C low CO concentrations.

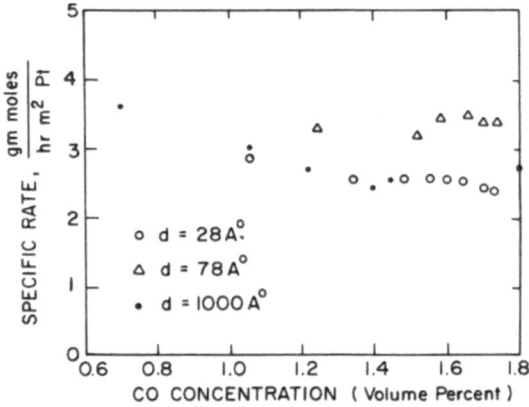

Fig. 7. Effect of particle size on specific rate of CO oxidation at 180°C. High CO concentrations.

In Table 1 there are summarized the kinetic parameters employed in the equation (1). In addition the activation energies E where available as well as the initial Pt area A_o and average initial size d_o are listed.

Specific Reaction Rates

For the catalysts containing 0.036% Pt the rates of CO oxidation were measured in the Notre Dame gradientless reactor[4]. The global rate under the conditions of excess of O_2 was found to conform with the equation

$$R = \frac{k[CO]}{(1+K_1[CO])^2} \tag{2}$$

where [CO] is the CO concentration and K_1 a constant. Hence at small values of K_1 [CO] (low [CO] or/and high temperatures)

$$R = K_1 [CO] \tag{3}$$

while at high values of K_1 [CO]

$$R = K_1/[CO] \tag{4}$$

Rate versus CO concentration for fresh catalyst and two levels of sintering of that same sample are set forth in Fig. 5. The telling influence of sintering upon R is evident.

However when the rate is expressed on a specific metal area basis, i.e., per M^2 of Pt, as in Fig. 6, a low [CO] indicates demanding or structure sensitive behavior. In contrast at higher [CO] structure insensitivity is manifested as shown in Fig. 7.

Discussion

As mentioned above sintering kinetics in air and argon conform to the empirical equation (1). As can be seen from Table 1, the rates of metallic particles growth in air are greater than in argon, although the activation energy of the process in argon seems to be lower than that in air.

It is difficult to identify the mechanism which prevails. The curves of the area change with time for the catalyst containing 0.036% Pt sintered in air seem to fit equation (1) with n = 2, except perhaps at 800°C and for longer sintering times where better fit is obtained with n = 3. An exponent lower than 3, according to prevailing theories would indicate the random walk of particles on the surface of the support[5]. Indeed, the particle size estimated from surface area measurements for low Pt-content catalyst (0.036% Pt) are very small, less than 50 Å which is the upper limit of the feasibility of this mechanism. Therefore it is not inconsistent with this theory that at the temperatures lower than 800°C and for short times of annealing at the latter temperature, particle migration-collision-coalescence are the most likely mechanisms of

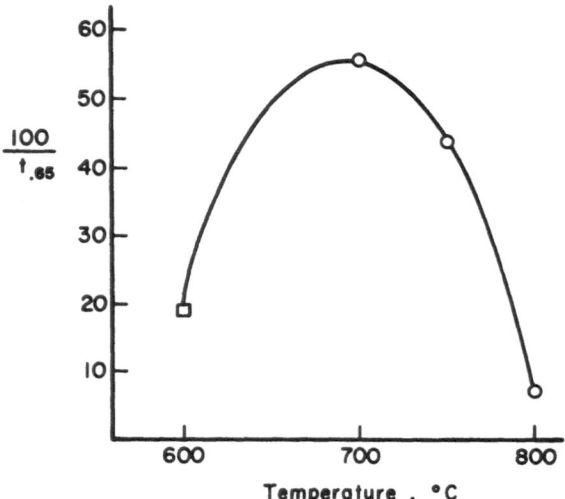

Fig. 8. The relative rate of sintering of a catalyst heated in
nitrogen atmosphere, as a function of temperature.

sintering, which at longer times and high temperatures changes into
Ostwald ripening, (n = 3). Similar experiments with a catalyst
with higher Pt-content (0.046% Pt) yield n = 3. From the start the
average particle size of this catalyst is of the order of magnitude
of 50 Å and then grows larger. Therefore one is tempted to assign
Ostwald ripening-interface reaction controlled mechanism to this
process.

However this interpretation breaks down in the case of sintering of catalyst of low metal-content (0.036% Pt) in Ar. Although
the particles are very fine the kinetics can be described by
equation (1) with n = 3. Still the predominant mechanism could be
that involving particle migration because the n values for this
mechanism can range from 2 to 7. Clearly we are getting deeper and
deeper into pure speculation. It should be remembered that the
assignment of the values of the exponent n to various mechanisms
operating in Ostwald ripening makes sense, only when the particle
distribution is quasi-stationary[6]. But the approach to this state
is a process requiring relatively long time[7], hence there is a
small chance to ascertain the mechanism of sintering of dispersed
fine particles from the experiments as those described in the
present report.

Let us now turn to a truly remarkable effect displayed during
sintering of Pt catalyst (0.036% Pt) in N_2 atmosphere. Convention-

al plot of A/A_o as a function of time is given in Fig. 4. It indicates that the process of sintering is the slowest at highest ($800^\circ C$) and lowest ($600^\circ C$) temperatures. This effect is even more clearly shown in Fig. 8, where relative rates of sintering, defined as the reciprocal time required to reach A/A_o = .65, is plotted against sintering temperature, resulting in a C curve. Such a curve is characteristic of kinetics involving nucleation. The possibility of a growth mechanism controlled by nucleation of new crystallographic planes on a particle has been suggested before[5], however the smallness of the particles used in these experiments seem to rule it out. Other possible explanation is the modification of the surface of the support by reaction with N_2. Indeed it has been demonstrated that $\alpha-Al_2O_3$ is able to dissolve considerable amount of nitrogen[8]. A compound Al_3O_3N having spinel structure has been identified. Therefore if such a reaction could be induced on the surface of support it may change the morphology of the surface by formation of micropores, thus facilitating the nucleation of new Pt particles from the metal atoms migrating on the support surface. In consequence the sintering would be accompanied by redispersion which of course would slow down the process. If the dominant mechanism is that of particle migration, micropores would obviously impede their movement. Clearly more study is needed to resolve this problem.

References

1. J. Sinfelt, Catalysis Reviewes 3 (1967).

2. J. Benson and M. Boudart, J. of Catalysis 4, 704 (1965).

3. M. Boudart, Adv. in Catalysis 20, 153 (1969).

4. J. J. Carberry, Ind. Eng. Chem. 56, No. 11, 39 (1964).

5. P. Wynblatt and T. M. Ahn, "Crystallite Sintering and Growth in Supported Catalysts," this volume.

6. G. C. Kuczynski, "Statistical Approach to the Theory of Sintering," this volume.

7. R. D. McKellar and C. B. Alcock, "Particle Coarsening in Fixed Salt Media," this volume.

8. I. Adams, T. R. AuCoin and G. A. Wolff, J. Elecrochem. Soc. 109, 11, 1050 (1962).

ZEOLITE CHEMISTRY AND REACTIONS

Donald W. Breck

Union Carbide Corporation

Tarrytown, New York

The zeolites are complex, crystalline, inorganic macromolecular compounds. Their chemical and structural aspects are involved singly or cooperatively in a diverse assortment of chemical reactions and transformations. This paper is an attempt to illustrate some of the unusual chemical behavior of these materials.

STRUCTURE

Zeolites are framework aluminosilicates based on an infinitely extending, three-dimensional network of oxygen tetrahedra containing either aluminum or silicon.[1] The framework contains channels and interconnected voids which are occupied by the cations and water molecules. The cations are mobile and may be exchanged by other cations. The intracrystalline zeolite water may be removed continuously and reversibly, leaving a stable, host structure permeated by uniform micropores. In many zeolites cation exchange and/or dehydration does produce structural perturbations which affect the subsequent adsorption properties.

Zeolite Structural Formula

$$M_{x/n} [(AlO_2)_x (SiO_2)_y] \cdot w \, H_2O$$

M is exchange cation of valence n; w is number of water molecules; x + y is the total number of tetrahedra per unit cell.

Unlike the feldspars which have dense structures, zeolite structures are open and contain large cavities. The cavities may be interconnected in one, two or three dimensions. Metal ions which are needed for charge compensation occupy sites in the channel or adjacent to the cavities and are generally available for exchange by other cations.

Zeolite framework structures consist of simple polyhedra, each polyhedron in itself is a three-dimensional array of individual tetrahedra in a simple geometric form. These arrangements of polyhedra are referred to as secondary building units. Examples include single rings of four or six, cubic arrays, and hexagonal prism arrays or double six rings.

In the structure of the synthetic zeolite A, units of double four rings or cubes are joined together[2]. This cubic arrangement produces a larger polyhedron known as the truncated octahedron. (Figure 1) Internally, in the middle of the cube there results another type of polyhedron referred to as the truncated cube octahedron. This internal cavity is 11 A in diameter and is entered through six circular apertures formed by a regular ring of eight oxygens giving a free diameter of 4. 2 A. It is these apertures which produce the molecular sieve character. (Figure 2) In the crystal these cavities are stacked in a continuous three dimensional pattern forming a system of unduloid-like channels with 4. 2 A periodic restrictions. Zeolite A has the following chemical composition of the unit cell:

$$Na_{12} [(AlO_2)_{12} (SiO_2)_{12}] \cdot 27 H_2O.$$

Eight of the twelve sodium ions lie in the center or near the center of the six rings in the hexagonal faces and four occupy positions adjacent to the larger eight-membered rings. Dehydration leaves these four cations in positions near the eight-membered ring where they protrude into the apertures and may interfere with the movement of molecules. When replaced by other cations, such as calcium, the adsorption behavior changes. Since one calcium ion replaces two sodium ions, the cations are removed from the eight-membered rings and the zeolite has a larger effective pore size. Replacement of the sodium by larger univalent ions, such as potassium, restricts the eight-membered ring even more and produces a zeolite of a smaller pore size.

A second structure which illustrates the structural principles is that of the faujasite-type structure which is found in the synthetic zeolites X and and Y.[3] This framework is based upon a different configuration of the truncated octahedra. Eight units are arranged in a diamond-type, tetrahedral configuration which produces eight super-cages of about 13 A in diameter. The cage is entered through four, 12-membered rings which

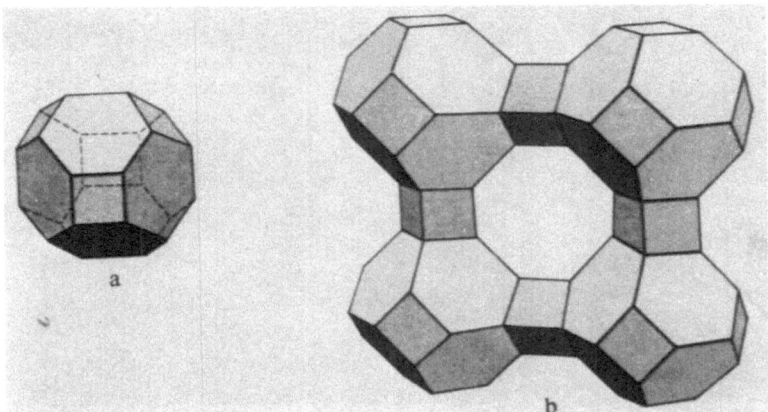

Figure 1   The truncated octahedron (a) and (b) the array of truncated octahedra in the framework of zeolite A.   The linkage is shown via the double 4-rings.

produce a pore size of about 8 A. (Figure 3) The cation positions are even more complex since there are five types of positions. This complex cation siting accounts for the variation in the adsorptive and catalytic properties as a result of cation exchange.

Zeolites A and X and Y have three dimensional channel systems. The zeolite mordenite is different; its structure is more complex and provides for a one-dimensional channel system. The main channels do not intersect but run parallel to one particular crystallographic direction; they have a cross-sectional diameter of about 6 x 7 A.

The total void volume in zeolites is usually calculated from the amount of adsorbed water on the assumption that the water is present as the normal liquid.[4] It is also possible to calculate the micropore volume from the crystal structure if known.

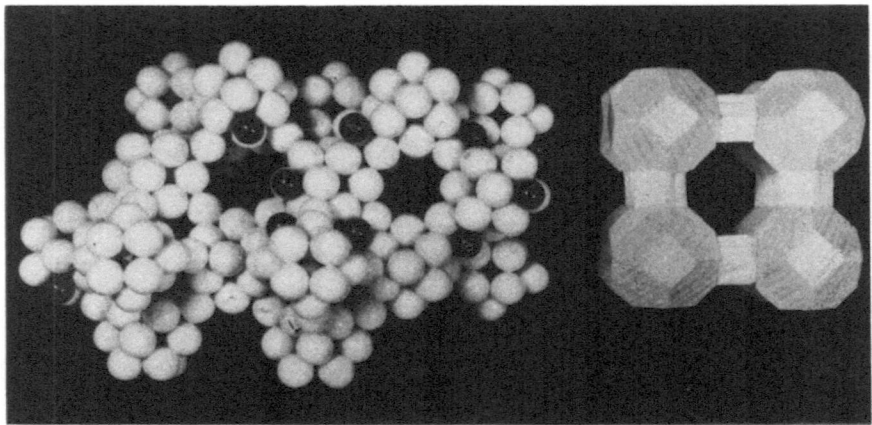

Figure 2   Model of the crystal structure (unit cell) of zeolite A showing
the cubic array of truncated octahedra, r.h., surrounding one adsorption
cavity.   A packing model at the left shows the apertures into the large
cages and some of the exchange cations, Na$^+$, near the main 4.2 A
apertures. The truncated octahedra, each of which consists of 24 octahedra, are joined by cubic units of 8 tetrahedra.

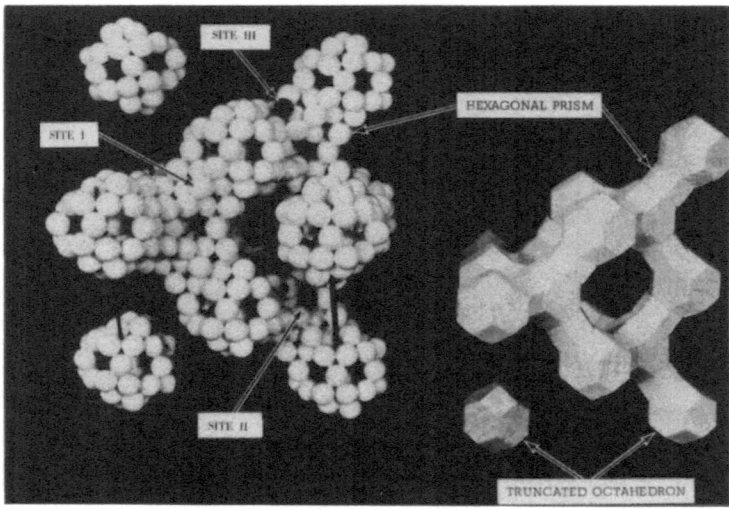

Figure 3   Model of the crystal structure (unit cell) of zeolites X and Y.
On the right is shown the tetrahedral arrangement of truncated octahedra
surrounding one large cavity.   On the left is shown the packing model. The
12-member ring is visible as well as the smaller 6-rings.   Cation positions
for 3 types of Na$^+$ ions in zeolite X are illustrated.

Total pore volume, V_p, $= X_s/d_a$

Where d_a is the density of the adsorbed phase (water for example)
and X_s is the maximum amount adsorbed in the crystal in g. per g.
of dehydrated crystal.

Measured void volumes can then be compared with the structural-derived
volumes. This has been done in several instances and the two values
correlated by a relationship of the type shown in Figure 4.

ZEOLITE SYNTHESIS

In a classical sense it is difficult to predict the chemical behavior of
zeolites because very little thermochemical data is available. There are
some basic free energy data, etc., available on one or two zeolites but we
cannot predict chemical reactions since necessary data are lacking.

Zeolites initially are formed hydrothermally as metastable phases
and, if left too long in their synthesis magma, will undergo various transformations to more stable phases. Most of the synthetic zeolites do not
exist as true equilibrium phases but, when properly removed from their
synthesis media, they exist metastably like diamond for prolonged periods
of time. However, exposure of these materials to reactive conditions may
provoke subsequent chemical transformations which are undesirable.[5]

Alkali aluminosilicate gels are prepared from aqueous solutions such
as sodium aluminate, sodium silicate, and sodium hydroxide. Sodium
zeolites result when the gels are crystallized at temperatures ranging from
room temperature to 150°C at atmospheric or autogenous pressure. This
method is best suited for the alkali metals since they form soluble hydroxides, aluminates, and silicates and it is possible to prepare very homogeneous gel mixtures. The gel structure is produced by the polymerization
of the aluminate and the silicate anions.[6] The composition and structure
of this hydrous polymer gel appears to be controlled by the size and structure of the polymerizing species. Differences in the chemical composition
and the molecular weight distribution of the starting species in the silicate
solutions lead to differences in gel structures and have been found to produce major differences in the zeolites phases.

During the crystallization of the gel the sodium ions, aluminate and
silicate components apparently undergo a rearrangement into the ordered
crystalline structure. This comes about by a depolymerization of the gel,
due to the hydroxyl ions present in the reaction mixture as sodium hydroxide,

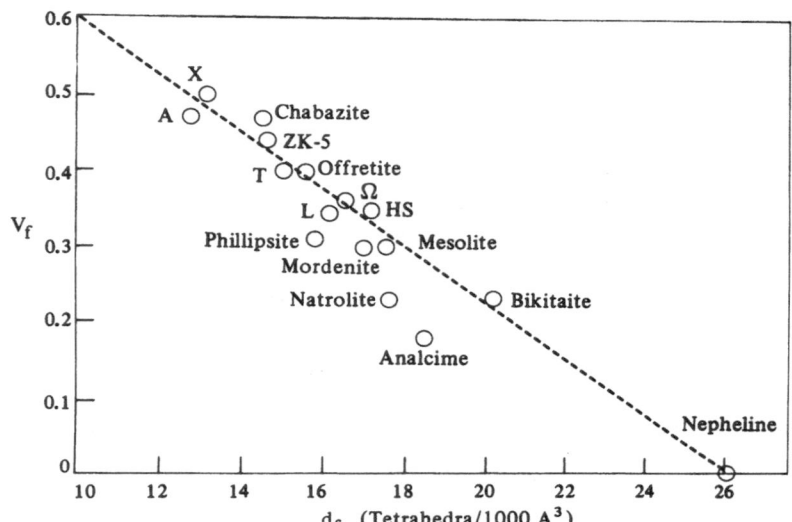

Figure 4   Relation between the measured void volume, expressed as the
void fraction V_f and the framework density, d.   The dashed line connects
the points corresponding to $V_f = 1.0$ at d = 0 and $V_f = 0$ at d = 26.   The line
is therefore expressed by $V_f = -d/26 + 1$.   The points corresponding to
natrolite and analcime deviate because the water molecules in these structures are tightly bound to cations and framework atoms indicating a
"feldspathoid" character.   The point representing sodalite hydrate corresponds to no occluded NaOH which is normally present.   Typically, synthetic
sodalite hydrate, or basic sodalite, contains occluded NaOH, and the void
fraction is accordingly much less, about 0.19 cm^3/cm^3.

Zeolite Crystals and Solution

$$NaOH\,(aq)\ +\ Na\,Al\,(OH)_4\,(aq)\ +\ Na_2SiO_3\,(aq)$$

$$\downarrow\ T_1\ \simeq\ 25\,^{\circ}C$$

$$[Na_a(AlO_2)_b\,(SiO_2)_c\cdot NaOH\cdot H_2O]\ gel$$

$$\downarrow\ T_2\ \simeq\ 25\ to\ 175\,^{\circ}C$$

followed by the formation of nuclei as small crystallites which are the
basis for the zeolite crystals. These nuclei, as in any crystallization
process, undergo growth during the crystallization period. Large numbers
of the crystallite nuclei are formed from the super-saturated gels and the
final product consists of a finely divided white powder of very small crystals in the order of only a few microns in size.

About 100 species of zeolites have been prepared in a pure state as a
result of the controlled variation of parameters such as the initial composition of the gel, crystallization temperature and type of reactant. Some of
these synthetic species appear to be structurally related to mineral zeolites;
others appear to have no known analog in the group of mineral zeolites.

It is not possible to treat these reacting systems in a phase-equilibrium
sense. Reaction diagrams similar to normal phase equilibrium diagrams
have been employed. One, illustrating the relation between composition and
some zeolite phases produced in the system Na_2O-Al_2O_3-SiO_2-H_2O at 100°C
is shown in Figure 5. In this instance three zeolites result: synthetic
zeolite type A, zeolite type P, and zeolite type X. Different zeolite species
are produced when potassium aluminosilicate gels are crystallized. Thus
the type of hydrated cations present in the starting gels is important. This
may be due to the size of the hydrated cation which acts as a template for
aluminosilicate polyhedra in the ordered crystal structure.

Crystallization of the hydrous gel exhibits an induction period which
can be determined by following the formation of the crystals as a function
of time. The crystallization curve shown in Figure 6 is sigmoid in shape.
The formation approximately follows first order kinetics and appears to be
autocatalytic. Prolonged exposure of zeolites to various conditions may
produce a series of transformations down the free energy "hill". Instead
of the free energy difference, the real driving force may be a tendency for
a reaction to take place that results in products of the highest entropy.

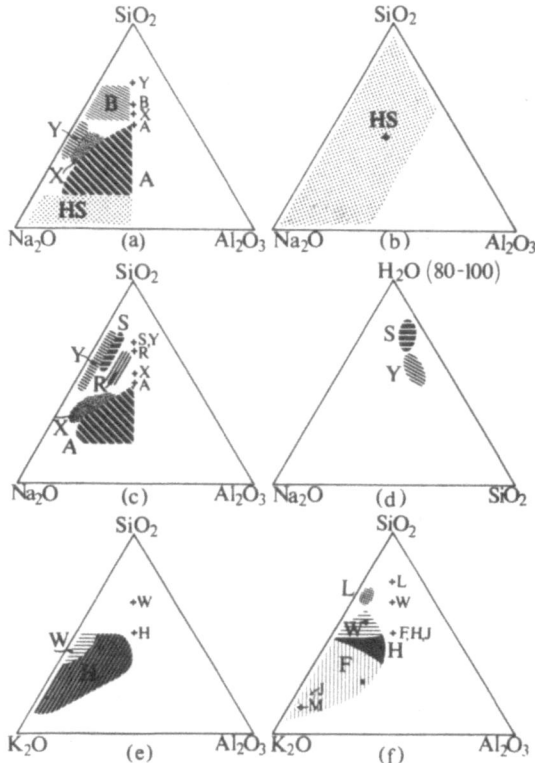

Figure 5   Reaction Composition Diagrams
(a) Projection of the Na$_2$O-Al$_2$O$_3$-SiO$_2$-H$_2$O systems at 100°C. H$_2$O content of gels is 90-98 mole %. Areas identified by letters refer to compositions which yield the designated zeolite. The points marked with (+) show typical composition of zeolite phase. Compositions in mole %. Sodium silicate used as a source of SiO$_2$.
(b) Same as (a) with 60-85 mole % H$_2$O in the gel.
(c) Same as (a). Colloidal silica used as source of SiO$_2$.
(d) Effect of water content in gel on synthesis of zeolites Y and S. Colloidal silica employed at 100°C. Al$_2$O$_3$ content is 2 - 10 mole % of anhydrous gel composition.
(e) Projection of the K$_2$O-Al$_2$O$_3$-SiO$_2$-H$_2$O systems at 100°C. Water content is 95 - 98 mole %.
(f) Same as (e). Water content is 80 - 92 mole %.

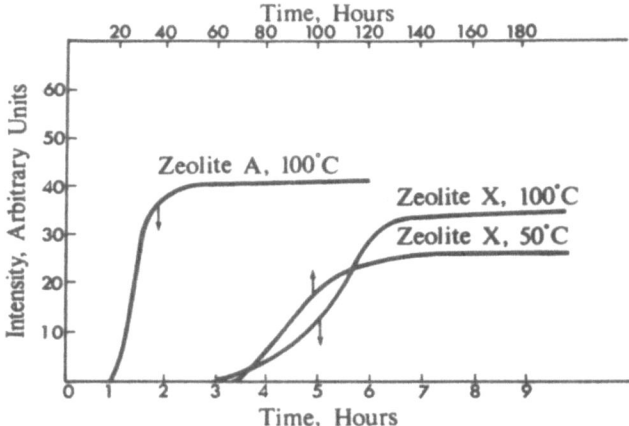

Figure 6   Crystallization of zeolite A and zeolite X at 50°C and 100°C, as a function of time. Intensity, in arbitrary units, indicates the degree of crystallization as determined by X-ray powder methods.

ZEOLITE WATER

One of the first subjects of the study of zeolites was concerned with the zeolite water and the way in which removal and readsorption altered the zeolite. Most of these early studies were confined to rather limited experimental measurements such as changes in weight and changes in optical characteristics. Prior to the advent of modern X-ray crystallography it had been shown that the presence of water in the zeolite is a stabilizing factor. As shown by Barrer the removal of water from the zeolite raises the chemical potential. (Figure 7) Consequently, the adsorption of water or other polar molecules such as ammonia lowers the chemical potential by a substantial amount. [7] This same effect is produced when nonvolatile salts are occluded in zeolite structures such as in the feldspathoids.

HYDROTHERMAL BEHAVIOR

The stability of zeolites and their structural changes when exposed to water vapor at elevated temperatures and elevated pressures is important

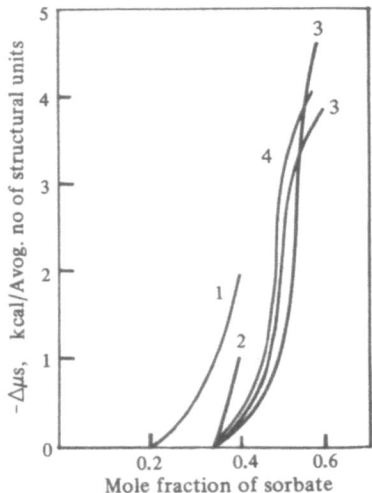

Figure 7   Lowering of the chemical potential, $-\Delta\mu S$, of zeolite X by the adsorption of H_2O and NH_3.   The structural unit is equal to 1/192 part of the unit cell which contains 192 tetrahedra   (1) NH_3 on NaX;   (2) NH_3 on NaX;   (3) H_2O on CaX;   (4) H_2O on Ca, CsX.

in understanding mechanism and conditions of their transformation as well as their use in high temperature, high water vapor environments.   These hydrothermal conditions will promote successive transformations to other structures.   One method which has been used to study zeolite behavior under these conditions is admittedly rather drastic.   It consists of heating the zeolite in a sealed container in a high pressure vessel at a constant applied pressure of water of several thousand atmospheres and following the transformations that occur.   We found that transformations of this type are time dependent.   As illustrated, temperature-time stability limits of several synthetic zeolites are shown.   (Figure 8)   These diagrams are not phase diagrams since the boundary has been approached from only one direction and in all of these cases the reaction is not reversible.[5]   Under some conditions where excess water is present there may be some preferential dissolution of the starting zeolite.   For example, zeolite A converts to an analcime type zeolite which has less alumina and soda.   The cation has an effect.   Whereas sodium X is converted to the higher density analcime, the magnesium exchanged form is instead transformed to a layer structure of the smectite type.

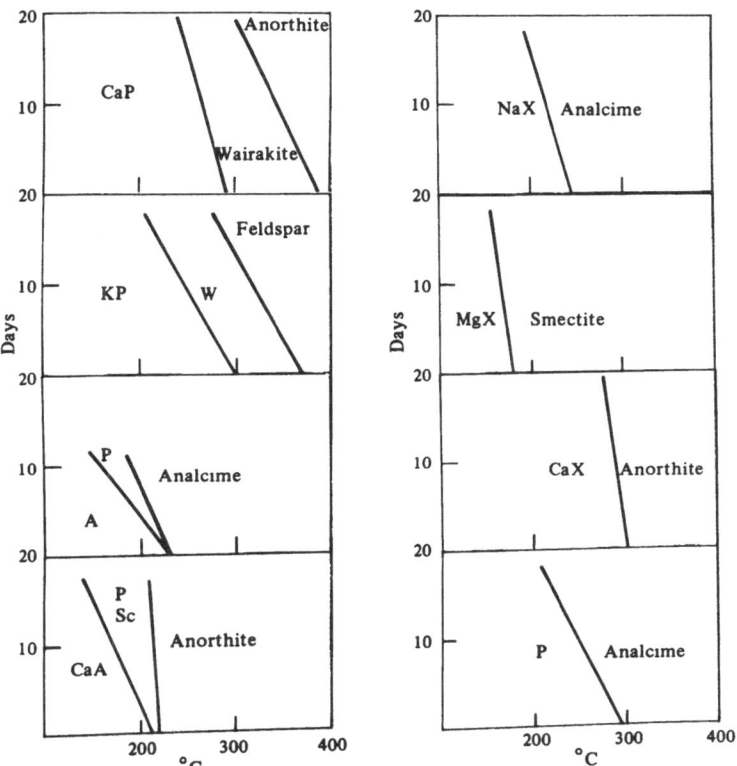

Figure 8   Time temperature diagrams showing hydrothermal transformation of some zeolites at 20,000 psi pressure of H_2O.   Compositions of starting materials are as follows (in moles of oxides):

(a) Zeolite X, NaX $Na_2O \cdot Al_2O_3 \cdot 2.4SiO_2 \cdot 4.3H_2O$
(b) Calcium X $CaO_{1.0} \cdot Al_2O_3 \cdot 2.4SiO_2 \cdot nH_2O$
(c) Magnesium X $MgO_{0.81} \cdot Na_2O_{0.19} \cdot Al_2O_3 \cdot 2.4SiO_2 \cdot nH_2O$
(d) Zeolite P $Na_2O \cdot Al_2O_3 \cdot 5SiO_2 \cdot 4.3H_2O$
(e) Calcium P $CaO \cdot Al_2O_3 \cdot 4.9SiO_2 \cdot 5.5H_2O$
(f) Zeolite A, NaA $Na_2O \cdot Al_2O_3 \cdot 1.9SiO_2 \cdot 3.7H_2O$
(g) Potassium P $K_2O_{0.88}Na_2O_{0.12} \cdot Al_2O_3 \cdot 4.8SiO_2 \cdot 2.8H_2O$
(h) Calcium A $CaO_{0.96}Na_2O_{0.04} \cdot Al_2O_3 \cdot 1.9SiO_2 \cdot 5H_2O$

RECRYSTALLIZATION

When heated to elevated temperatures after dehydration, zeolites generally break down or recrystallize forming an amorphous solid followed by recrystallization to some non-zeolitic species. Some examples of this type of anhydrous recrystallization are:

Zeolite A $\xrightarrow{800°C}$ β-Cristobalite-type

Zeolite X $\xrightarrow{1000°C}$ Carnegeite-type

$Mg^{ex}X$ $\xrightarrow{1500°C}$ glass $\xrightarrow{1000°C}$ cordierite

Zeolite Y $\xrightarrow{1000°C}$ glass

The magnesium exchanged form of zeolite X is transformed to a glass which undergoes the usual subsolidus recrystallization to cordierite at 1000°C.

Information on the thermal stability in air of various synthetic zeolites has been accumulated. (Figure 9) The typical type of behavior is also illustrated for the zeolite Y. All of the zeolites studied exhibit this type of decomposition curve as evidenced by a catastrophic loss of crystalline structure over a narrow temperature interval. The weight loss curve exhibits a change in weight over this same temperature interval in which the crystal structure is destroyed. For the polyvalent ions such as calcium, the weight loss curve has steps which are attributed to loss of water from hydroxylated cations.

At high pressures and temperatures zeolites may transform to denser aluminosilicates. For example, zeolite Y with a void volume of about 0.5 is transformed to various species depending upon the applied pressure and temperature. Ultimately, as illustrated, it can be transformed into jadeite, a high density aluminosilicate with a chain structure and the mineral basis for the gemstone jade. Depending upon the conditions it may also transform to zeolite P or a feldspar type of material. The data indicate that the zeolite does not convert directly to the high density material but must first pass through the intermediate forms of zeolite P.[5]

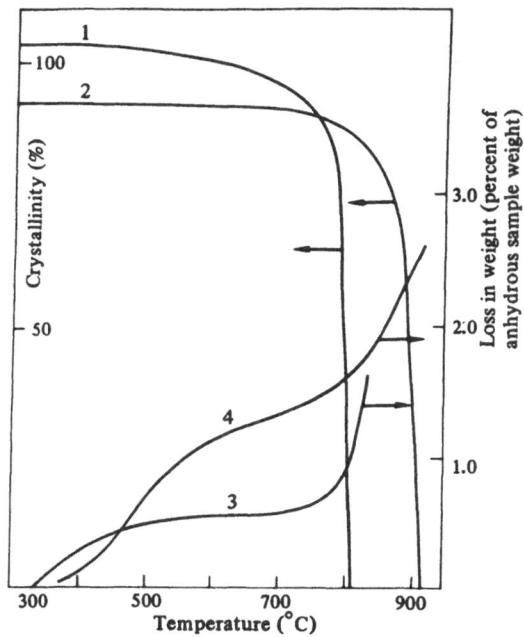

Figure 9   Thermal stability in air of zeolite Y. Heated for 16 hours after an initial dehydration in air at 350°. Si/Al ratio = 1.75; Ca exchange = 0.91 eq. fraction. (1) NaY - crystallinity; (2) CaY - crystallinity; (3) NaY - loss in weight (% w/w); (4) CaY - loss in weight (% w/w).

Zeolite Y, d_f = 1.25 $Na_{64}[(AlO_2)_{64}(SiO_2)_{128}] \cdot 260\ H_2O$

$$\xrightarrow[\text{15 kilobars}]{\text{300°C}} \text{(a)}\ 12[Na_{5.33}(AlO_2)_{5.33}(SiO_2)_{10.66} \cdot 15\ H_2O] + 80\ H_2O$$
Zeolite P, d_f = 1.57

$$\xrightarrow[\text{10 kilobars}]{\text{400°C}} \text{(b)}\ 4\ [Na_{16}(AlO_2)_{16}(SiO_2)_{32}] \cdot 16\ H_2O] + 196\ H_2O$$
Analcime, d_f = 1.85

$$\xrightarrow[\text{20 kilobars}]{\text{500°C}} \text{(c)}\ 16\ [Na_4Al_4(Si_2O_6)_4] + 260\ H_2O$$

$$\xrightarrow[\text{15 kilobars}]{\text{700°C}} \text{(d)}\ 8\ [Na_4Al_4(Si_3O_8)_4] + 4[Na_8Al_8Si_8O_{32}] + 260\ H_2O$$
Albite   Nepheline, d_f = 2.6

The relative "openness" of the structure is indicated by the framework density, d_f, which is in units of g/cc and is a value calculated from the tetrahedra per unit cell as given by the expression:

$$d_f = \frac{1.66\,[59x + 60y]}{V}$$

V is the unit cell volume in A^3.

ION EXCHANGE

Cations in zeolites exchange with other cations in aqueous solution under mild conditions. The cation exchange behavior depends upon the nature of the cation species; that is, the size and charge of the ion or complex cation as the case may be, and the type of zeolite structure. A wide range of ion exchange character has been observed. In aqueous suspension a synthetic zeolite exhibits a basic character that is due to limited carbon hydrolysis. The pH of such a suspension is generally above 7.

If the pH of a zeolite water suspension is deliberately reduced by the addition of a aqueous acid to a pH of below 5, aluminum ions are removed from the framework. However, a hydrogen or proton form of the molecular sieve zeolites can be prepared to a limited extent by first exchanging the alkali metal ions with ammonium and subsequently heating the ammonium form at 400°C to decompose the ammonium cation.

The behavior of the zeolite toward various cations in exchange is illustrated by ion exchange isotherms of the type given in Figure 10. Here the equivalent fraction of cation in the zeolite is plotted against the same quantity in the equilibrium solutions. The data given are for zeolite type X. These few examples illustrate a wide range in ion exchange selectivity for various cations; from the very great affinity for silver ion (for example, with an equivalent fraction of 0.5 in the zeolite the silver ion is almost undetectable in the equilibrium solution) to the very weak exchange exhibited by lithium. In these cases the competing cation is sodium. Similar studies have been made for exchange equilibria involving many other cation pairs. In the case of potassium ion the selectivity does not deviate too much from a factor of 1. At low concentrations the zeolite is slightly selective for potassium but at higher solution concentrations the selectivity is less than 1.

Some zeolites such as mordenite resist acid treatment although the cations may be removed by strong acids to produce a hydrogen form.

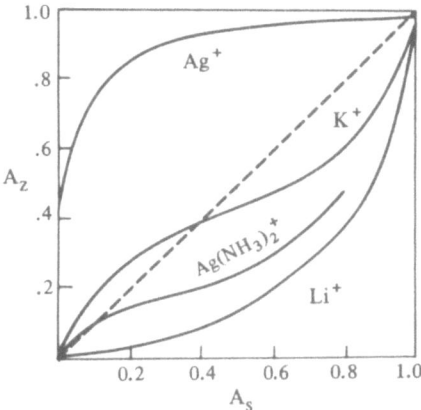

Figure 10   Exchange isotherms for univalent cation exchange in zeolite X at 25°C. $Na^+ \longrightarrow Ag^+$, 0.2 N; $Na \longrightarrow K^+$, 0.2 N; $Na^+ \longrightarrow Li^+$, 0.2 N; and $Na^+ \longrightarrow Cs^+$, 0.1 N.   The equivalent fraction of exchange ion in the zeolite at equilibrium for the reaction $Na^+_{zeolite} + A^+_{solution} = A^+_{zeolite} + Na^+_{solution}$ is given by A_Z and in solution by A_S.   The Cs^+ isotherm terminates at $Cs_Z \doteq 0.5$.

Under alkaline conditions many synthetic zeolites transform if exposed for prolonged periods of time.   Thus, zeolite A when exposed to sodium hydroxide solutions is converted to zeolite P.   If further exposed to more concentrated solutions conversion to the more stable sodalite hydrate, a feldspathoid, occurs.

Theoretical cation exchange capacities for the synthetic zeolites are quite high, 5.5 milliequivalents/gram of hydrated zeolite 4A and 4.7 milli-equivalent/gram of hydrated zeolite type X.

METALS AS GUEST PHASES

A dispersed metal phase may be produced within the pores of the zeolite crystal structures by the decomposition of an adsorbed, volatile, inorganic compound.   A typical example of this is illustrated for the case of nickel carbonyl.   Thermal decomposition of adsorbed nickel carbonyl produces a metal dispersion in the zeolite channels, although there is some migration to the exterior surfaces.   At room temperature dehydrated zeolite X adsorbs 0.30 cc per gram of nickel carbonyl expressed as the liquid. By repeated adsorption and subsequent decomposition, the zeolite channels can be nearly filled with nickel.

$$NaX + Ni(CO)_4 \longrightarrow NaX \cdot Ni(CO)_4$$

$$NaX \cdot Ni(CO)_4 \longrightarrow NaX \cdot Ni + 4CO$$

also $Fe(CO)_5$

Other volatile inorganic compounds such as other carbonyls behave in the same manner. Dispersed iron introduced in this way can be converted to intracrystalline clusters of alpha iron oxide. Similarly, some of the non-metallic elements such as sulfur and tellurium have been deposited within the zeolite cavities of zeolite X by vapor phase treatment and by ball milling.

Some reducible metal cations when exchanged into a zeolite may be reduced upon further contact with a metal vapor such as mercury. This is illustrated in the case of mercury exchanged zeolite X.

$$Hg\,X + Hg \xrightarrow[\text{300 torr}]{236°} Hg_2^{2+}X$$

The charge compensating cations in a zeolite may be removed by exchange with a reducible metal cation. As illustrated, exchange of the Na^+ by Ni^{2+}, followed by reduction with hydrogen, should result in nickel dispersed in the zeolite and, in order to maintain charge balance, a hydrogen on the framework. Several combinations of zeolites containing dispersed metal phases have been prepared including metals such as lead, copper and silver. Other chemical reactants such as carbon monoxide have been used.

$$NiX + H_2 \xrightarrow{\text{300-350°}} HY \cdot Ni$$

$$Cu^{++}X + CO \longrightarrow Cu^+, HX + CO_2$$

In one instance copper exchanged zeolite X was reduced by heating in carbon monoxide at 350°C. The question of maintaining charge balance during reduction by carbon monoxide is of interest. Under completely anhydrous conditions this can only be accomplished by the removal of an oxygen from the zeolite framework. However, it is more probable that some residual water produces hydrogen which acts as the reducing agent.

More exotic types of metal zeolite compositions are possible, particularly those of considerable interest as catalysts. The platinum group metals are introduced by cation exchange using an amine complex followed by hydrogen reduction. Dispersion of platinum is accomplished by thermal decomposition at 500°C to give platinum cations. It is concluded that the platinum cation occupies the surface sites in the large cages of the zeolite because it is too large to enter the structure to smaller sites.

Some reducible metal ions can be vaporized from the zeolite when heated in hydrogen. A mercury exchanged zeolite X lost the metal at 200°C and a cadmium exchanged zeolite X at 450°C. The formation of a copper nickel alloy in zeolite Y by reduction of the cations in hydrogen at 550°C was

$$HgX + H_2 \longrightarrow HX + Hg\uparrow$$

shown by a decrease in the ferromagnetism of the nickel.

Characterizing the dispersed metal phase in zeolites, particularly if the metal content is very low, has always been a problem. In one study using X-ray adsorption edge spectroscopy on a calcium Y (0.5 wt.% platinum) combined with line broadening and hydrogen adsorption it was concluded that 60% of the platinum consisted of 10 A crystallites which are small enough to fit inside the wide zeolite super cages. However, 40% consisted of 60 A crystals.

REFERENCES

1. D. W. Breck, Zeolite Molecular Sieves: Structure, Chemistry, and Use, Wiley-Interscience: New York, N.Y. 1974

2. T. B. Reed and D. W. Breck, J. Amer. Chem. Soc., 78 5972 (1956)

3. D. W. Breck and E. M. Flanigen, Molecular Sieves, Society of Chemical Industry, London, 1968, p 47

4. D. W. Breck and R. W. Grose, Adv. Chem. Ser., 1973 121, 319

5. Ref. 1., Chap. 6

6. Ref. 1., Chap. 4

7. R. M. Barrer, J. Phys. Chem. Solids, 16, 84 (1960)

THERMODYNAMIC ASPECTS OF SOLID STATE SINTERING

Joseph A. Pask and Carl E. Hoge

Inorganic Materials Research Division, Lawrence Berkeley
Laboratory and Department of Materials Science and
Engineering, College of Engineering, University of
California, Berkeley, California 94720

The kinetic approach, initially developed by Kuczynski,[1]
based on the use of a two-sphere model has led to an understanding
of the mass transport processes that can occur during solid state
sintering. The factors that lead to the grain growth that is
practically always observed during sintering, however, are not as
well understood. The following phenomenological analysis based on
a thermodynamic approach provides additional understanding of the
densification processes, driving forces for mass transport, and the
conditions under which grain boundary movement and consequently
grain growth occur.

Because of the complexity of real systems, model systems are
used to develop concepts and principles that play a role in sintering. The particles are considered to be single crystals and at
chemical equilibrium. The thermodynamic analysis thus deals with
interfacial energies and area changes, and is first applied to a
two-sphere model and then a many-sphere model.

TWO-SPHERE MODEL

The particles are assumed to be spherical, of uniform size and
with isotropic interfacial energies. On sintering, a thermodynamic
driving force is realized because of the decrease in free energy
due to the reduction of the surface area for the system, but the
formation and growth of the grain boundary or s/s interface which
is a positive contribution to the free energy of the system should
also be considered. As long as δG in Eq. 1 is negative, sintering

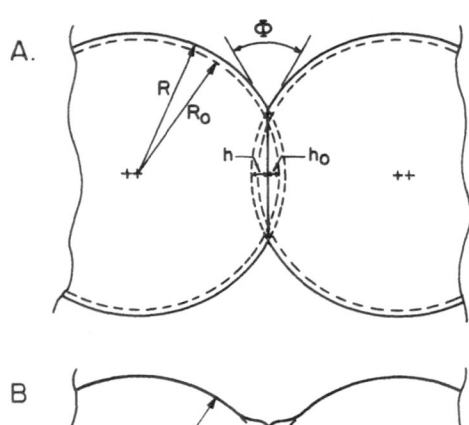

Fig. 1. Interpenetration at two sphere contact and distribution of mass in (a) minimum free energy configuration and (b) neck region.

$$\delta G = \delta \int \gamma_{sv} dA_{sv} + \delta \int \gamma_{ss} dA_{ss} \qquad (1)$$

continues; when δG is zero, the system is at equilibrium.

Figure 1 shows ideal two-sphere models (a) without and (b) with a neck. Shrinkage is realized because of mass transport from the grain boundary to the free surfaces. In the first case the material distribution over the free surfaces is faster than the mass transport to the neck region resulting in the lowest surface area for the system at every instant. It can be shown[2] that if the system is at thermodynamic equilibrium at any point in densification, i.e. when $\delta G = 0$ according to Eq. 1, the areas can be determined and the γ_{ss}/γ_{sv} ratio calculated. This ratio also corresponds to that calculated from the dihedral angle according to Eq. 2. The fractional linear shrinkage in this case

$$\gamma_{ss}/\gamma_{sv} = 2 \cos \Phi/2 \qquad (2)$$

is equivalent to h_o/R_o, as seen in Fig. 1

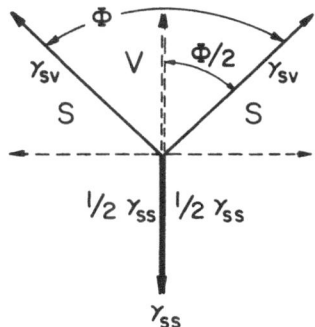

Isotropic Interfacial Energies

Fig. 2. Solid/vapor/solid dihedral angle showing sources of tensile stresses at the grain boundary due to the components of the surface tensions perpendicular and parallel to the grain boundary.

In the second case the mass transport from the grain boundary to the neck region is faster than its movement from the dihedral angle region resulting in the classical neck region. With single crystal spheres a grain boundary is present which forms an equilibrium dihedral angle with the free surfaces. The surface area, however, is not at a minimum and the reverse curvature provides a driving force for mass transport.[3] The resulting non-equilibrium perturbations imposed on the dihedral angle are counteracted by diffusion from the grain boundary, thus effectively maintaining an equilibrium dihedral angle until the equilibrium geometric configuration of the first case is attained.

In both cases, with the stated conditions, the grain boundary grows but is not able to move because any movement would cause an increase in its area without any compensating negative free energy contribution which is energetically unfavorable.

From a geometric viewpoint, using the conditions of case (a) the sphere centers move toward each other with an experimental dihedral angle starting at zero and increasing until the equilibrium angle is reached as determined by the γ_{ss}/γ_{sv} ratio for the system. There thus exists a driving force for the angle to move to its equilibrium value.[4] Figure 2 shows the balance of forces schematically; it emphasizes the fact that for a given grain the vertical component of its surface tension is balanced by its interfacial tension in contact with the adjoining grain, i.e. energetically it favors a grain contact in preference to vapor. The

sum of the interfacial tensions for the two adjoining grains constitutes the normal grain boundary energy.

MANY-SPHERE MODEL

The same ideal conditions as before are assumed. Analyses
have been carried out for simple cubic (SC), face-centered cubic
(FCC), body-centered cubic (BCC), and diamond cubic (DC) packings
of uniform size spheres.[2]   Only one packing, SC, is discussed
here to emphasize the significant points.

Figure 3(a) illustrates schematically the packing along (100)
and (110) faces. For a given sphere, shrinkage occurs equally at
all six contact points and the mass from the grain boundary regions
is distributed uniformly as in Fig. 2(a) in order to maintain at
every instant the lowest free energy for the system and to realize
the relationship between interfacial areas and γ_{ss}/γ_{sv} for the
system as stated in Eq. 1; during this process the experimental
dihedral angle increases from 0° and the γ_{ss}/γ_{sv} ratio decreases
from 2.0. When the angle reaches 90°, the (100) faces have
densified (as seen in Fig. 3(b)); a closed pore exists in the
center of each cube as seen along the (110) face and it is a
terminal point for grain boundaries. At this stage the porosity
is 3.6% and the linear shrinkage is 18.4%. Complete densification
results in a linear shrinkage of 19.6% and a geometrically determined maximum dihedral angle of 109°.

Several points should be emphasized. The mass transport
mechanisms are the same at every contact point and similar to that
for the two-sphere model until closed pores form. Mass transport
is not symmetrical in the last stage since material has to move out
of the densified (100) faces in order to maintain uniform three-
dimensional shrinkage. Also, the grain boundaries are energetically
pinned and cannot move during the open pore period. Any perturbations on the 90° S/S/S dihedral angles that form at this point
would cause grain boundary motion and grain growth since 120° S/S/S
dihedral angles are at the lowest energy state.

Another significant point is that the dihedral angle of 109°
and a γ_{ss}/γ_{sv} ratio of 1.416 become the critical values for this
packing. If the ratio for a real system is less than 1.416, then
the system has no thermodynamic barriers to reach theoretical
density. On the other hand, if the real system ratio is greater
than 1.416, then there will be a thermodynamic end point density.

Data for SC and the other packings are listed in Table 1.
Open pores are maintained and grain boundaries cannot move up to
theoretical densities of 91.2 to 96.5% depending on the type of
packing. Linear shrinkage vs experimental γ_{ss}/γ_{sv} ratios are

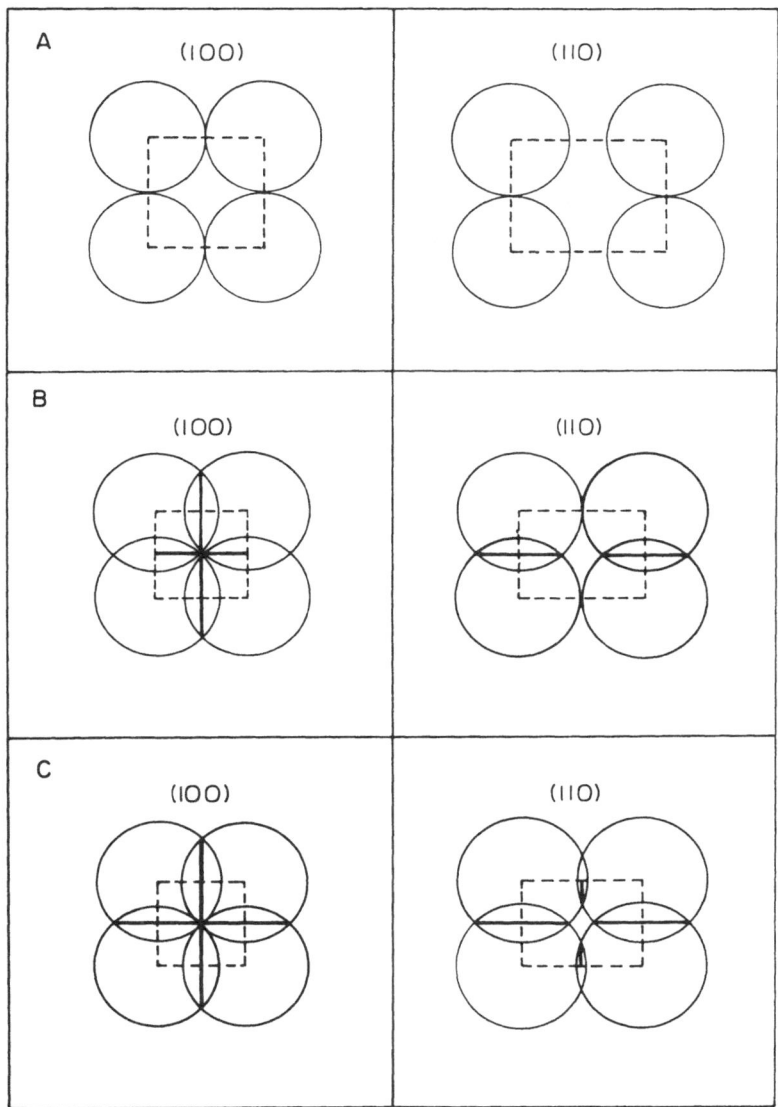

Fig. 3. Densification steps on sintering spheres in simple cubic packing: (a) at start, (b) complete densification on (100) face, and (c) continued densification along (110) face.

plotted in Fig. 4. For uniform size spheres the shrinkage in the
initial stage is essentially the same for all packings; the slight
deviations are due to the fact that as the number of contacts or
coordination number increases, more material is distributed in a
given time causing the radius of the spheres, R in Fig. 1, to
increase proportionately. Since the sintering mechanism is the
same at all points of contact, the horizontal axis in Fig. 4 could
represent time with zero at the right end. The rate of shrinkage
would be essentially the same for all packings in the open pore
stage, but the more dense packings would reach the closed pore
stage in less time.

The same analysis applies to a similar series of particles of
a uniform but different size. All the relationships are the same.
The scale of the time axis as discussed above, however, is dependent upon the size of the spheres: shorter for smaller particles,
longer for larger particles.

Disordered packings of uniform size spheres bring in some com-
plexities. As an example, particles in a compact with an unfired
theoretical density of 60% would have a statistical range of coordination numbers depending on the particle   distribution but
with an average of about 7 (as determined by inspection of Table 1).
The shrinkage curve as indicated in Fig. 4 would be interpolated
between that for SC (52% theoretical density) and BCC (68%); the
rate ,of shrinkage throughout the compact would be the same at the
start but the higher coordination number regions would reach the
closed pore stage in shorter times creating unbalances that could
lead to grain boundary motion. This situation emphasizes the
importance of attaining uniformity in disordered packings, e.g., an
agglomerate of lower bulk density than the continuous packing becomes a region with pores because of its inability to densify completely due to the constraint of the matrix.

The analysis of the relationship of the densification of different packings to the γ_{ss}/γ_{sv} ratio indicates that the most
favorable condition for a given system is the lowest γ_{ss}/γ_{sv} ratio
possible. Any additives or atmospheric conditions that reduce the
γ_{ss}/γ_{sv} ratio will enhance sintering, and also make it possible for
some systems to densify whose natural ratio is too high. Another
factor to consider is the experimental dihedral angle. Since a
driving force exists for a dihedral angle to reach its equilibrium
value, any conditions that would tend to keep the dihedral angle
below its equilibrium value during the densification process, as
described above in the non-neck-forming model system, will increase
the rate of sintering. If the dihedral angle is continuously maintained at equilibrium due to the formation of a neck, the driving
force for sintering is the reverse curvature in the neck region
and becomes the controlling step in the sintering process.

Table 1. Parameters for solid phase sintering models

	DC	SC	BCC	FCC
Coordination number	4	6	8	12
Fractional initial void volume:	0.68	0.48	0.32	0.26
At point of second neighbor contacts:				
Linear shrinkage, h_o/R_o	0.277	0.184	0.102	0.084
Φ (dihedral angle)	104.4	89.6	59.6	59.6
γ_{ss}/γ_{sv}	1.226	1.416	1.734	1.734
Fractional void volume	0.101	0.036	0.062	0.035
At theoretical density:				
Linear shrinkage of unit cube	0.316	0.196	0.121	0.095
Φ (dihedral angle)	109		71.5	109
γ_{ss}/γ_{sv}	1.074	1.161	1.625	1.161

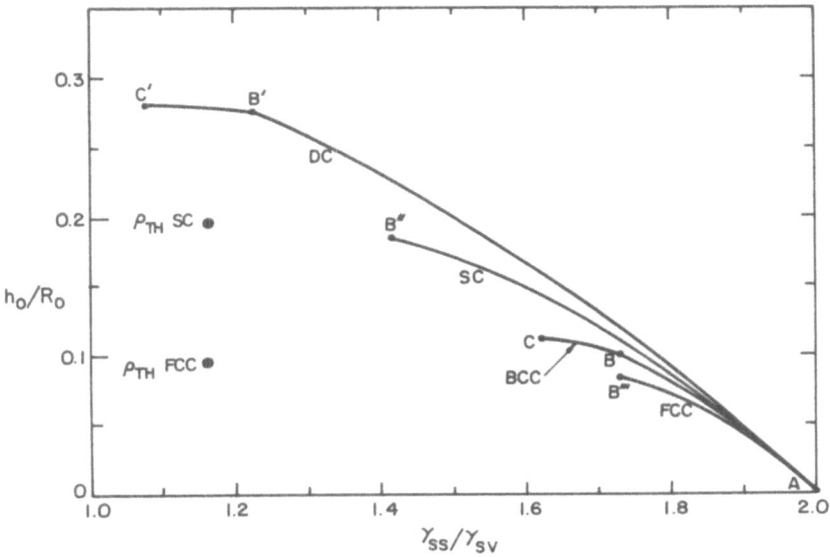

Fig. 4. Fractional shrinkage vs critical γ_{ss}/γ_{sv} ratio for different arrays of uniform size spheres.

GRAIN BOUNDARY MOTION

In model systems as presented, grain boundaries increase in
area at contact points between spheres but complete densification
can occur without any grain growth in the absence of grain boundary
motion. Grain growth definitely does not occur in the open pore
stage; in the closed pore stage unbalances can occur which result
in grain boundary motion and grain growth. In real systems, however, it is observed that some grain growth occurs during the open
pore stage. It critically affects the sintering process in the
sense that growth modifies the sintering parameters and could lead
to entrapment of pores within grains. Since grain growth is
dependent on grain boundary motion, it is important to explore
some of the factors that play a role in this process.

Anisotropy of Interfacial Energies

Appearance of flat faces, sharp edges and corners,and facets
on crystals is a reflection of the existence of a significant
difference in surface energies for different crystallographic
faces. The particle shape favors the face that has the lowest
surface energy since the free energy for the system is then at a
minimum. This situation can lead to a high γ_{ss}/γ_{sv} ratio which
would be detrimental to sintering.

Anisotropy also affects the dihedral angle. The S/V/S angle
in Fig. 2 is symmetrical with equal γ_{sv} values and a grain boundary
energy that has equal amounts of excess internal energy on either
side of the S/S interface. In a symmetrical S/S/S dihedral angle
the γ_{ss} values are equal and the angles are 120°. In either case
if an unbalance of interfacial energies occurs at triple points,
the interfaces will bend in order to realize a balance of forces
and thus equilibrium angles.

Curved grain boundaries can be shown thermodynamically to have
a tendency to move towards their center of curvature. A straight
grain boundary, however, can also exhibit movement if the excess
energies on both sides of the interface are not equal. The crystal
oriented with a higher energy surface at the interface will have a
driving force to grow at the expense of the other. Or, if not
constrained, the orientation of the boundary can change to a lower
energy state.

Movement of Grain Boundaries from Pores

Closed pores when first formed are always associated with
grain boundaries. A pore at a triple point can be shown to be
pinned from energy considerations. Boundary rearrangement, however, could result in a configuration with a pore on a planar
boundary.

If the boundary remains flat, it can be shown thermodynamically that the pore irregardless of the magnitude of its dihedral angle remains pinned since the energy necessary to form a continuous boundary that replaces the pore exceeds the free energy gain on spheroidization of the released pore.[2] On the other hand, if the boundary is curved, the additional free energy gain on shortening of the boundary after breaking away makes it possible for the event to happen. It can be shown that boundaries with pores with dihedral angles up to about 73° can break away if their curvature is greater than about 26° to 36°; for pores with dihedral angles greater than about 73°, the boundaries with curvatures greater than 36° can break away.

It is possible that a boundary in a certain configuration may initially have low or no curvature, but it may curve due to aniso-tropy until it reaches sufficient curvature to break away from the pore. Or, it may pull a pore along leading to coalescence. It thus is apparent that any additives to a system that will tend to reduce anisotropy at the interfaces is beneficial in keeping the pores on the grain boundaries.

Range of Particle Sizes

All real particle compacts have a range of particle sizes. Two spherical particles of different sizes on sintering will form a curved boundary with its center of curvature in the direction of the smaller sphere because of the unsymmetrical dihedral angle. When the grain boundary grows to a large enough size, it will be able to move out causing the smaller grain to coalesce with the larger. This process is facilitated if a neck forms between the particles and also if the difference in particle sizes is large.

During this stage the average size of the particles is in-creasing, and continuous and open pores still remain. Consequently, the kinetics of sintering are also being continuously reduced. No closed pores will form in this period as long as all of the pores remain associated with grain boundaries.

It should be pointed out that the thermodynamic requirements for densification as represented by Eq. 1 are not affected by this grain growth. The δG change is associated only with the grain boundary area growth and decrease of free surface area that occur as part of the densification process. Any grain growth that occurs in densified regions is an incidental process relative to sinter-ing.

This analysis again emphasizes the need for uniformity and homogeneity in the compact. For example, if agglomerates of lower density are dispersed in a powder that packs more densely, the

powder could sinter to theoretical density and form a matrix that
would restrain the agglomerates from sintering completely, resulting in an end point density for the compact. This result is not
truly representative for the material since it is the result of
poor processing.

CONCLUSIONS

Grain growth is extrinsic relative to sintering. Its occurrence affects sintering kinetics because of the resulting modification of sintering parameters. In an ideal system densification
can occur without any grain growth, at least up to the closed pore
stage.

Desirable conditions for sintering based on this analysis are:

(a) Uniform size particles as a first approximation--in any
case, complete homogeneity regardless of the type of packing.

(b) Highest green density--results in less shrinkage and a
larger critical γ_{ss}/γ_{sv}.

(c) As low a γ_{ss}/γ_{sv} ratio for the system as possible--
achieved with additives or controlled sintering conditions.

(d) A minimum amount of anisotropy of interfacial energies--
also achieved with additives or controlled sintering conditions.

(e) Experimental conditions that will tend to maintain a small
experimental dihedral angle.

REFERENCES

1. George C. Kuczynski, "Self-Diffusion in Sintering of Metallic
    Particles," Metals Transactions of AIME, pp. 169-177, Feb. 1949.

2. Carl E. Hoge and Joseph A. Pask, "Thermodynamics of Solid State
    Sintering," in "Physics of Sintering, Vol. 5," edited by
    M. M. Ristić, Boris Kidrić, Institute of Nuclear Sciences,
    Beograd, Yugoslavia, pp. 109-142, Sept. 1973.

3. Conyers Herring, "Surface Tension as a Motivation for Sinter-
    ing," pp. 143-179 in the Physics of Powder Metallurgy, ed.
    W. E. Kingston, McGraw-Hill Book Co., New York, 1951.

4. Carl E. Hoge and Joseph A. Pask, "Dihedral Angle as a Motivation
    for Sintering," to be published.

CONTRIBUTIONS OF GRAIN BOUNDARY AND VOLUME DIFFUSION TO SHRINKAGE

RATES DURING SINTERING

R.L. Eadie and G.C. Weatherly

Department of Metallurgy and Materials Science
University of Toronto
Toronto, Ontario

I INITIAL STAGE MODEL

A. Introduction

Previous sintering models (4),(6) that attempted to calculate the shrinkage rates associated with atom fluxes from grain boundary to neck have treated volume diffusion and grain boundary diffusion separately. In the present model the uniform shrinkage condition at the grain boundary is applied to the sum of the volume and grain boundary fluxes rather than to each flux separately. The assumption of the earlier models that each flux provided uniform shrinkage at the grain boundary cannot be justified on physical grounds. The present model also differs from previous ones in that the surface curvature of the neck surface is inserted in the problem in functional form; by this we mean that correct account is taken of the way in which surface curvature varies at the neck region. The ability to insert an arbitrary form and value for the surface curvature give this proposed solution a good deal of flexibility.

B. Diffusion Equation

We shall consider a line of spheres and later show that the restriction to this geometry can be removed and the solution applied to a 3D compact of sintering spheres. The diffusion equation as developed by C. Herring (3) is given by equation (1):

$$J = - \frac{D_V \nabla\mu}{\Omega_0 kT} \tag{1}$$

where D_V is the self diffusion coefficient of pure solid, μ is the difference in chemical potential of atoms and vacancies and kT has its usual meaning. Anywhere inside the material the flux must satisfy the condition:

$$\text{div } J = 0 \tag{2}$$

Hence combining equations (1) and (2) the partial differential equation for the diffusion is:

$$\nabla^2\mu = 0 \tag{3}$$

This is Laplace's equation. The initial stage geometry is approximated by cylinder as shown in figure 1, and we shall therefore work in cylindrical co-ordinates, r and z.

C. Boundary Conditions

1) <u>At the surface of the cylinder, r=x.</u> At the junction of the surface and the grain boundary (z=0), the chemical potential is given by:

$$\mu(x,0) = \gamma\Omega_0 \ (1/x - 1/\rho) \tag{4}$$

where γ is the specific surface free energy and Ω_0 is the atomic volume.

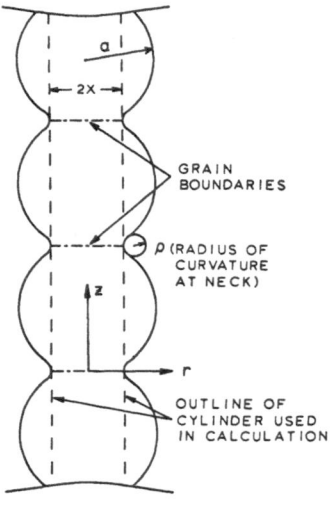

Fig. 1. A row of sintering spheres, radius a, and the cylinder used to approximate them.

A reasonable estimate of the form of the function μ (x,z) can be made from the following observations:

i) Somewhere between $z=0$ and $z=a$ the chemical potential assumes the value associated with the sphere viz. $2\gamma\Omega_0/a$.

ii)The value of z at which μ (x,z) first becomes essentially the same as $2\gamma\Omega_0/a$ will be called the decay distance. All published photomicrographs of sintering necks imply decay distances of order ρ.

iii) Surface diffusion will act to remove any large gradients in surface curvature along the surface. Hence the increase in μ from $\gamma\Omega_0$ $(1/x - 1/\rho)$ to $2\gamma\Omega_0/a$ at the surface of the cylinder shown in Fig. 1 should be fairly smooth and continuous.

For these reasons we chose to model $\mu(x,z)$ by an exponential function which approaches $2\gamma\Omega_0/a$ asymptotically. A simple function such as that given in equation (5) can be used:

$$\mu(x,z) = \mu_0 + 2\gamma\Omega_0/a - \frac{\gamma\Omega_0}{\rho} \exp\ (-F2z) \qquad (5)$$

The value of F2 is a variable in the solution. Values of F2 selected to illustrate the solution were $2/\rho$, $1/\rho$, $1/2\rho$; these give a decay distance of order ρ. The second exponential term added in reference 1 only affects the solution by 1% and need not be considered here. The solution presented elsewhere (1) illustrates the principle that the form of $\mu(x,z)$ can be varied by adding further terms of this type.

2) At the grain boundary (z=0). We shall assume that the grain boundary is a perfect source or sink for vacancies and that it is the only grain boundary in the neck region. If continuity is to be maintained across the grain boundary matter must leave or enter the boundary uniformly. i.e., the flux out of each incremental area of grain boundary is a constant. If we consider an incremental annular ring as shown in figure (2) we can write:

$$2(2\pi rdr\ J_v) - 2\pi r\ \delta J_b(r) + 2\pi(r+dr)\delta J_b(r+dr) = K* \ 2\pi rdr \quad (6)$$

for the shrinkage rate K* per unit area of boundary. Now substituting for J_v and J_b from equation (1) we obtain:

$$2D_v \frac{\partial\mu}{\partial z} + \delta\ D_b\ (\frac{\partial^2\mu}{\partial r^2} + \frac{1}{r}\frac{\partial\mu}{\partial r}) = K \qquad (7)$$

where $K = \Omega_0$ kTK*.

A further condition is required at the grain boundary to evaluate K: this comes from the requirement that there be no net force on the grain boundary.

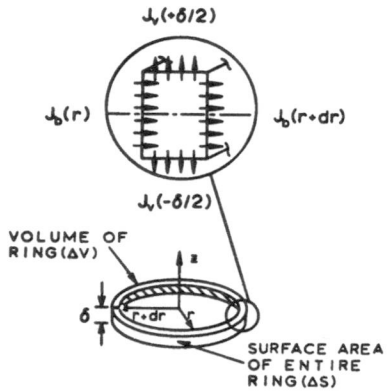

EXPLODED VIEW OF CROSS SECTION
OF RING SHOWING DIFFUSION
FLUXES IN GRAIN BOUNDARY

VOLUME OF
RING(ΔV)

SURFACE AREA
OF ENTIRE
RING(ΔS)

ANNULAR RING OF GRAIN
BOUNDARY

Fig. 2. Annular ring of grain
boundary at the neck, used to
calculate uniform shrinkage
condition.

i.e. $\iint \mu ds = 2\pi \, x\gamma\Omega_0$ (8)

The right hand side of equation (8) could be modified to take
account of the dihedral angle between the grain boundary and the
surface. If this term were to be introduced it would reduce the
right hand side by perhaps 5% (the dihedral angle is about 160°
for most metals). This term has been left out of the present
solution since equation (8) is accurate enough for metals. If
smaller dihedral angles are observed (for example in liquid phase
sintering or with ceramics) then the dihedral term must be
included.

D. Solution

The solution can be obtained in analytical form, as described
fully in reference 1. Rather than present the full analytical
solution here we shall enumerate some important conclusions which
can be drawn from the form of the solution.
i) Provided a > 10ρ the symmetry provided by the row of spheres
is not important and the solution can be applied to random necks
in a 3-dimensional compact.
ii) The value of the arbitrary constant F_2 (see equation 5) does
not affect shrinkage rate substantially. A 400% increase in F_2
produces at most a 33% decrease in shrinkage rate.
iii)The relative contributions of grain boundary and volume
diffusion to shrinkage are extremely sensitive to the decay
distance of $\mu(x,z)$. It would therefore be unrealistic to use

sintering data to determine either volume or grain boundary
diffusion data, except where experimental conditions preclude one
of these mechanisms throughout the experiment. This point is
considered further, later in this paper.

iv) The shrinkage rate $\underline{dL}$ is given by:
   dt

$$\frac{dL}{dt} = \frac{F\Omega_0 D_v}{2\pi x^3 kT} \left\{ \sum_n \frac{\coth(X_n a/x)}{X_n^3 \left(1 + \frac{\delta D_b X_n}{2 D_v x}\right) \coth(X_n a/x)} \right\}^{-1} \qquad (9)$$

where:

$$\Omega_0 F = 2\pi x \gamma \Omega_0 - \iint_{g.b.} \mu b \, dS \qquad (10)$$

L is the centre to centre distance of the sintering spheres and
Xn is the nth zero of the zeroth order Bessel function. The
dependence of the shrinkage rate on the neck geometry and the
diffusion coefficients is entirely expressed in this equation.
Note that the function $\mu(x,z)$ enters the solution through the
integral $\iint \mu b \, dS$ which is fully defined in reference 1.

II EXTENSION TO INTERMEDIATE STAGE

In developing an intermediate stage model some approximations
have to be made with the pore geometry. One of the most commonly
used assumes that the pore structure and neck geometry are not
significantly different in different regions of the compact.
Using this approximation the pores can be modelled as tori, of
tube radius ρ, surrounding circular grain boundary regions, of
radius x (7). In this structure surface transport mechanisms are
no longer important because gradients in surface curvature are
small. As a result changes in pore geometry are dictated entirely
by the fluxes of atoms from grain boundary sources to sinks at
surfaces of negative curvature. Figure 3 shows the idealized
structure we used to model the geometry of the intermediate stage.
In Figure 4, the intial stage geometry is shown in overlay on the
intermediate stage geometry.

On examining Figure 4 it is clear that provided a suitable
$\mu(x,z)$ is chosen the initial stage model should also serve for the
intermediate stage. This conclusion is supported by the observation
that the shrinkage rate changes rather slowly with changes in
$\mu(x,z)$. The changes in shrinkage rate as $\mu(x,z)$ is varied are
directly proportional to the variation of F in equation (9). F as
defined in equation (10) may be thought of as an

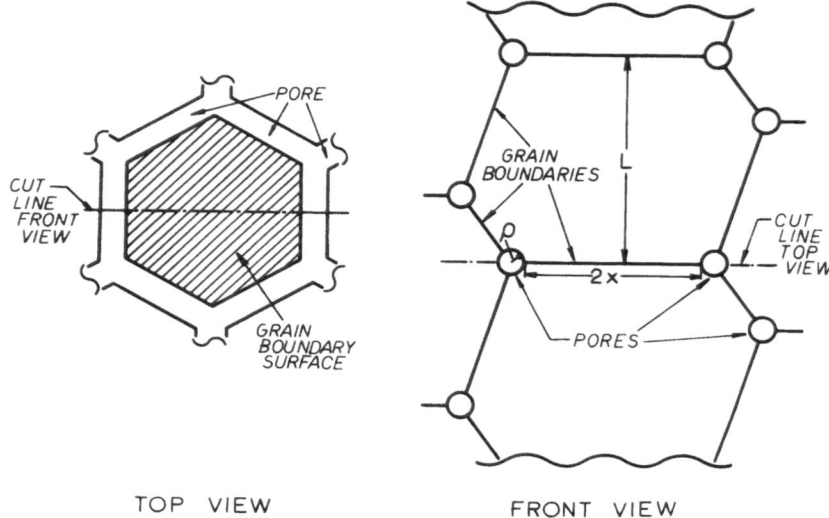

Fig. 3. Idealized geometry of pores in the intermediate stage. Cross-sectional views of structure consisting of equal sized tetrakaidecahedra after Coble (7).

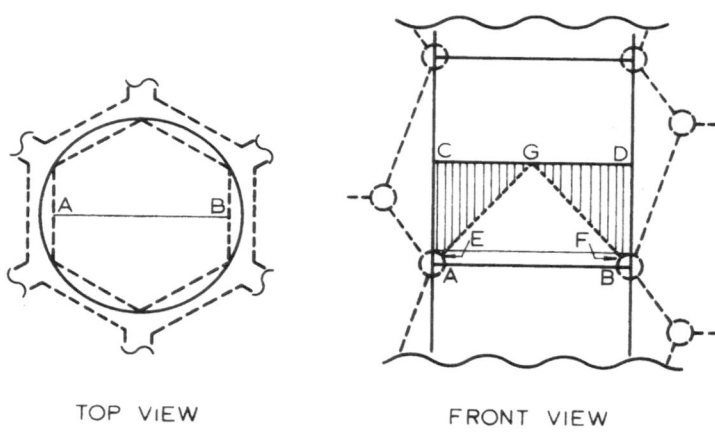

Fig. 4. Geometry used in calculation of shrinkage rates. The structure is shown overlaid on the tetrakaidecahedra structure. Symmetry would indicate the volume AGB as the appropriate one. The added volume EGG and FDG is not significant for flux from a source along AB to sinks at AE and BF.

effective sintering force. The relative accuracy of F can be established by examining two limiting cases. An upper limit occurs when $\mu(x,z) = \gamma\Omega_0 \ (1/x - 1/\rho)$ for all values of y. A lower limit is established as $\mu(x,z) = \gamma\Omega_0 \ (1/x - 1/\rho)$ for $z \leq \rho$ with flux restricted to the slice of cylinder within a distance ρ of the grain boundary. The total variation in F for these two limiting cases for silver was only 40%.[8] In fact, if the original initial stage model with a decay constant F2 = $1/2\rho$ is used, a good estimate of intermediate stage shrinkage is obtained, since the F obtained lies between these limiting values.

III   APPROXIMATIONS TO THE RATE OF SHRINKAGE EQUATION

Since $a \geq x$ and the minimum value of X_n is 2.405, hence $1.02 \geq \coth (\bar{X}_n a/x) \geq 1.00$. Thus the shrinkage rate may be written to a good approximation as:

$$\frac{dL}{dt} \simeq \frac{F\Omega_0 D_v}{2\pi x^3 kT} \left[\sum_n \frac{1}{X_n^3 + \frac{\delta D_b X_n^4}{2 D_v x}} \right]^{-1} \tag{11}$$

Provided $\delta D_b X_n \gg 2D_v x$, the X_n^3 term in the summation is negligible relative to $\delta D_b X_n^4/2 D_v x$. Since $\left[\sum_n \frac{1}{X_n^4} \right]^{-1} = 32$, the shrinkage rate becomes:

$$\frac{dL}{dt} = \frac{8 \ F\Omega_0 \ \delta D_b}{\pi \ x^4 \ kT} \tag{12}$$

If $F/\pi x^2$ is set equal to $-\gamma \ (1/\rho + 1/x)$, then equation (12) is identical to the solution to the shrinkage rate for grain boundary diffusion acting alone which was obtained by Johnson (4). This value for F is suitable for grain boundary diffusion but, as we shall see, also for volume diffusion control - a rather remarkable result!

If $2 \ D_v \ x \gg \delta D_b X_n$, then the X_n^3 term dominates the summation in (11). Since $\left[\sum_n \frac{1}{X_n^3} \right]^{-1} \simeq 12$, equation (11) becomes:

$$\frac{dL}{dt} \simeq \frac{6 \ D_v \Omega_0 F}{\pi kT x^3} \tag{13}$$

An upper limit for the case of volume diffusion control can be calculated by setting $\mu(x,z) = \gamma\Omega_0 \ (1/x - 1/\rho)$ and with the boundary condition at the grain boundary, $\partial\mu/\partial z$ = a constant. The method of solution to this problem is similar to the method followed in reference (1). The solution is identically equal to equation (13) provided $F/\pi x^2$ again = $-\gamma \ (1/\rho + 1/x)$!

Recalling our earlier observation that this upper limit value of F is only 40% greater than the lower limit value, this value of F can be used with some confidence during intermediate stage sintering. Note that the limiting values of F are identical both for the combined flux equation (11), and the volume diffusion equation (13).

These results are conveniently summarized in Figure (5) which shows the neck size and temperature regimes where the various solutions, i.e. equation (11),(12) or (13), can be safely used. The data used in these calculations were those for Ag (1).

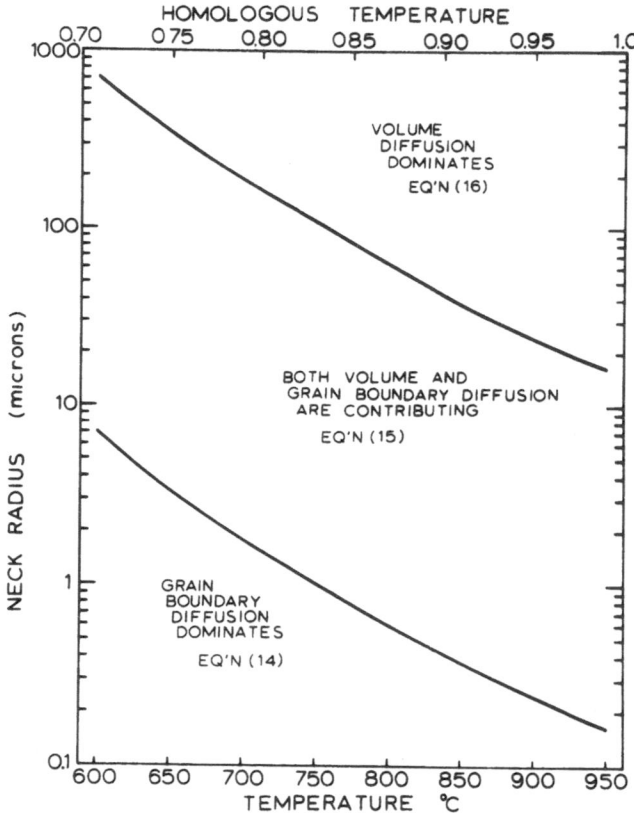

Map showing regions of applicability of approximations to the shrinkage rate equation for pure silver.

For small neck sizes and low temperatures, grain boundary diffusion dominates. The appropriate equation here is:

$$\frac{dL}{dt} \simeq - \frac{8\Omega_0 \, \delta \, D_b \, \gamma}{kTx^2} \, (\frac{1}{\rho} + \frac{1}{x}) \tag{14}$$

For an intermediate region as shown in Figure (5), the equation combining grain boundary and volume diffusion should be used. If one is satisfied with an upper limit value for F* the appropriate equation is:

$$\frac{dL}{dt} \simeq - \frac{\gamma \Omega_0 D_v (1/\rho + 1/x)}{2kTx} \left[\sum_n \frac{1}{X_n^3 + \frac{\delta D_b X_n^4}{2 D_v x}} \right]^{-1} \tag{15}$$

For larger neck radii and high temperatures, volume diffusion is dominant. The value of F , $\pi x^2 \, \gamma \Omega_0 \, (1/x + 1/\rho)$, is known to be an upper limit. However, the approximation we made in the summation term to obtain equation (13) tends to be an underestimate. Hence the approximation errors subtract. With this value of F the equation becomes:

$$\frac{dL}{dt} \simeq - \frac{6 \, D_v \, \gamma \Omega_0 \, (1/x + 1/\rho)}{kTx} \tag{16}$$

Calculations on copper using the diffusion coefficients quoted by Ashby (5) gave rise to approximately similar regions on this type of map.

From Figure (5) it can be concluded that for many metals, equations incorporating either volume diffusion or grain boundary diffusion alone are not useful for describing intermediate stage sintering unless:
i) The homologous temperature is less than .75 and the neck radius less than 5 microns (then grain boundary diffusion dominates).
ii) The homologous temperature is greater than .95 and the neck radius is greater than 20 microns (volume diffusion dominates).

ACKNOWLEDGEMENTS

     The authors would like to thank Prof. K.T. Aust, the Dept. of Trade, Industry and Commerce of Canada, and the National Research Council of Canada for financial support during the course of this work. R.L. Eadie gratefully acknowledges the J. Edgar McAllister Graduate Bursary from the University of Toronto.

*This estimate could be 20% too high.

REFERENCES

1. R.L. Eadie, D.S. Wilkinson, and G.C. Weatherly, Acta Met., 22, 1185, (1974).

2. B.H. Alexander and R.W. Balluffi, Acta Met., 5, 666, (1957).

3. C. Herring, "The Physics of Powder Metallurgy", Chap. 8, McGraw Hill, (1951).

4. D.L. Johnson, J.Appl.Phys., 40, 192, (1969).

5. M.F. Ashby, Acta Met., 22, 275, (1974).

6. T.L. Wilson and P.G. Shewmon, Trans. Met. Soc. A.I.M.E., 236, 48, (1966).

7. R.L. Coble, J.Appl.Physics, 32, 787, (1961).

8. R.L. Eadie and G.C. Weatherly, Scripta Met., 9, 285, (1975).

A KINETIC MODEL FOR THE REDUCTION IN SURFACE AREA DURING INITIAL

STAGE SINTERING

R. M. German* and Z. A. Munir[+]

*Sandia Laboratories, Livermore, CA 94550

[+]University of California, Davis, CA 95616

INTRODUCTION

Recent morphology calculations[1-3] have demonstrated the usefulness of geometric parameters other than neck size in monitoring sintering progression. These calculations have provided the first precise relationship between neck size and the specific surface area based on a minimum surface energy consideration in a sphere-to-sphere configuration. Also, shrinkage in particle size or interparticle distance has been directly related to the neck size for the surface- and bulk-transport mechanism,[1,2] respectively. Similar relations have been found from an analysis of the sintering of wires.[3]

Although similar attempts have been reported by numerous authors, these have suffered from imprecise approximations of the neck geometry. The present analysis uses the previously reported morphology calculations to introduce kinetics into the specific surface area reduction. Through this technique, identification of the characteristic rate-controlling sintering mechanism is possible for the initial sintering stage. In a second presentation,[4] the validity of this approach is demonstrated through analysis of published results.

BACKGROUND

The classical approach to studying sintering kinetics is to monitor the neck size growth between spheres under isothermal conditions. Kuczynski[5] was the first to provide the characteristic neck growth equation:

$$\left(\frac{x}{a}\right)^N = Bt \tag{1}$$

where $\frac{x}{a}$ is the ratio of the interparticle neck radius to the sphere radius; N is a characteristic mechanism exponent; t is the isothermal sintering time; and B is a constant containing such parameters as material transport properties, temperature, and particle size. Subsequently, numerous modifications and extensions of this kinetic relationship have been provided. Notable among these are the investigations of Rockland,[6,7] Kingery and Berg,[8] Kuczynski,[5,9] Johnson and Cutler,[10] and Coble.[11] For the bulk-transport processes interparticle distance decreases with increasing neck growth. However, sintering controlled by a surface-transport mechanism does not contribute to shrinkage, and neck size has been the only available measure of the transport kinetics.

The realization that surface area is the driving force for sintering has led to consideration of various mathematical relations between surface area and time. Schlaffer, Adams, and Wilson[12] determined that an empirical equation of the form

$$\frac{dS}{dt} = -kS^n \tag{2}$$

where S is the surface area and k and n are constants, adequately represented the initial decrease in specific surface area for alumina and silica gels. However, at extended sintering times, their mathematical model did not correlate with their experimental data. From density changes during the second stage of sintering, Rhines and DeHoff[13-15] have given similar expressions for the decrease in surface area with time. Analysis of Cu and Ni sintering data led them to formulate a final mathematical relation equivalent to Eq. 2. However, data reported by these authors, as well as Watanabe and Masuda,[16] can only partially substantiate such an approach.

An alternative approach to the kinetics of surface area reduction has been a geometric correlation between the neck size and surface area. Jellinek and Ibrahim[17] approximated the change in surface area resulting from neck growth by an equation of the form

$$\frac{\Delta S}{S_0} = k_1 \left(\frac{x}{a}\right)^2 + k_2 \left(\frac{x}{a}\right)^3 \tag{3}$$

where ΔS is the change in surface area from an initial value of S_0, and k_1 and k_2 are constants for a given particle packing. Application of this relation to Cu spheres by Nyce and Shafer[18] showed a reasonable agreement with experimental data. Similarly, Prochazka and Coble[19,20] have geometrically approximated the surface area reduction for the sphere-sphere case during surface-diffusion controlled sintering. Their resulting expression is similar to Eq. 3. Also, Prochazka and Coble[20] have integrated the sphere-sphere computer simulation profiles of Nichols and Mullins[21] to arrive at a modified geometric relation. In such cases, time dependence is introduced through the appropriate expression corresponding to the form shown in Eq. 1.

In all of the above cases, approximate relations between neck size and surface area have been developed, based on assumed simple geometries. Most recently, German and Munir[1,2] have provided precise morphology calculations of the interparticle neck geometry using a minimum surface energy neck configuration. It was shown that a modified catenary represented a given neck profile with a minimum surface area.[1] This model was used to determine specific surface area variation with neck size and particle coordination for both surface- and bulk-transport mechanisms during the initial stage of sintering.[1,2] An important consideration in such an analysis is the number of particle contacts per particle. From a practical standpoint, estimation of the coordination number N_c is possible through determination of the powder-packing fraction if monosized spheres are considered. Fig. 1 shows the relation for both regular and random packing geometries.[15,19,22-25]

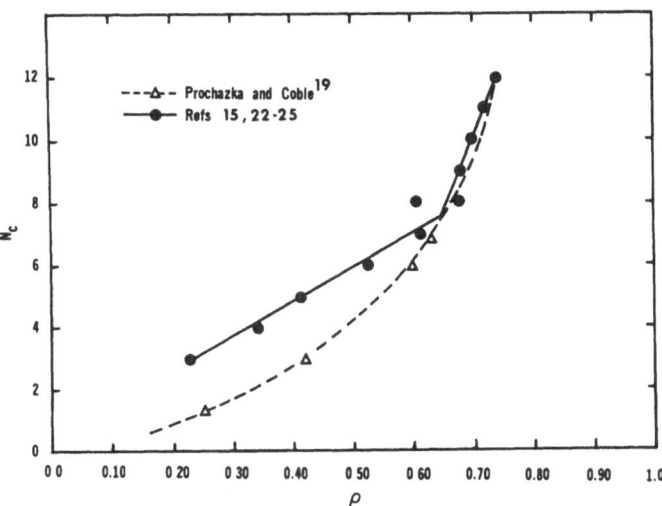

Fig. 1. Packing density dependence on particle coordination; solid line – regular packing, dashed line – random packing.

KINETICS OF SURFACE AREA REDUCTION

    With respect to the decrease in surface area ΔS, the most
significant parameter involved is the neck size, with coordination
having a secondary influence. For a first approximation, $\Delta S/S_0$
can be considered to be a function of x/a only. Shown in Figs. 2
and 3 are the log-log plots of the normalized surface area reduc-
tions $\Delta S/S_0$ versus the normalized neck size x/a obtained from the
results of the earlier studies.[1,2] The resulting straight lines
have slopes which increase with increasing coordination. Fig. 4
shows the variation in slopes with coordination number for both
the bulk- and surface-transport cases. Slope values of approximately 2 correspond to the geometric values used by Jellinek and
Ibrahim[17] and Prochazka and Coble.[20] Mathematically, the surface
area reduction can be expressed as a function of neck size as
follows:

$$\frac{\Delta S}{S_0} \;=\; K \left(\frac{x}{a}\right)^m \tag{4}$$

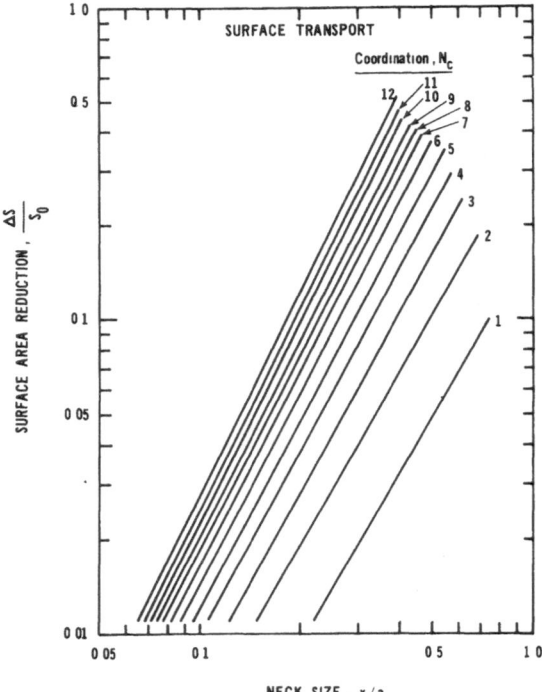

Fig. 2. Variation in $\Delta S/S_0$ with x/a and N_c for surface-transport
         controlled sintering.

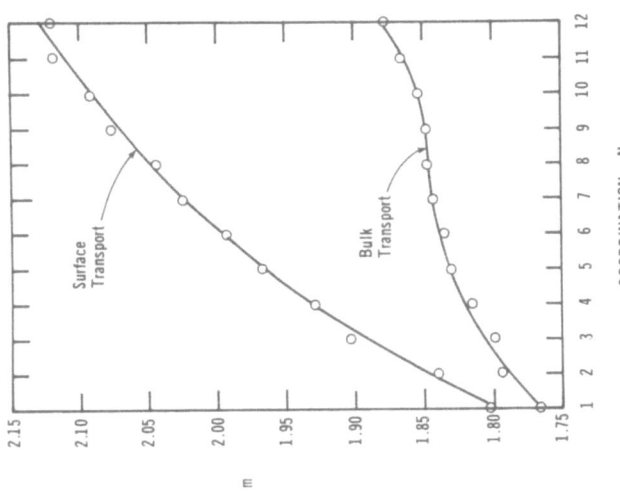

Fig. 4.   Dependence of the slope (m of Eq. 5) of Figs. 2 and 3 with coordination (particle packing) and mechanism.

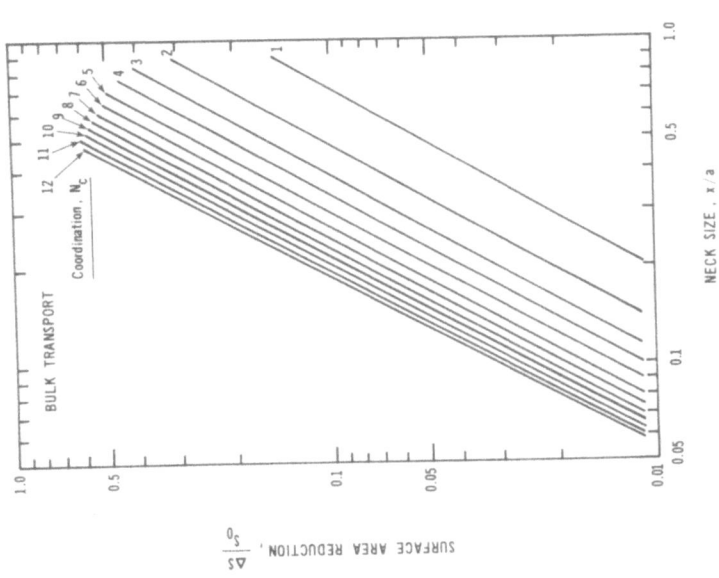

Fig. 3.   Variation in $\Delta S/S_0$ with x/a and N_C for bulk-transport controlled sintering.

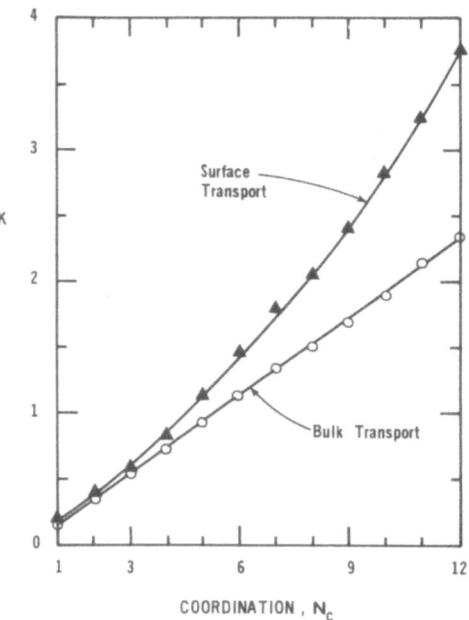

Fig. 5. Increase in the preexponential factor K with coordination
N_c for both surface- and bulk-transport processes.

where K and m are both mechanism- and coordination-dependent constants. The dependence of m is shown in Fig. 4. The variation of
K with coordination for the two categories of mass transport processes is shown in Fig. 5. For loose powder compacts, values of
K near unity would suffice for both forms of mass transport.

The final link between specific surface area reduction and
isothermal sintering time for monosized spheres is established
from a combination of Eqs. 1 and 4. This relation can be expressed
as follows:

$$\frac{\Delta S}{S_0} = K(Bt)^{m/N} \tag{5}$$

Eq. 5 is generally valid up to $\Delta S/S_0 \approx 50\%$ for most monosized
powder compacts. Thus identification of the dominant sintering
mechanism, be it a surface- or bulk-transport process, can be
accomplished through the surface area reduction kinetics. Such
a procedure is more general than the previous shrinkage techniques,
while being relatively more precise and less tedious than the
classical neck growth measurements. Finally, it should be pointed
out that the specific surface area can be determined by a variety
of techniques and thus provides an averaging of all interparticle
necks.

PRACTICAL APPLICATIONS

Most powder compacts have interparticle coordinations between 6 and 8. As shown in Fig. 1, the interparticle coordination can be estimated from the packing fraction. If one assumes an N_c value of 7, generally characteristic of loose monosized spherical particle compacts,[14],[18] from Fig. 4 we see that the appropriate m values are 1.84 and 2.02 for bulk- and surface-transport, respectively. Thus, experimental identification of the rate-controbling sintering mechanism results from determining the slope of log $(\Delta S/S_0)$ versus log (t) data. Inherently, it is assumed that shrinkage or densification checks have accompanied such measurements. The occurrence of shrinkage indicates that a value of m = 1.84 is appropriate. Hence, from the slope of the experimental log $(\Delta S/S_0)$ versus log (t) data the mechanism characteristic exponent N can be extracted. Identification of the rate-controlling mechanism results from a comparison of the experimental N value with those values predicted theoretically.

The application of this approach to identification of the rate-controlling mechanism is simply an alternative to the classical neck growth measurements. Combined with shrinkage and possibly scaling[26] experiments, this technique offers an extremely sensitive measure of the sintering process. Such combined measurements would provide cross-checks on the adequacy of a single mechanism model. Differing results derived from surface area and shrinkage or scaling experiments could be indicative of a multiple mechanism process. In the subsequent presentation,[4] the results of Eq. 5 are applied to several sets of experimental data to demonstrate the usefulness of the technique. The rate-controlling mechanism is identified from surface area reduction data and theoretical neck growth laws. Comparison of these results with expected mechanism dominance substantiates the proposed approach.

CONCLUSIONS

The classical approach to determining the rate-controlling mechanism during initial stage sintering is extended to include the parameter of specific surface area. The results of this extension indicate that specific surface area reduction kinetics can lead to identification of the controlling mechanism. Combining these results with shrinkage or scaling measurements provides a means of assessing the action of multiple mechanisms.

ACKNOWLEDGMENTS

The authors are grateful to D. R. Adolphson of Sandia Laboratories for his support in the preparation of this presentation. This work was supported by the U.S. Atomic Energy Commission under Contract Number AT-(29-1)-789.

REFERENCES

1. R. M. German and Z. A. Munir, "Morphology Changes During Surface-Transport Controlled Sintering," accepted, Met. Trans., 1975.

2. R. M. German and Z. A. Munir, "Morphology Changes During Bulk-Transport Sintering," submitted, Met. Trans., 1974.

3. R. M. German and Z. A. Munir, "The Geometry of Sintering Wires," accepted, J. Mat. Sci., 1975.

4. R. M. German and Z. A. Munir, "The Identification of the Initial-Stage Sintering Mechanism: A New Approach," published in these proceedings.

5. G. C. Kuczynski, Trans. AIME, vol. 185, p. 169, 1949.

6. J. G. R. Rockland, Acta Met., vol. 14, p. 1273, 1966.

7. J. G. R. Rockland, Acta Met., vol. 15, p. 277, 1967.

8. W. D. Kingery and M. Berg, J. Appl. Phys., vol. 26, p. 1205, 1955.

9. G. C. Kuczynski, Advan. Colloid Interface Sci., vol. 3, p. 275, 1972.

10. D. L. Johnson and I. B. Cutler, J. Amer. Ceram. Soc., vol. 46, p. 541, 1963.

11. R. L. Coble, J. Amer. Ceram. Soc., vol. 41, p. 55, 1958.

12. W. G. Schlaffer, C. R. Adams, and J. N. Wilson, J. Phys. Chem., vol. 69, p. 1530, 1965.

13. F. N. Rhines, R. T. DeHoff and R. A. Rummel, Agglomeration, W. A. Knepper (editor), Interscience, New York, p. 351, 1962.

14. F. N. Rhines, R. T. DeHoff and J. Kronsbein, A Topological
     Study of the Sintering Process, Final Report, U.S. Atomic
     Energy Commission Contract Number AT-(40-1)-2581, University
     of Florida, Gainesville, 1969.

15. R. T. DeHoff, R. A. Rummell, H. P. LaBuff and F. N. Rhines,
     Modern Developments in Powder Metallurgy, vol. 1, H. H.
     Hausner (editor), Plenum Press, New York, p. 310, 1966.

16. R. Watanabe and Y. Masuda, Trans. Japan Inst. Metals, vol. 13,
     p. 134, 1972.

17. H. H. G. Jellinek and S. H. Ibrahim, J. Colloid Interface Sci.,
     vol. 25, p. 245, 1967.

18. A. C. Nyce and W. M. Shafer, Inter. J. Powder Met., vol. 8(4),
     p. 171, 1972.

19. S. Prochazka and R. L. Coble, Phys. Sintering, vol. 2(1), p. 1,
     1970.

20. S. Prochazka and R. L. Coble, Phys. Sintering, vol. 2(2),
     p. 15, 1970.

21. F. A. Nichols and W. W. Mullins, J. Appl. Phys., vol. 36,
     p. 1826, 1965.

22. L. C. Grayton and H. J. Fraser, J. Geol., vol. 43, p. 785,
     1935.

23. E. Manegold, R. Hofman and K. Solf, Koll. Z., vol. 56,
     p. 142, 1931.

24. E. Manegold and W. Von Engelhardt, Koll. Z., vol. 62, p. 285,
     1933.

25. E. Manegold and W. Von Engelhardt, Koll. Z., vol. 63, p. 12,
     1933.

26. C. Herring, J. Appl. Phys., vol. 21, p. 301, 1950.

THE IDENTIFICATION OF THE INITIAL-STAGE SINTERING MECHANISM:

A NEW APPROACH

R. M. German* and Z. A. Munir[+]

*Sandia Laboratories, Livermore, CA 94550
[+]University of California, Davis, CA 95616

INTRODUCTION

In the preceding paper we have provided a relationship between neck size and surface area reduction based on the morphological changes accompanying the initial stage sintering of monosized spheres. It was shown that such a relation is dependent on both mechanism and particle coordination (powder packing fraction) in the following manner:

$$\frac{\Delta S}{S_0} = K(Bt)^{\frac{m}{N}} \quad (\frac{\Delta S}{S_0} \leq 0.5) \tag{1}$$

Here $\Delta S/S_0$ is the normalized specific surface area reduction; t is the isothermal sintering time; B is a constant containing material properties, particle size, and temperature; and K, m, and N are constants for a given set of conditions dictated by the controlling mechanism. From theoretical models, N is expected to have a characteristic value for each mass transport mechanism. Examination of Eq. 1 reveals that a log ($\Delta S/S_0$) versus log t plot will show a characteristic inverse slope of N/m if a single mechanism is operative. Since m can be determined from simple compact density determinations,[1] calculation of N is straightforward. Thus, $\Delta S/S_0$ has the same fundamental role in sintering kinetics as neck size data, without the inherent errors associated with measurements of the latter parameter.[2-4]

PREDICTED SLOPES

Sintering phenomena can be classified as either adhesion or densification processes.[5] The exponential factor in Eq. 1 has been shown to be 2.02 for the surface-transport processes and 1.84 for the bulk-transport processes in normal loose powder compacts.[1] Selection of the appropriate m value is then determined, in general, by the presence or lack of densification. Numerous authors have provided theoretical examination of the appropriate N values. Based on a review of the models, assumptions, and criticisms of each treatment, Table I was constructed.[5-13] It shows the best N values according to current theories and the expected N/m values for each individual mechanism. Comparison of experimental $\Delta S/S_0$ data within this format leads to identification of the dominant initial stage sintering mechanism.

Table I

RELATION BETWEEN MECHANISM AND $\dfrac{N}{m}$

Mechanism	Densification	Neck Size Exponent, N	Approximate $\dfrac{N}{m}$
Viscous Flow or Plastic Flow	yes	2	1.1
Evaporation-Condensation	no	3	1.5
Volume Diffusion	yes	5	2.7
Grain Boundary Diffusion	yes	6	3.3
Surface Diffusion	no	7	3.5

RESULTS

Several authors have monitored the surface area reduction during sintering.[2-4,14-21] Unfortunately, not all reports contain sufficient experimental data to be analyzed by the present model. Some of the investigations containing data within the limits of the proposed model are summarized in Table II. Figs. 1 through 7 show plots of the log ($\Delta S/S_0$) versus log (t) of the data referenced in Table II. Shown with each data set is the inverse slope as determined by a least squares fit.

Table II

EXPERIMENTAL VERIFICATION OF THE MODEL

Material	Particle Size	T,K	Atmosphere	Inverse Slope*	Densification	Mechanism	Figure	Reference
Alumina Gel	54Å	849	steam	3.6	no	S.D.	1	18
Al_2O_3	0.26 μm	1023	air	3.3	no	S.D.	2	19
Silica Gel	90Å	1136	steam	2.8	yes	V.D.	3	18
"	"	1053	steam	2.7	yes	V.D.	3	18
"	"	951	steam	2.7	yes	V.D.	3	18
"	"	849	steam	3.5	no	S.D.	3	18
TiO_2	0.14 μm	1023	air	3.4	yes	G.B.D.	4	17
Cu	48 μm	1123	H_2	3.2	yes	G.B.D.	5	4
	115 μm	1223	H_2	2.5	yes	V.D.	5	4
B	9 μm	2155	vacuum	3.5**	no	S.D.	6	20
	9 μm	2175	vacuum	3.5**	no	S.D.	6	20
	9 μm	2222	vacuum	3.5**	no	S.D.	6	20
$As_{3/4}S_{1/2}Se_{5/2}$	<90 μm	431	He	1.0	yes	V.F.	7	21
	"	433	He	1.1	yes	V.F.	7	21
$As_{5/4}S_{5/2}Se_{1/2}$	"	448	He	1.0	yes	V.F.	7	21
$As_{3/4}Se_3$	"	421	He	0.98	yes	V.F.	7	21
	"	425	He	0.98	yes	V.F.	7	21
$As_{3/4}S\,Se_2$	"	430	He	0.93	yes	V.F.	7	21

S.D. = surface diffusion; V.D. = volume diffusion; G.B.D. = grain boundary diffusion; V.F. = viscous flow

*slopes from least-squares fit of data unless otherwise indicated

**imposed slopes, see figures

DISCUSSION

Examination of Table II shows a generally good agreement between the N/m values calculated from the experimental data and those predicted in this work (Table I). Cases where apparent uncertainties exist (e.g., surface diffusion, N/m = 3.5, versus grain boundary diffusion, N/m = 3.3) can be easily verified from observations of dimensional changes, if any, during sintering. For example, the Al_2O_3 data of Prochazka and Coble[19] gave an N/m ratio of 3.3. However, lack of densification rules out the possibility of grain boundary diffusion in favor of surface diffusion with its anticipated N/m ratio of 3.5. Indeed, Prochazka and Coble[19] concluded that surface diffusion was the dominant mechanism of material transport. Furthermore, Wilson and Shewmon[22] arrived at the same conclusion by analyzing the appropriate diffusion coefficients. Further substantiation for this conclusion comes from the results of Schlaffer, Adams, and Wilson[18] on the sintering of alumina gel.

While the identification of the dominant mechanism in the sintering of alumina gel was possible, no conclusions were made by Schlaffer, et al.[18] regarding the sintering of silica gel. The present analyses shed some light on the possible material transport processes involved. The sintering of the 90Å gel appears to be dominated by a volume diffusion mechanism, except at the lowest experimental temperature. At 849K, where no densification occurred, the evidence is in favor of a surface diffusion controlled sintering.

The TiO_2 results obtained by Vergnon, Astier, and Teichner[17] led them to conclude a Nabarro-Herring mechanism of mass transport. Data by Anderson[23] suggests that grain boundary diffusion may be rate-controlling in rutile, in agreement with the present analysis of Reference 17 results.

The sintering mechanism of Cu has long been disputed. Grain boundary and volume and surface diffusion mechanisms, as well as plastic flow, have all been reported for this material. Quite possibly the contradictory nature of these results stems from differences in the prevailing experimental conditions. Therefore it is not surprising to see the disagreement shown in Table II. Analysis of the Cu data of Nyce and Shafer,[4] shows the dominant mechanism for the finer powder, sintered at the lower temperature (see Table II), is grain boundary diffusion. On the other hand, compacts of larger particles maintained at a higher temperature sinter by a volume diffusion process. However, the possibility of concomitant surface diffusion makes such a definitive analysis unlikely.[5]

The conclusion arrived at by German, Mar, and Hastings[20] regarding the sintering of boron is in agreement with our analysis.

The dominant mechanism for the sintering of this element at temperatures in the neighborhood of 2200K is surface diffusion. Finally, the results of Carrell and Wilder[21] offer an interesting addition to the variety of sintering mechanisms assembled in Table II. The sintering of glassy phases in the As-S-Se ternary system has been shown to proceed by a viscous flow mechanism.[21,24] Our analysis of Carrell and Wilder[21] results gave inverse slopes N/m ranging from 0.93 to 1.1 depending on composition and other experimental variables. These slopes are in good agreement with the predicted value of 1.1 (Table I).

The basis for the present model is a monosized particle distribution and a singly operative mass transport mechanism. Typically, polydisperse systems are encountered in practical systems. Thus exact agreement between predicted and calculated N/m values is not to be anticipated. Furthermore, the study of late-stage sintering is not a valid application of this model. Our experience shows such sintering situations typically provide inverse slopes between seven and ten. It should be pointed out that temperature variations such as those shown in Figs. 3 and 6 provide a technique for determining the activation energy for the sintering process. The problem of multiple active sintering processes (a very realistic problem) would have the same effect on the time dependence of the surface area reduction as on neck size.[5] Such difficulties must be anticipated and require judicious treatment of experimental data.

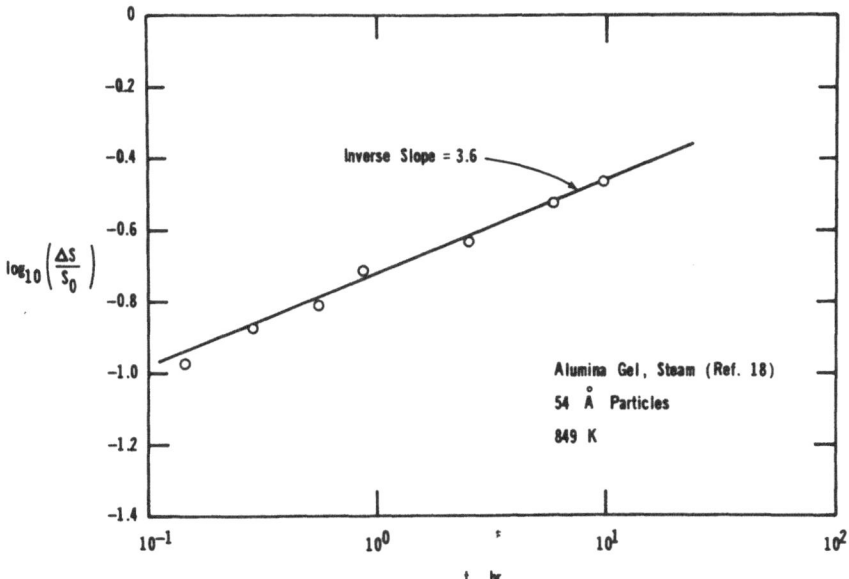

Figure 1

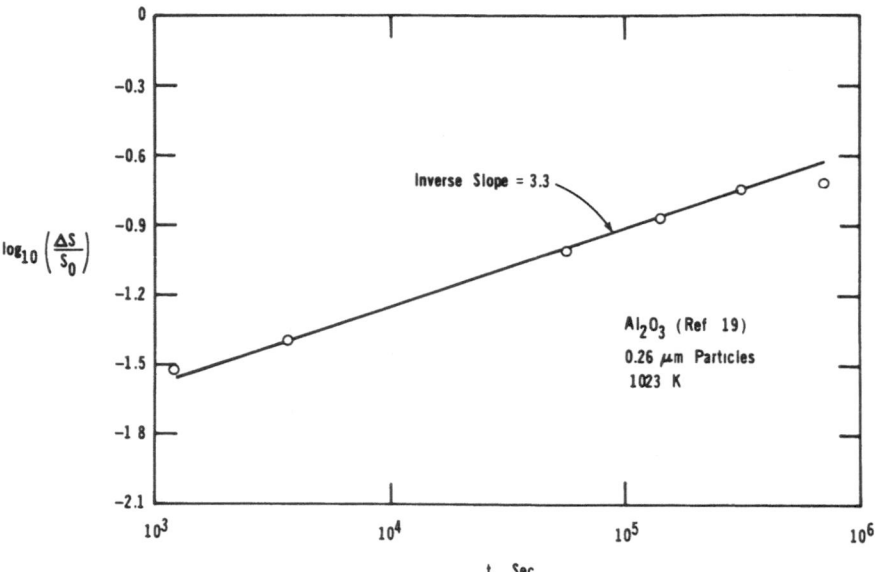

Figure 2

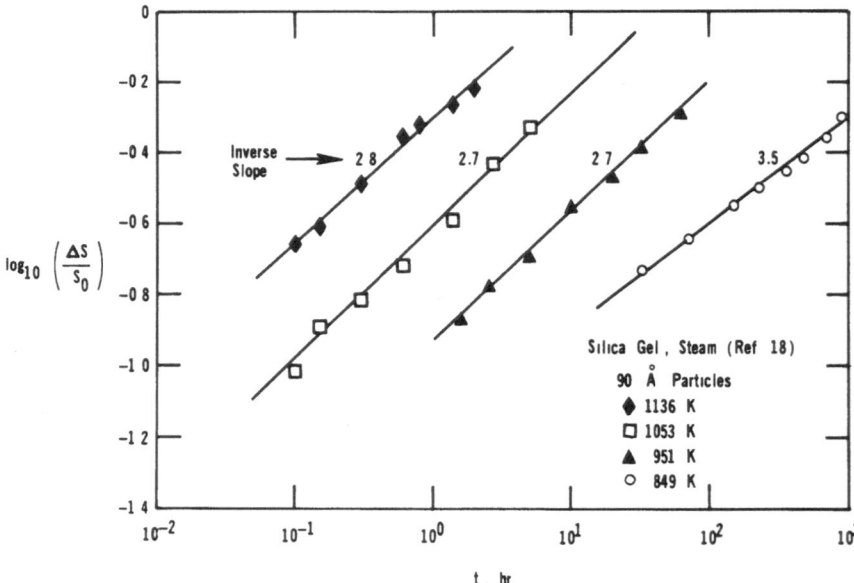

Figure 3

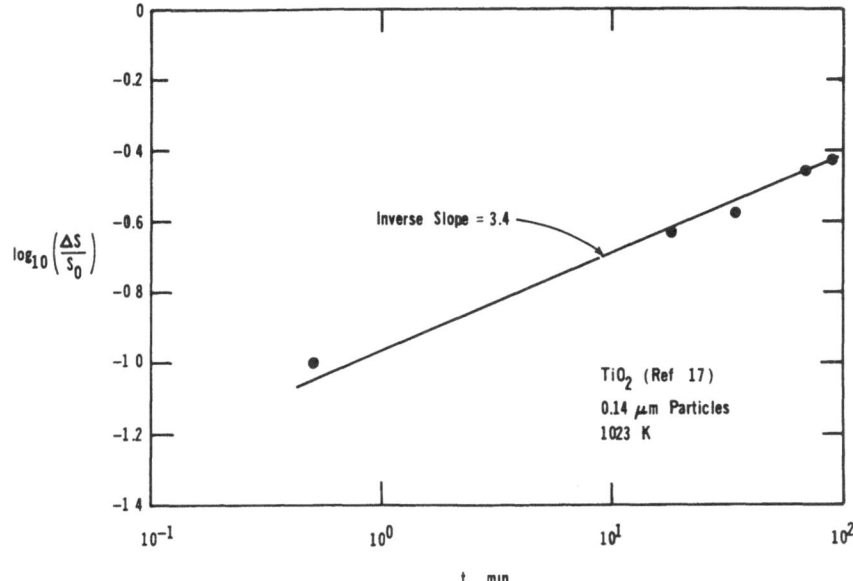

Figure 4

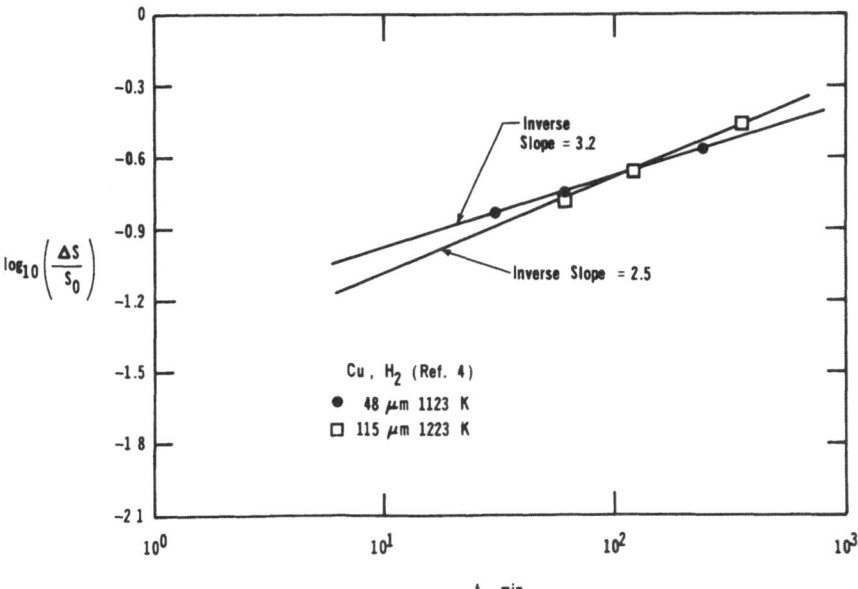

Figure 5

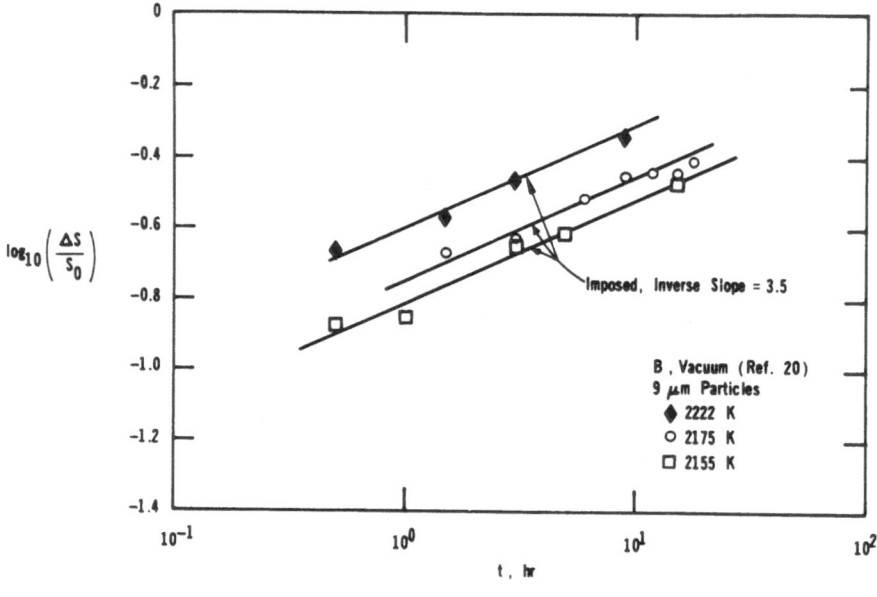

Figure 6

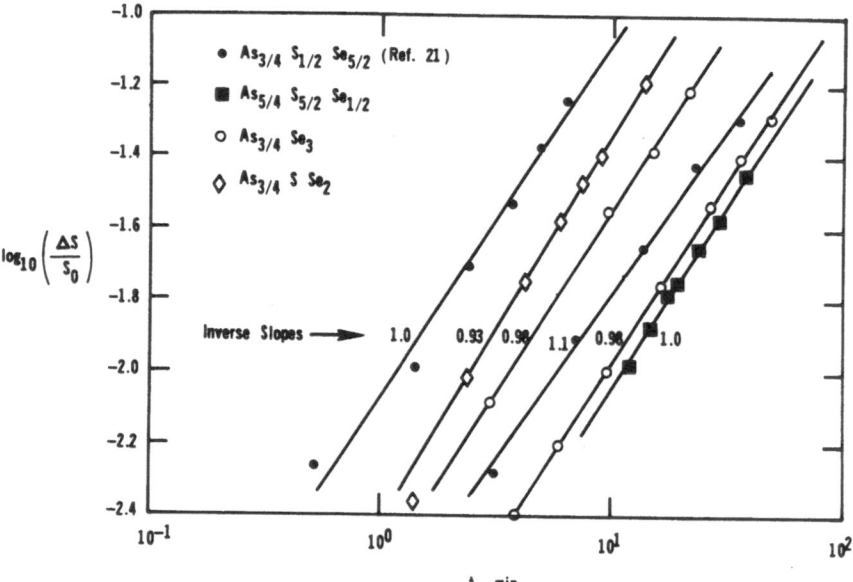

Figure 7

CONCLUSIONS

Examination of Table II shows that the predicted slopes are well matched by the experimental data. The addition of specific surface area measurements to other techniques of mechanism identification provide a new tool to the experimentalist. The variety of sintering techniques now include 1) neck size growth rate, 2) density and shrinkage determinations, 3) surface area reduction measurements, 4) nonisothermal sintering rate determinations, 5) activation energy checks and the Dorn technique, 6) scaling laws, 7) checks with known diffusivities, and finally, 8) intuition based on known physical parameters of a material. No single experimental approach is universally applicable to the determination of the rate-controlling mechanism. With regard to the specific surface area reduction kinetics, two positive factors are obvious. First, the measurements can be conducted nondestructively by a variety of techniques, which are far less laborious than neck size measurements. Also, each measurement gives an averaging of many neck formation processes. Thus, the surface area reduction measurement has the same general benefit that is associated with shrinkage measurements without being limited to bulk-transport mechanisms.

ACKNOWLEDGMENTS

The authors are grateful to D. R. Adolphson of Sandia Laboratories for his support in the preparation of this presentation. This work was supported by the U.S. Atomic Energy Commission under Contract Number AT-(29-1)-789.

REFERENCES

1. R. M. German and Z. A. Munir, "A Kinetic Model for the Reduction in Surface Area During Initial Stage Sintering," published in these proceedings.

2. F. N. Rhines, R. T. DeHoff, and R. A. Rummel, Agglomeration, ed. by W. A. Kanepper, Interscience, New York, 1962, p. 351.

3. R. Watanabe and Y. Masuda, Trans. Japan Inst. Metals, 1972, vol. 13, p. 134.

4. A. C. Nyce and W. M. Shafer, Inter. J. Powder Met., 1972, vol. 8, p. 171.

5. J. G. R. Rockland, Acta Met., 1967, vol. 15, p. 277.

6. J. G. R. Rockland, Acta Met., 1966, vol. 14, p. 1273.

7. G. C. Kuczynski, Trans. AIME, 1949, vol. 185, p. 169.

8. W. D. Kingery and M. Berg, J. Appl. Phys., 1955, vol. 26,
    p. 1205.

9. G. C. Kuczynski, Adv. Colloid Interface Sci., 1972, vol. 3,
    p. 275.

10. R. L. Coble, J. Amer. Ceram. Soc., 1958, vol. 41, p. 55.

11. D. L. Johnson and I. B. Cutler, J. Amer. Ceram. Soc., 1963,
    vol. 46, p. 541.

12. J. R. Moon, Powder Met. Inter., 1971, vol. 3, p. 147 and p. 194.

13. F. Thummler and W. Thomma, Met. Rev., 1967, vol. 12, p. 69.

14. R. T. DeHoff, R. A. Rummel, H. P. LaBuff, and F. H. Rhines,
    Modern Developments in Powder Metallurgy, vol. 1, ed. by
    H. H. Hausner, Plenum Press, New York, 1966, p. 310.

15. H. H. G. Jellinek and S. H. Ibrahim, J. Colloid Interface Sci.,
    1967, vol. 25, p. 245.

16. W. Beere, J. Mat. Sci., 1973, vol. 8, p. 1717.

17. P. Vergnon, M. Astier, and S. J. Teichner, Sintering and
    Related Phenomena, (Vol. 6 of Materials Sci. Research) ed. by
    G. C. Kuczynski, Plenum Press, New York, 1973, p. 301.

18. W. G. Schlaffer, C. R. Adams, and J. N. Wilson, J. Phys. Chem.,
    1965, vol. 69, 1530.

19. S. Prochazka and R. L. Coble, Phys. Sintering, 1970, vol. 2(2),
    p. 17.

20. R. M. German, R. W. Mar and J. C. Hastings, Amer. Ceram. Soc.
    Bul., 1975, vol. 54, p. 178.

21. M. A. Carrell and D. R. Wilder, J. Amer. Ceram. Soc., 1967,
    vol. 50, p. 604.

22. T. L. Wilson and P. G. Shewmon, Trans. TMS AIME, 1966, vol. 236,
    p. 48.

23. H. V. Anderson, J. Amer. Ceram. Soc., 1967, vol. 50, p. 235.

24. R. N. Lott and Z. A. Munir, J. Non-Cryst. Solids, 1972, vol. 7,
    p. 86.

EFFECT OF BeO INCLUSIONS ON THE RATE OF SINTERING OF Cu WIRES

S. Bahk, M.F. Ashby[*], J. Bevk and D. Turnbull

Harvard University
Division of Engineering and Applied Physics
Cambridge, Massachusetts 02138 U.S.A.

ABSTRACT

Single crystal Cu-Be alloys were made and drawn to wires of 0.005" diameter. The wires were annealed and internally oxidized to precipitate BeO particles. (The BeO particle size, determined by transmission electron microscopy was 300 ~ 400 Å). The volume fraction range of BeO particles investigated was from 0.09% to 0.93%. Generally, the neck size of sintered wires was found to decrease with increasing volume fraction of BeO particles. The interface reaction is proposed to control the rate of the sintering process. The effects of cold work on the sintering behavior of Cu-BeO alloys are also discussed.

INTRODUCTION

Shrinkage and neck growth during sintering of a powder aggregate are a result of mass transport which occurs by a number of physical processes. The main objective of sintering studies has been to predict the time dependence and relative contribution of each of these mechanisms to the overall process. While there is little doubt that glasses sinter by viscous flow, there is still considerable controversy about the importance of various mechanisms in sintering of crystalline solids.

Even in a pure, one-component system, at least six distinguishable mechanisms contribute to material transport. Other

[*]Now at University of Cambridge, University Engineering Department, Cambridge CB2 1PZ, ENGLAND.

factors, such as applied pressure or a difference in diffusion coefficients in multi-component systems further complicate the matter.

In view of these considerations, it is not surprising that sintering behavior of multi-component, multi-phase systems has been almost entirely ignored. A few scattered attempts reported in the literature utilized second-phase particles either as inert markers to monitor mass flow on the macroscopic scale (as in the work by Seigle and Brett)[1]or to study the influence of the dispersed second phase upon the sintering rate through dislocation-inclusion interaction (as in work by Early et al)[2] The main objective of both of these studies was to distinguish the plastic flow from the Nabarro-Herring mechanism. The conclusions were diametrically opposite, mainly because of the oversimplified models used commonly to describe a process as complicated as sintering. Early et al. explained the observed decrease of sintering rate in Cu powders by a model originally developed to account for the steady-state creep behavior of dispersion-strengthened alloys. Yet another interpretation of their results was given by Ashby[3] who proposed the interface reaction (the absorption or emission of vacancies) as the rate-controlling mechanism.

One purpose of this work is to study the effect of inclusions on the sintering rate in a more quantitative way. In particular, the particle size and spacing should have a pronounced effect on the sintering rate. We are not only interested in the presence or absence of inclusions in the neck region of the sintered product (using inclusions merely as markers) but also in the kinetics of sintering (the rate of neck-growth, shrinkage rate, etc.) as a function of volume fraction of the dispersed phase. This goal might seem ambitious at this point since even the sintering processes of pure metals are not yet fully understood. However, if one can estimate the effect of inclusions on individual mechanisms, then one should be able, in principle, to extract the contributions from the individual mechanisms and thus gain more understanding of the sintering process as a whole.

Our preliminary work has brought to our attention certain problems which one normally does not encounter in sintering single-phase materials. They stem from the inhibited dislocation movement during recovery processes and lead to shape instabilities and in turn accelerated neck growth during the early stages of sintering. These effects are discussed in the first part of the paper; in the second, we present some preliminary results and conclusions.

EXPERIMENTAL PROCEDURE

Cu-Be single crystals, 2 mm in diameter, were grown in graphite crucibles. During the initial stage of this investigation, these alloys were first internally oxidized and then drawn to 0.005"(127μm) diameter wire without intermediate annealing or heat treatment. Because the stored energy of cold work in these alloys results in shape instabilities at elevated temperatures (see the following section) the procedure was later reversed. The Cu-Be alloys were first drawn to the final size and then internally oxidized.

Four different alloys were prepared, with volume fraction of BeO particles between 0.09 and 0.93%. The BeO particle size, determined by transmission electron microscopy was 300 ~ 400 Å for all alloys. Three wires of each Cu-BeO alloy as well as pure Cu were twisted on the lathe and a slight tension (16 g) was applied to the wires during winding. Before winding, the wires were rinsed in dilute hydrochloric acid, water and acetone. After sintering the samples were mounted vertically in 'epoxy', polished and examined under the microscope. The neck sizes were measured as a function of time, temperature and composition. The results are listed in Table I.

RESULTS AND DISCUSSION

Some Consequences of Stored Energy of Cold Work in Cu-BeO Alloys

Stored energy of cold work has little or no effect on sintering of pure metals. At normal sintering temperatures, the stored energy of cold work has already been released by processes of recovery and recrystallization before any appreciable shrinkage can occur. The fact that grinding powders often results in higher

Table I: Sintering data for pure Cu and Cu-BeO 3-wire compacts at 1050°C.

Time (hrs)	Average Neck Width at 1050°C (cm × 10^{-3})		
	pure Cu	Cu - 0.09 v/o BeO	Cu - 0.93 v/o BeO
13	3.82	2.59	2.13
20	4.09	2.65	2.20
40.5	4.72	2.75	2.33
70	5.19	2.84	2.38

density products has been attributed to breaking up of particle aggregates rather than to increased driving force due to the stored energy of cold work.

This is no longer true in certain oxide-dispersion-strengthened alloys. Because the oxide particles very effectively pin disloca- tions, cell walls and sub-boundaries, the recrystallization tem- perature can be much higher than in a pure metal. If the metal containing inclusions is heated rapidly enough, a substantial amount of stored energy can be retained up to the onset of melting, when its release should cause a depression of T_M.

To check this hypothesis, differential thermal analysis was carried out on three different wires: pure copper, copper con- taining 0.93% and 0.09% BeO by volume. All wires were first ini- tially oxidized and then drawn from 2 mm diameter down to 127 μm without intermediate annealing; thus, they suffered the same amount of plastic deformation (99.6% reduction in cross-sectional area). In addition to these strained wires, melting points of pure Cu single crystal and unstrained copper with BeO particles were also measured for comparison.

The measurements have shown that the melting point of the unstrained internally oxidized alloys is within the experimental error ($\pm$ 0.12°C) of that of pure copper. This is not surprising since the matrix after internal oxidation is essentially pure Cu. Similarly, cold work did not alter the melting points of pure Cu or Cu-0.09 vol % BeO. Both materials recrystallized completely before melting – a fact confirmed by metallographic examination. The cold worked samples containing the higher volume fraction of BeO, on the other hand, showed consistent lowering of the melting point by 4.2°C. Clearly, these samples did not fully recrystallize; a fraction of the stored energy of cold work was retained up to the melting point. Melting of course releases this energy.

The magnitude of the depression, ΔT_M, can be estimated from simple thermodynamic considerations[4]; it is related to the excess enthalpy, ΔH, due to the cold work by:

$$\Delta H \geq \Delta H_{S-L} \frac{\Delta T_M}{T_M} \qquad (1)$$

where H_{S-L} are heat of crystallization at the melting point for the stress free, undeformed material. Substituting the measured value of ΔT_M into Eq. 1, we obtain $\Delta H \geq 9.4$ cal/gm. atom. From the pub- lished[5] value of the stored energy of cold work in pure copper which underwent the same amount of plastic deformation ($\Delta H = 20.0$ cal/gm. atom for the value of true strain, $\bar{\epsilon} = 2.77$), one can con-

clude that at least half of the energy of cold work was retained up
to the melting point and that the recrystallization temperature
cannot be far below T_M. The measured values of ΔT_M and ΔH are
of course not unique but depend somewhat on the scanning rate
(in this case, ~5°C/min.).

Our observations of depression of T_M are consistent with
recently published work by Hansen and Bay[6] who investigated
the effect of $A\ell_2O_3$ particles on the recrystallization temperature, T_r, of $A\ell$. A rapid increase of T_r for submicron particle
spacings in $A\ell$-based alloys is in qualitative agreement with our
own work. The average particle spacing for the two copper wires
containing BeO particles is estimated to be $0.26 \pm 0.05\mu m$ and
$0.85 \pm 0.1\mu m$ for higher and lower volume fraction, respectively.
Even in the latter case, the recrystallization temperature most
likely increased (see the following section) but the wire was fully
recrystallized before the melting point was reached and hence no
depression of T_M was observed.

A direct consequence of the increase in T_r is shape instability
at elevated temperatures. The internally stored energy of cold work
can lead to the creation of additional surface area whenever a
decrease in the total strain energy exceeds the increase in the
surface energy. We observed this phenomenon in Cu-0.93 vol % BeO
alloy annealed either in an enert atmosphere or vacuum at temperatures between 750 and 1000°C, for times up to 200 hours. Figure 1(a)
shows a typical example of a wire treated in this way. The initial
circular cross-section is unstable on heating and changes, probably
by surface and volume diffusion to a periodically perturbed, cog-
like profile. Etching reveals the outer part of the teeth to be
particle-free [Fig. 1(b)]. The average tooth spacing is consistent
with a calculated value assuming an idealized sinusoidal shape of
the wire surface.[4] This "humping" phenomenon led to an accelerated neck growth during the early stages of the sintering experiments. The results were very irreproducible and these experiments
were not pursued any further. Instead the oxidation-drawing procedure was reversed.

Sintering Results

Some of the three-wire compacts of different composition,
sintered at 1050°C, are shown in the photomicrographs of Fig. 2
and the time dependence of neck sizes is presented in Fig. 3 where
$\ell n \frac{x}{a}$ was plotted against $\ell n t$. Each experimental point on the plot
is the average of at least 16 readings and the error bars indicate
the standard deviation. Occasional values which were much smaller
than the average value and which obviously resulted because of the
lack of the initial contact, were discarded.

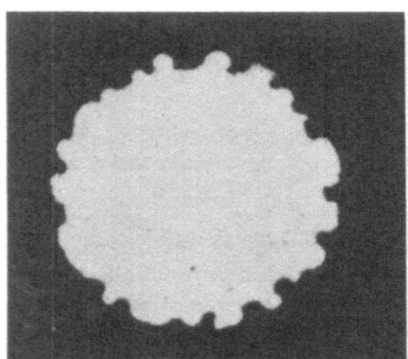 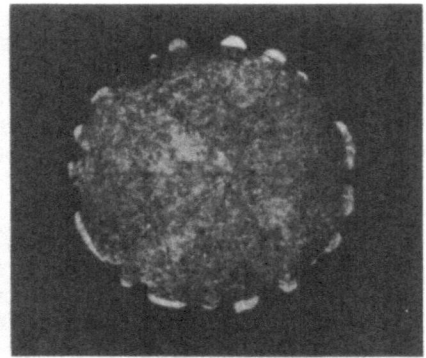

Figure 1(a)   Copper containing .93% BeO by volume, wire-drawn
               from 2 mm to 127 μm and annealed at 1000°C for
               118 hours. 1(b) The same, etched.

As one can see clearly in Figs. 2 and 3, the presence of
BeO particles in the Cu matrix inhibits the sintering process.
Consequently, at any given time the neck sizes of Cu-BeO alloys are
considerably smaller than those of pure Cu; also, as shown in
Fig. 3, the slopes of $\ln\frac{x}{a}$ vs. $\ln t$ for the Cu-BeO alloys are much
smaller than the slope for pure Cu (~1/5) indicating a decrease in
sintering rate. Similar behavior has been observed at temperatures
950 and 1000°C for all compositions although only results of 0.09
and 0.93 vol. % Cu-BeO alloys, sintered at 1050°C, are reported in
this paper.

The results presented in this section should be regarded as
only preliminary. Because of the initial problems with the experimental procedure, the scatter in neck sizes was substantial and it
was impossible to measure the shrinkage or the void area with
meaningful accuracy. The sintering times we have used so far are
insufficient to effect measurable shrinkage. The shrinkage of pure
copper compacts, corresponding to the same neck sizes as in Cu-BeO
alloys would be less than 1% so it is clear that longer times and
higher precision are necessary to resolve the question of shrinkage
of the Cu-BeO system. Because of the lack of precision, we attribute no particular significance to the fact that the two lower
curves in Fig. 3 are almost exactly parallel. It is clear, though,
that their slope (~1/18) is much smaller than that for the pure Cu
and that none of the sintering mechanisms studied in pure metals can
account for this low value.

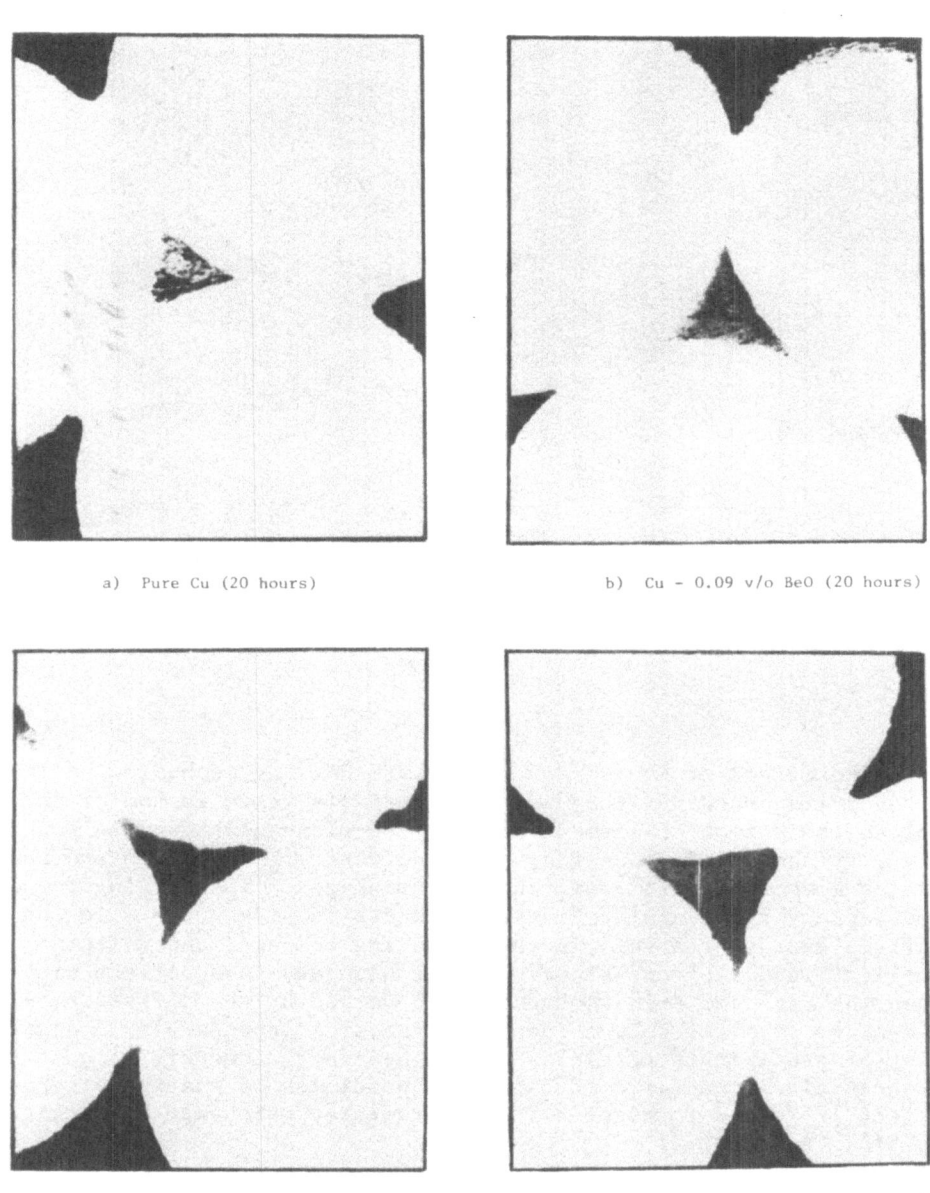

a) Pure Cu (20 hours)   b) Cu - 0.09 v/o BeO (20 hours)

c) Cu - 0.09 v/o BeO (70 hours)   d) Cu - 0.93 v/o BeO (13 hours)

Figure 2   Three-wire compacts of various composition, sintered at 1050°C.

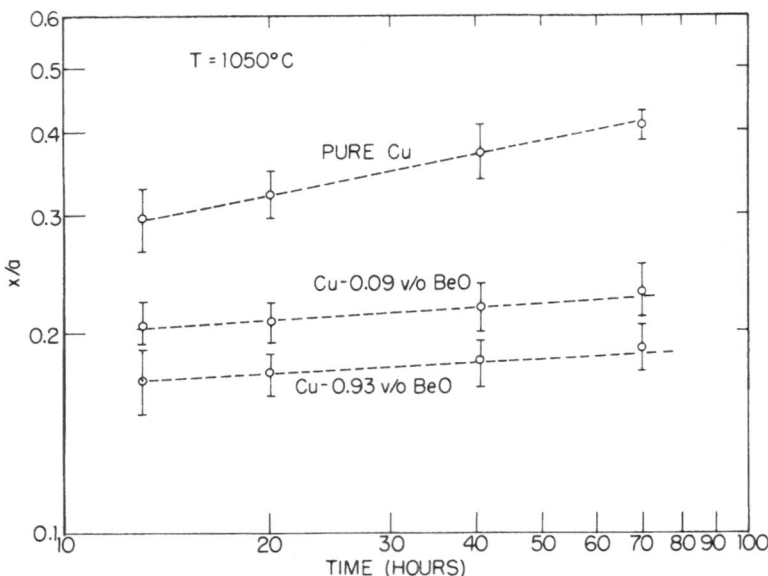

Figure 3   Plot of $\ln\frac{x}{a}$ vs. $\ln t$ for pure Cu and two Cu-BeO alloys, sintered at 1050°C.

        The effect of inclusions on Nabarro-Herring mechanism has been already discussed by Ashby.[3]   Very briefly, this mechanism in-volves the motion of line defects in the plane of the grain bound-ary. By absorbing vacancies, these defects move by climb and the grains move together (resulting in shrinkage). The driving force for vacancy flow in sintering experiments is $\sim\gamma/R$ where γ is the surface energy of a void in the sintering compact, and R its smallest radius of curvature. If the boundary is a perfect sink, then the rate of this mechanism is governed by the diffusion of vacancies from the source (neck) to the sink (boundary). A disper-sion of precipitate particles in the boundary, however, affects this process significantly. In bypassing particles of spacing λ, the defect is forced to adopt a local curvature, which may be as high as $2/\lambda$. A potential

$$\Delta\mu \simeq \frac{2E\Omega}{b\lambda}$$

is required to do this. Here E is the energy of a defect (regarded as a flexible line) per unit length and the other symbols have their usual meaning. The Nabarro-Herring mechanism will contribute to the sintering process only if the driving force, $\sim\gamma/R$, exceeds the threshold stress $2E/b\lambda$; the sintering rate will of course decrease. Thus, the mechanism is no longer diffusion but interface-reaction controlled.

The mechanism involving neck dislocations as sinks of vacancies and the grain boundary diffusion mechanism will be affected in a similar manner. These are the only three mechanisms which lead not only to neck growth but also to densification. A discontinuity in the shrinkage process should therefore serve as an indication that the above three mechanisms ceased to operate. The threshold stress could then be estimated from the curvature measurements. We suspect that all our sintering experiments were terminated after shrinkage had already stopped. More accurate measurements of shrinkage in the future experiments will certainly resolve this question.

There is sufficient evidence that mechanisms which involve free surface as a sink of vacancies are also strongly affected by dispersed particles. The slopes on $\ln\frac{x}{a}$ vs. $\ln t$ plot for our Cu-BeO alloys are much lower than predicted for the above mechanisms. Brett and Seigle[1] reported (see especially Fig.3ab) that the central pores in their Cu-$A\ell_2O_3$ compacts did not spheroidize even after prolonged sintering (589 hours). They shrunk very little and developed facets. These observations cannot be explained by changes in various diffusion coefficients and free surface energy only. These changes would only affect the rate of neck growth or pore rounding but would not stop it. In addition, the slope of $\ln\frac{x}{a}$ vs. $\ln t$ would remain the same. A decrease in slope can be explained by a surface potential barrier. The exact mechanism is not well understood but it may involve the interaction of precipitate particles with the surface steps. This model would also account for Brett and Seigle's observations. It is clear that in the final sintering stages one cannot apply the usual simple geometrical relationship between the neck curvature and neck size and therefore the published sintering equations for pure metals are in this case no longer valid.

In summary, the size and spacing of precipitate particles are important factors in sintering processes. Densely distributed particles will inhibit sintering or completely stop it while scarcely scattered particles will have little effect or might even enhance it (through a decrease in grain boundary energy, possible enhancement of volume diffusion). Since kinetics of recovery and recrystallization is closely related to sintering processes, changes in sintering rate should be also reflected in the change of recrystallization temperature.

S. BAHK, M.F. ASHBY, J. BEVK, AND D. TURNBULL

ACKNOWLEDGMENTS

This work was supported in part by the National Science Foundation under Contract DMR-72-03020 and by the Division of Engineering and Applied Physics, Harvard University.

REFERENCES

1. J. Brett and L. Seigle, Acta Met. 14, 575 (1966).

2. J.G. Early, F.V. Lenel, and G.S. Ansell, Trans. Met. Soc. AIME 230, 1641 (1964).

3. M.F. Ashby, Scripta Met. 3, 837 (1969).

4. S. Bahk and M.F. Ashby, Scripta Met. 9, 129 (1975).

5. H. Wenzl, Z. Angew. Phys. 15, 286 (1963).

6. N. Hansen and B. Bay, Scripta Met. 8, 1291 (1974).

SHRINKAGE AND REARRANGEMENT DURING SINTERING OF GLASS SPHERES

H.E. Exner and G. Petzow

Max-Planck-Institut für Metallforschung
Institut für Werkstoffwissenschaften
Stuttgart, Germany

INTRODUCTION

In the field of sintering, shrinkage of amorphous materials is regarded as being well understood. At high temperatures, inorganic glasses behave as ordinary Newtonian liquids. Due to the action of surface tension, viscous flow takes place during sintering of a system of glass particles resulting in neck growth and approach of particle centers, i.e. shrinkage. For spherical particles of equal size, the kinetics of initial stage sintering were analytically described by Frenkel (1,2) as early as thirty years ago. He, as well as later authors (3-5), assumed that the shape of the spheres changes in the neck region only, and that, at a given time of sintering, the curvature over the neck regions is constant (Fig. 1). This model leads to the following equation for the neck radius x as a function of sintering time t:

$$\left(\frac{x}{a}\right)^2 = \frac{3}{2} \cdot \frac{\gamma}{\eta} \cdot \frac{1}{a} \cdot t \tag{1}$$

where a is the particle radius, γ and η are the surface tension and the viscosity, respectively, at the sintering temperature. This equation was confirmed by Kuczynski (3) for pairs of glass spheres.

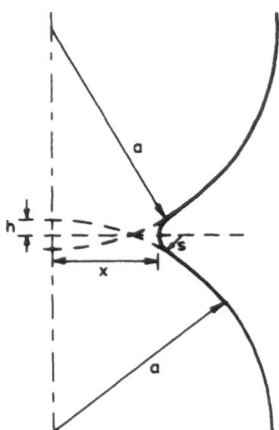

Fig. 1 The parameters of the two-sphere model.

From the neck growth, equation (1), an equation (2) for the rate of approach of the centers of two spheres during sintering can be derived making the usual assumption $h/a \approx s/a \approx x^2/4a^2$:

$$\frac{h}{a} = \frac{3}{8} \cdot \frac{\gamma}{\eta} \cdot \frac{1}{a} \cdot t \qquad (2)$$

In the original work of Frenkel, a numerical factor 3/4 has been calculated on the assumption $s/a=x^2/2a^2$ which does not account for center-to-center approach. This fact is of little importance unless absolute values of γ and η are known.

The equations for the rate of approach of the centers of two adjacent spheres has been generalized to describe the rate of linear shrinkage of an array of spherical particles by writing

$$\frac{\Delta l}{l_o} = \frac{h}{a} = \frac{3}{8} \cdot \frac{\gamma}{\eta} \cdot \frac{1}{a} \cdot t \qquad (3)$$

Excellent agreement with this equation was claimed in experiments on the shrinkage of spherical glass powder arrays for shrinkage values up to 8 % (5-7).

However, in an earlier paper (8) we have pointed out that asymmetric neck growth and formation of new contacts during sintering makes the quantitative description of the shrinkage of irregularly packed particles by relations derived for the two particle model quite doubtful. The purpose of this paper is to present additional evidence that the agreement between experimental results and the shrinkage equation (3) for particle systems may be fortuitous.

EXPERIMENTAL

Spherical glass powder, made by S. Lindner, Warmen-
steinach/Bayreuth, Germany, was used (main components
SiO_2, CaO, K_2O, Na_2O, additional metal oxides 10 wt %).
The fraction +355 to -400 µm was obtained by careful
sieving. Non-spherical particles were eliminated by
rolling the powder down a sheet of paper several times.
All spheres used were cleaned with alcohol and water.

Sintering experiments were made on arrays of two
and three spheres and on planar arrays of larger number
of spheres. The arrays were arranged on graphite substrates and sintered at 750°C. No reaction or adhesion
of the glass spheres to the graphite was observed. In
order to compare the arrays after sintering for different intervals of time they were photographed in
transmitted light by transferring them from the graphite
substrate to a glass slide. The first of the photographs
representing a presintered condition was taken after
heating the array just long enough so that the spheres
adhered to each other sufficiently for safe handling.
Sintering was carried out in dry argon, normal air (dew
point approximately 15°C) and wet air (dew point 95°C).

For the two and three sphere arrays, spheres as
perfect as possible were selected under a stereo-
microscope. The ratio of neck to sphere radius x/a and
the relative center-to-center approach h/a were measured. In addition, the changes in the angle between the
particles were determined for the three sphere models.
Representative photographs of these arrays are shown in
Fig. 2.

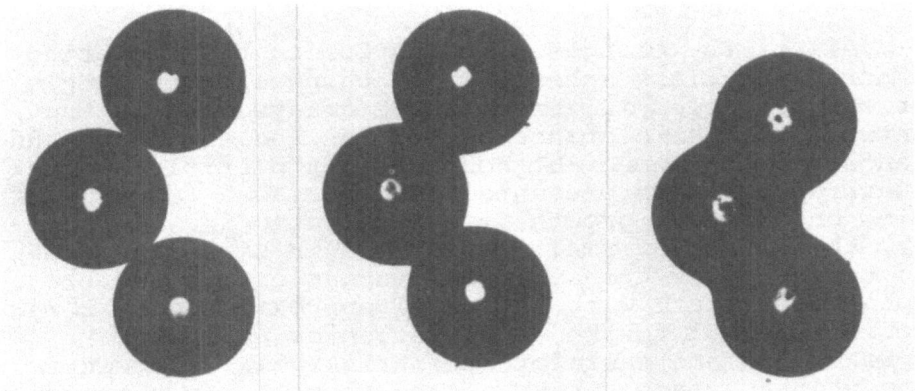

Fig. 2 Three particle array made of 400 µm glass
        spheres. Sintered at 700°C in argon for 1,2 and
        7 minutes.   ├────────┤ 0.5 mm

    An example of shadowgraphs of a planar array of
spheres at different stages of sintering is shown in
Fig. 3. Values for x/a and for the relative center-to-
center approach h/a of adjacent spheres were determined. In addition, the linear shrinkage of the planar
arrays was measured by comparing the distances between
the center of two spheres lying several sphere diameters apart. Finally, the change of number of contacts per particle due to the formation of new contacts during shrinkage was counted in identical fields
of view, i.e. for identical spheres. For each atmosphere, a total number of approximately 500 spheres in
three independently sintered samples was considered to
avoid systematic errors.

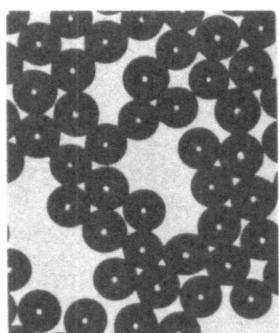 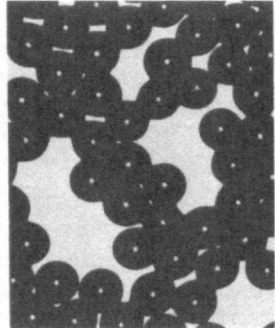 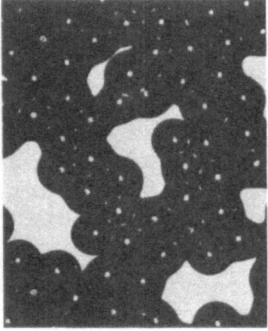

<u>Fig. 3</u> Planar array of glass spheres sintered at
        700°C in argon for 1,2 and 5 minutes.┝━━━━━━┥
                                              1 mm

                            RESULTS

    All of the changes occurring during the sintering
of arrays of glass spheres (i.e. changes of the angle
between particles in three sphere arrays, and of the
center to center distance as well as the shrinkage and
changes in the number of contacts per particle in the
planar arrays) are presented here not as a function of
time, but of neck growth, i.e. the ratio x/a. In this
way, the effect of small variations in sintering times
and temperatures upon the results was eliminated.
Emphasizing that, with the usual approximations, $x^2/4a^2$
should be equal to the relative center-to-center
approach h/a and therefore to shrinkage $\Delta l/l_o$, center-
to-center approach shrinkage and increase in the number
of contacts per particle were plotted vs $(x/a)^2$. This
type of plot suppresses the early stages of sintering
where only little shrinkage occurs.

Neck Geometry

The changes in neck geometry when two spheres coalesce as a function of the stage of neck growth are shown in Fig. 4. To compare the real geometry with the model assuming a constant neck and particle curvature in the neck region (Fig. 1), neck shadowgraphs at given stages of neck growth (i.e. at given x/a-values) are superimposed upon drawings of the hypothetical neck geometry. The figure shows that the particle surfaces near the neck are deformed in regions much larger than the radius of curvature, that the neck is sharper than given by the tangent-circle approximation and that this effect is atmosphere dependent. For sintering experiments in wet air the deviations are small. For those in argon the effect is more pronounced and for dry air the deviations from the ideal geometry are of the same order as those observed by Geguzin and Kruzhanov (9,10) for large epoxy spheres in water.

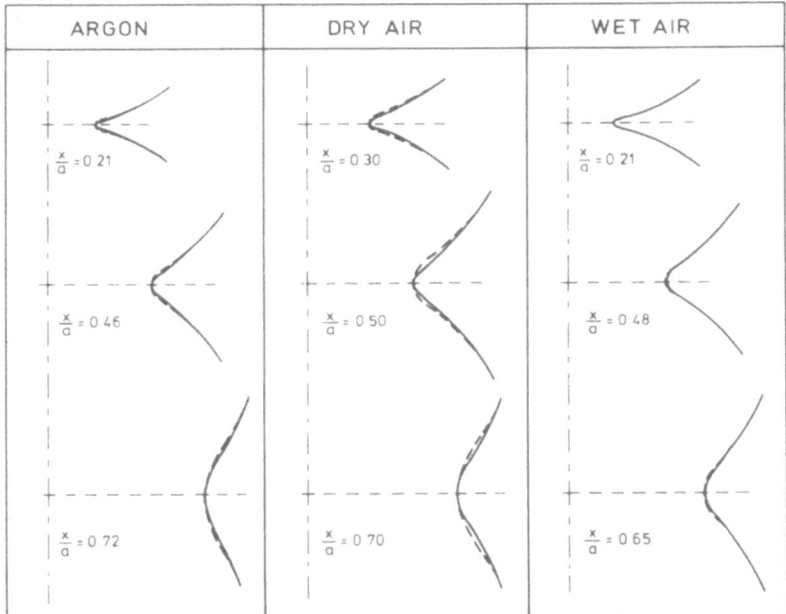

<u>Fig. 4</u> Real and ideal neck geometry of 400 µm glass spheres sintered in different atmospheres at various sintering stages. (Solid lines represent the contours of shadowgraphs, dashed lines are best fits of tangent circles for given neck and particle radius).

Changes in the Angle between Particles
in Three Particle Arrays

As reported in an earlier paper (8), the angle between three glass spheres increases during sintering. Figure 5 shows that the sintering atmosphere has a strong effect on the angle change: In wet air, it is approximately proportional to the relative neck radius x/a. In dry air, there is a reduction of the angle up to x/a = 0.4, followed by a pronounced increase. In argon, the angle changes to a larger degree (note the different scale!), the rate increasing with increasing x/a. In all these atmospheres, the effects are most pronounced for small angles (60 to 100 degree range) and least pronounced for large angles (140 to 180 degree range). It must be noted that the large scatter of results and the small number of time consuming experiments (limited to approximately 25 models for each atmosphere) does not allow a more quantitative evaluation of angle changes as a function of initial angle and neck radius.

Center-to-Center Approach and Shrinkage

Figure 6 shows the shrinkage occurring between pairs of spheres calculated from the distance of adja-

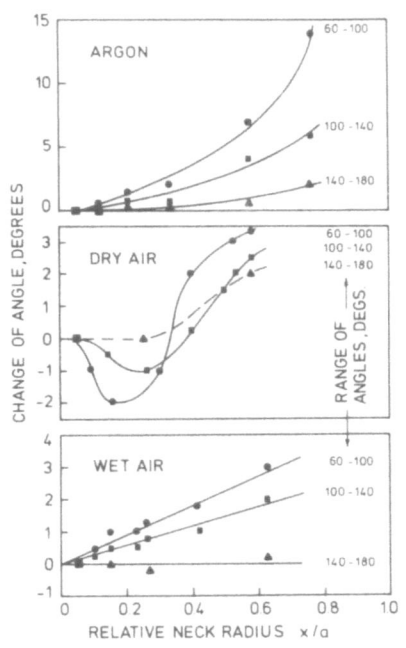

Fig. 5 Change of angle between center-to-center lines in three particle models during sintering in different atmospheres as a function of x/a.

cent sphere centers before and after sintering. There
is no systematic difference between the shrinkage behavior of three-particle models and that of contacting
spheres in planar array up to 25 % reduction in the
center-to-center distance (corresponding to a relative
neck radius x/a of approximately 0.85). Up to 4 %
shrinkage or x/a = 0.4 there is only a slight deviation
from the straight line plotted according to the approximation h/a = x^2/4a^2. This agrees with the limits estimated by Kuczynski (11) and Geguzin (10) for this approximation (x/a = 0.33). At higher values of x/a, a
marked deviation from the straight line occurs. Better
agreement is obtained when it is assumed that the neck
volume V_N is equal to the volume of penetration of the
two spheres V_P. This curve has been calculated by computer iteration and is also given in Fig. 6. Still the
experimental values are too high for a given x/a. This
behavior is easily explained by the fact that the particle surfaces near the neck are deformed in regions
much larger than the radius of curvature and that the
neck is sharper than given by the tangent-circle approximation. This was clearly demonstrated in Fig. 4.

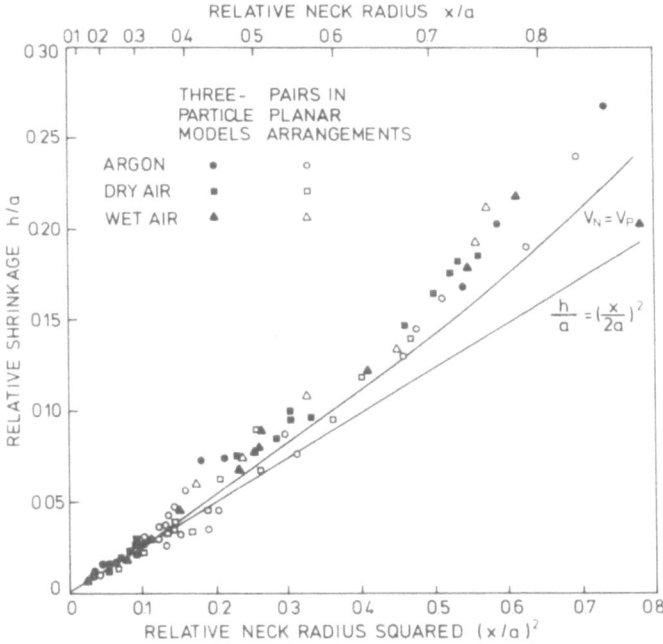

Fig. 6 Relative shrinkage calculated from the approach
of centers in three-sphere models and in neighbouring pairs of spheres in planar arrays.

Shrinkage of planar arrays (changes of distances between the centers of two spheres lying several sphere diameters apart) is shown in Fig. 7. For the sintering runs in argon and wet air, the experimental points fall on a curve which is curved upward. Considering the experimental errors, the straight line given by equation (3) is quite satisfactory. Comparing Figs. 6 and 7 the shrinkage of the planar arrays proves to be markedly less than the center-to-center approach of adjacent pairs of spheres in the same arrangement. This effect is revealed even more clearly by the shrinkage values of the arrangements sintered in normal air. Though the necks grow up to 75 % of the particle radius, shrinkage is limited to less than 3 % compared to a center approach of adjacent spheres of approximately 18 % in these arrays. If the viscosity η were calculated from these results on the basis of neck growth (equation 1) on the one hand and on the basis of shrinkage (equation 3) on the other hand, very different values would be obtained.

Changes in Particle Positions

In previous papers (8, 12) we have shown that the position of spherical particles of copper relative to each other may be shifted when arrays of three or more particles are sintered. Similar observations have been made by other authors (13-16) and very clearly

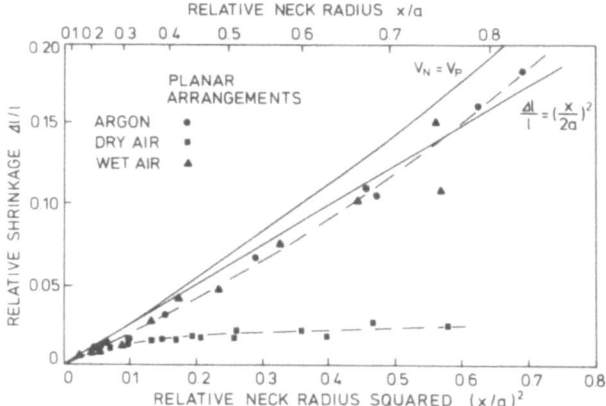

Fig. 7 Relative shrinkage calculated from the approach of centers of spheres in extreme corners of planar arrays.

demonstrated in a movie taken in a hot stage scanning
microscope (17). Our explanation for this effect is the
simultaneous transport of material from the surface and
the grain boundaries of the crystalline particles resulting in asymmetric neck growth for asymmetric
arrangements of contacts (8, 12).

For viscous materials, rearrangement of particles
during sintering has not yet been reported to our
knowledge. Figure 8 clearly shows that the position of
some spheres relative to each other have shifted pronouncedly and pore growth has resulted. Rearrangement
has been observed in all planar arrangements, i.e. in
all sintering atmospheres used.

 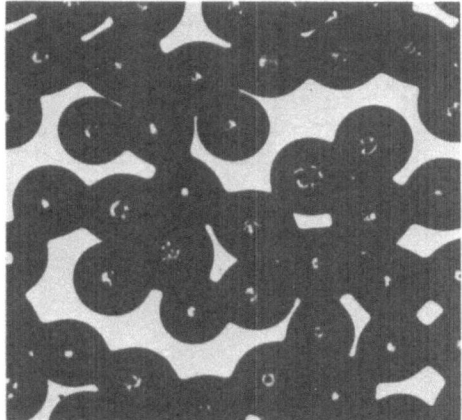

Fig. 8 Pore opening in a planar array of spherical
       glass particles during sintering in dry air
       for 9 minutes. ┝━━━━┥ 0.5 mm

Formation of New Contacts

The formation of additional contacts during
sintering is another reason to be considered why the
center-to-center approach of adjacent spheres and the
shrinkage measured on an irregularly packed planar
array of spheres differ from each other.

For three and four particle arrays, the relationship between shrinkage and the formation of new contacts has been discussed in a previous paper (8).
Typical examples for the formation of new contacts in
planar arrays are shown in Fig. 9. If no rearrangement
of particles occurred, a linear relationship could be

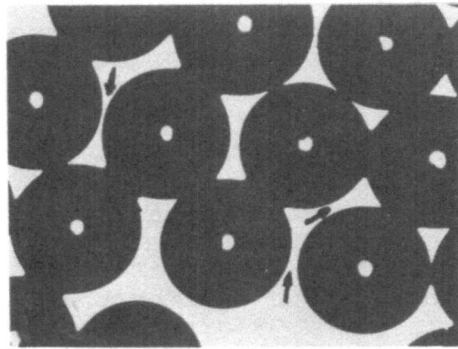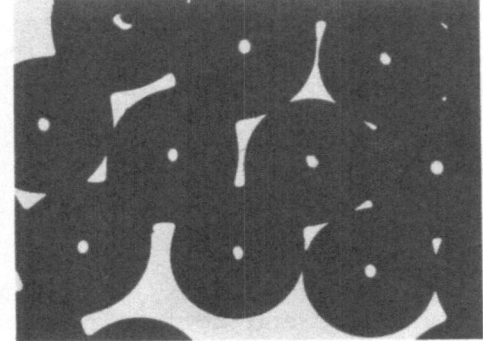

Fig. 9 Formation of new contacts during sintering in
       argon for 1 minute·
       The arrows indicate some new contacts. |———————————|
                                              0.5 mm

expected between shrinkage and the number of new contacts due to the fact that the spaces between the
spheres should be randomly distributed in a random array. In Fig. 10, the number of contacts formed during
sintering of planar arrays is plotted vs. $(x/a)^2$.

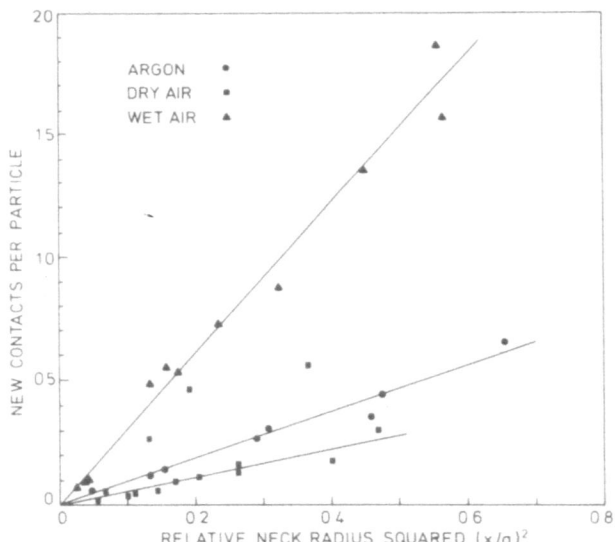

Fig. 10 Increase of average number of
        contacts per particle in planar
        arrays.

Straight lines are obtained though the scatter is
quite large in some cases. It is interesting to note
that at a given x/a-ratio the number of new contacts
formed is strongly dependent on the sintering atmosphere. The slopes of the lines in Fig. 10 are approximately 0.6, 1 and 3 (corresponding to approximately 0.7, 1.3 and 5.2 percent increase in the
average number of contacts per particle for each percent of center-to-center approach) for normal air,
argon and wet air, respectively. These numbers may be
somewhat influenced by differences in the initial number of contacts of the arrangements: For the experiments in different atmospheres the exact number of the
average varies slightly (3.32$\pm$0.25 in argon,
2.90$\pm$0.22 in dry air and 2.83$\pm$0.05 in wet air, respectively, where the standard deviations were calculated
from the averages of three independently sintered samples).
No corrections for this difference was applied since
only qualitative conclusions were intended.

Figure 11 shows the distribution of contacts per
sphere for increasing values of x/a. During sintering
in dry air, the distribution curves are only slightly
shifted towards higher contact numbers. In argon, the
shape of the distribution curve changes pronouncedly.
In wet air, the effect is by far the strongest. The
high frequency of spheres with five and six contacts
indicates that dense regions have formed.

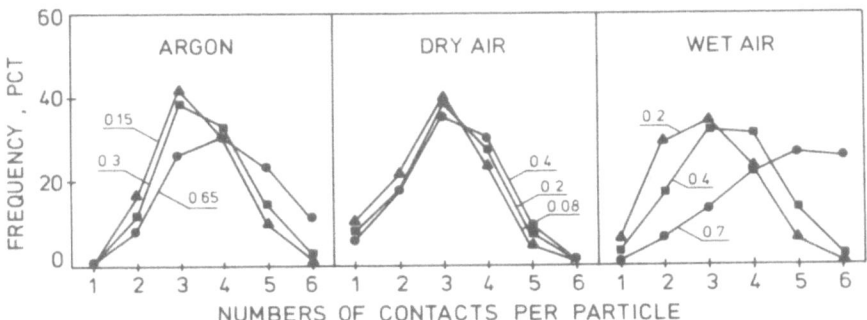

Fig. 11 Distribution of contacts per sphere
(parameter $(x/a)^2$) in planar arrays
sintered in different atmospheres.

DISCUSSION

The sintering behavior of spherical glass particles takes place by viscous flow and has frequently been described quantitatively assuming that

1) The geometry of the neck formed between two particles conforms to the tangent circle model of Fig. 1.

2) The behavior of irregularly packed arrays of spheres can be directly derived from that of the two sphere model.

This investigation has shown that both of these assumptions are incorrect and that the degree to which the behavior deviates from these models is a function of the sintering atmosphere. We cannot at present offer a final explanation for the pronounced influence of the atmosphere on the neck geometry. Qualitatively, it may be speculated that the water vapor and oxygen influence the viscosity of a relatively thin surface layer only. A similar argument has been used to explain other observations on the deformation of glass (18). At any rate, the different neck geometry readily explains the higher shrinkage occurring between adjacent pairs of spheres at large values of x/a in all atmospheres compared to that expected from the tangent circle approximation.

Turning now to the sintering of arrays of more than two spheres, the deviation from the two particle model towards lower shrinkage values is caused by the interference of neighboring necks. It produces asymmetric neck growth, rearrangement of particles in regions of small packing density and consequently opening of large pores and reduced shrinkage. This rearrangement of particles may also interfere with formation of new contacts during sintering and thereby counteract its effects. Since the newly formed contacts have small radii of curvature and put additional forces upon the necks initially present or subsequently formed, they promote shrinkage and accelerated pore closure. Shrinkage varies strongly throughout an irregularly packed array of glass spheres from a maximum value in relatively dense areas after formation of new contacts to positive values of length change, i.e. swelling in regions where pores open up.

The influence which neighboring necks exhibit upon each other and therefore the degree of rearrangement depends upon the sintering atmosphere. Since the sintering atmosphere also determines the degree to which the particle geometry is deformed in the two particle model (Fig. 4), these two phenomena must be directly related with each other, in the sense that minimum deformation of particle geometry should cause minimum influence of neighboring necks and minimum rearrangement. In fact, during sintering in wet air where the particle geometry of the neck region deviates the least from the tangent circle model, the angle changes in the three particle arrays (Fig. 5) are also the smallest. This fits with the observation that the overall shrinkage of planar arrays sintered in wet air (Fig. 7) is much nearer the center-to-center approach of adjacent particles than for planar arrays sintered in dry air. Finally, more new contacts are formed in planar arrays sintered in wet air with increasing neck growth than in planar arrays sintered in dry air.

The situation is not as clear cut for arrays sintered in argon. There is no direct relationship between the degree of neck deformation and angle change. Also, with the large angle changes observed during sintering of three sphere arrays one might have expected smaller over all shrinkage in planar arrays. Actually, the argon sintered arrays exhibit as much shrinkage as those sintered in wet air.

What is needed, is an exact description of the geometry of the region influenced by capillary forces as well as a quantitative description of rearrangement and its effects on shrinkage. Due to the statistical nature of rearrangement and the numerous parameters influencing it, the analytical treatment of the sintering behavior of irregularly packed glass powders in the later stages is a very difficult problem. Agreement between experimental shrinkage data and theoretical shrinkage equations derived from the two particle model is relevant for the early stages only. The agreement observed for the later stages in planar arrays in this work as well as in three dimensional stacking of glass spheres (6,7) must be the result of the combined effects of contact geometry, rearrangement due to asymmetric neck growth and new contact formation.

ACKNOWLEDGEMENT

    The authors wish to thank Prof. F.V. Lenel for his
help in preparing this manuscript and for stimulating
discussions. We greatly appreciate the efforts of Prof.
Kuczynski in initiating our work with glass spheres
during his visit at our laboratory and in encouraging
its progress. Thanks are also due to Prof. E.E. Underwood
and Prof. P.D. Ownby for valuable comments and Mr. H.
Kummer for the careful experimental work.

REFERENCES

1. Ya.T. Frenkel, J. Phys. USSR 9, 385 (1945).
2. Ya.T. Frenkel, Zh. Experim. Theor. Fiz. 16(1)29(1946)
3. G.C. Kuczynski, J. Appl. Phys. 20, 1160 (1949).
4. G.C. Kuczynski, Trans. AIME 185, 169 (1949).
5. W.D. Kingery, M.Berg, J. Appl. Phys. 26, 1205 (1955).
6. I.B. Cutler, J. Amer. Ceram. Soc. 52, 11 (1969).
7. R.E. Heinrichsen, I.B. Cutler, Proc. Brit. Ceram.
   Soc. 12, 155 (1970).
8. H.E. Exner, G. Petzow, H. Kummer, P. Wellner, H.L.
   Lukas, in Proceedings of the Fourth International
   Conference on Powder Metallurgy in CSSR, Vol. 3,
   (1974) p. 47.
9. Ya.E. Geguzin, V.S. Krushanov, Soviet Powder Met.
   and Met. Ceram. 367 (1972); translated from
   Poroshkovaya Metallurgiya (5) 33 (1972).
10. Ya.E. Geguzin, Physik des Sinterns, VEB Deutscher
    Verlag für Grundstoffindustrie, Leipzig (1973);
    Translation of Fizika Spekania, Nauka, Moscow (1967)
11. G.C. Kuczynski, in Powder Metallurgy for High-Per-
    formance Applications, J.J. Burke and V. Weiss
    (Editors), Syracuse University Press, Syracuse (1972)
    p. 101.
12. H.E. Exner, G.Petzow, P. Wellner, in Sintering and
    Related Phenomena, G.C. Kuczynski (Editor), Plenum
    Press, New York, London (1973), p. 358.
13. L.K. Barrett, C.S. Yust, Trans. Met.Soc. 239, 1167
    (1967).
14. A. Nayala, L. Mansour, J. White, Powder Met. 6, 108
    (1963).
15. P.C. Eloff, F.V. Lenel, in Modern Developments in
    Powder Metallurgy, H.H. Hausner (Editor, Plenum

Press, New York, London (1971) p. 291.

16. P.C. Eloff, Ph.D. Thesis, Rensselaer Polytechnic
    Institute, Troy (1969).

17. R.M. Fulrath, Movie presented at the Third Inter-
    national Conference on Sintering and Related
    Phenomena, June 5-7, 1972 at Notre Dame, Indiana.

18. W.A. Weyl, E.C. Marboe, Sprechsaal 108, 79 (1975).

EVIDENCE FOR ENHANCED GRAIN BOUNDARY HETERODIFFUSION

G.H. Gessinger and Ch. Buxbaum

Brown Boveri Research Centre and

Division EKR, Brown Boveri & Cie, Baden,Switzerland

ABSTRACT

Above a critical temperature the electron emission from thoriated tungsten cathodes begins to drop off because thorium atoms adsorbed to the wire surface evaporate faster than they can diffuse to the surface. In this temperature regime the diffusion of thorium through the grain boundaries of tungsten and W_2C is rate controlling.

This paper describes the observation that small additions of platinum to the wire surface increase the diffusion of thorium through the gràin boundaries of tungsten and W_2C. Consequently the temperature limit above which electron emission slows down, can be increased from 1950° K to 2150° K, and the maximum emission current is increased from 3 to 7,5 amps/cm^2.

The process is compared with the activated sintering of tungsten by the addition of platinum to the tungsten grain boundaries.

INTRODUCTION

The possibility of lowering the sintering temperature of tungsten by the addition of small amounts of activating agents has for many years been attracting the attention of investigators. It has been found that the use of certain transition metals as activators has been particularly promising. The best known example for activating the sintering of tungsten is through the addition of nickel[1], but other group VIII transition elements show this effect as well[2].

295

Only iridium additions[3] seem to decelerate the rate of sintering. Several mechanisms have been proposed to account for this unusual effect[4-9]. All models try to give an explanation of the higher diffusional mobilities of atoms, either in terms of the electronic structure of atoms of the metals[5] or in proposing specific diffusion paths which would lead to the high rate of densification[4,5,8,9]. The most plausible model describes the enhanced sintering rate by an enhancement of the tungsten grain boundary diffusion, if this grain boundary is covered by a thin film ($\sim$ 1 monolayer) of the transition metal. Such grain boundaries therefore assume an important carrier function.

It was the purpose of our investigation to show if tungsten grain boundaries enriched by a transition metal would show this carrier function also for small amounts of a second type of impurity atoms.

A convenient and sensitive experiment to demonstrate such an effect is the measurement of thermionic emission in the thoriated tungsten cathode. It is well known[10] since 1913 that under the presence of dissolved carbon or an outer tungsten carbide layer partial reduction of ThO_2 to thorium takes place in a thoriated tungsten wire which has been heated to temperatures above 1300° K in high vacuum. The thorium then diffuses to the wire surface via diffusion through the tungsten and tungsten carbide. Thorium atoms arriving at the surface are spreading to form a monoatomic layer which helps to lower the work function for thermionic electron emission. So long as at least one monolayer of W exists on either the W_2C or W-surface, saturation electron emission as a function of temperature is governed by the Richardson-Dushman equation.

$$i_e = AT^2 \exp\left(-e\phi/kT\right) \qquad\qquad (1)$$

where i_e is the current density, A is a constant, $e\phi=\psi$ is the work function, k is the Boltzmann constant and T is the sample temperature. Above a certain temperature the loss of Th by evaporation will exceed the diffusional flux of Th to the free surface. This means that the surface coverage Θ will start to decrease from unity with increasing temperature until an essentially Th-free tungsten surface is obtained at very high temperature. Thus the electron emission will no longer obey the original Richardson equation for Th-covered W-surface, but will gradually drop to the level of emission from pure tungsten. In this transition region diffusion of thorium to the W surface is the major rate-controlling step. Thus it should be possible to correlate aspects of this curve to the diffusion coefficient of thorium through tungsten. The scarce information available on the Th-W phase diagram[11] suggests that thorium diffusion must preferentially proceed via the tungsten grain boundaries, since no solubility of Th in W has been found. If thorium is now chosen as the impurity, then any carrier effect of a transition metal in the W or W_2C-grain boundaries should result in a change in the transition tempe-rature, above which the surface coverage will be less than a mono-layer of thorium.

One convenient transition metal is platinum, since the melting point of platinum is comparatively high, and no low melting point intermetallic compounds are formed[11].

EXPERIMENTAL

A thoriated tungsten wire (1.5 wt% ThO_2, 1 mm ϕ) was obtained from Philips. After carefully cleaning the wire surface, the wire was galvanically coated by a 5 μm layer of Pt. The layer thickness was determined by weighing the wire before and after applying the surface coating. The W_2C-layer was formed by carburizing the wire at 2070° K in a benzole/hydrogen mixture. The carburizing tempera-ture was high enough to allow some of the Pt diffuse into the grain boundaries of tungsten and W_2C. The thermionic emission was measu-red in a high vacuum glass tube (10^{-8} torr) under an applied pulse voltage with a gradient of 10^5 volt cm^{-1}. The cathode was at first activated by heating for 10 min at 2020° K; this treatment helped to reduce a sufficient amount of ThO_2 to Th and furthermore secured rapid outward diffusion of Th and formation of the mono-layer. The saturation electron emission current was then measured as a function of temperature from 1700° K to 2350° K.

RESULTS

Figure 1 shows the electron emission current densities i_e as a function of temperature for 2 cathodes that had both been coated with platinum and subsequently carburized. The figure also shows the emission characteristics of a Pt-free carburized thoriated-tungsten cathode. Figure 2 shows the logarithm of the current den-sity plotted as a function of the inverse temperature for the Pt-coated cathode (Richardson-lines).

From both figures it can be seen that at temperatures from 1700° K to 1950° K the emission current can be described by the same law as for the Pt-free wires. Between 1950° K and 2050° K there is a transition to a slightly lower level of emission, which continues to rise above 2050° K along a curve parallel to the ori-ginal one (up to 1950° K). Above 2150° K the current increases

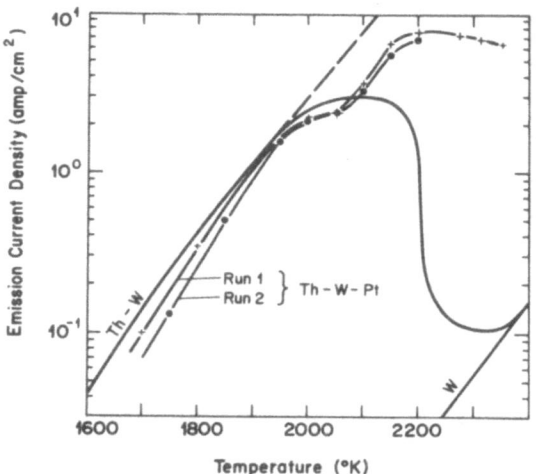

Fig. 1. Logarithm of emission current density vs. temperature for W-ThO$_2$-W$_2$C and W-ThO$_2$-W$_2$C-Pt cathodes.

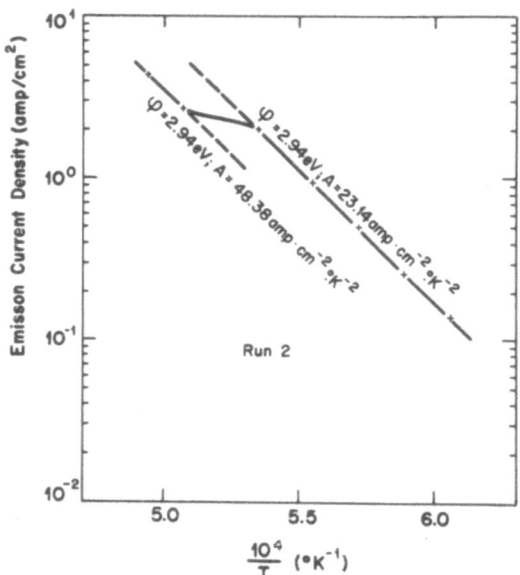

Fig. 2. Logarithm of emission current density vs. inverse temperature for run 2 (W-ThO$_2$-W$_2$C-Pt).

slower with increasing temperature and finally begins to drop. The maximum current density that could be measured was 7.5 amp.cm^{-2} at 2200° K, whereas Pt-free cathodes at this temperature are in the stage of socalled emission-collapse, where the emission drops to about 0.15 amp.cm^{-2}.

THE MODEL

Most parts of Fig. 1 and Fig. 2 can be explained by the following model, which is shown schematically in Fig. 3.

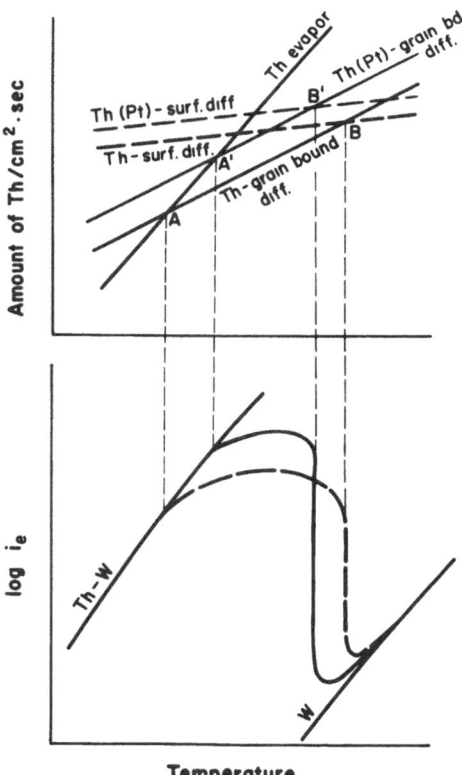

Fig. 3. Schematic correlation between evaporation, grain boundary diffusion, surface diffusion and emission current.

In Fig. 3a the amount of Th per unit time and unit area is plotted
as a function of temperature for the three possible rate controlling
mechanisms:
1. Rate of Th-diffusion along W-grain boundaries.
2. Rate of spreading of Th by surface diffusion on the tungsten
   surface.
3. Rate of evaporation of Th-atoms from W-surfaces.

In Fig. 3b the corresponding emission current is plotted. The
important phases of emission activation are designated by A and
B, which have the following meaning:
1. At $T \leqslant T_A$, the rate of diffusion is faster than the rate
   of evaporation. With a full surface coverage with Th-atoms
   the emission current can be described by the Richardson-
   Dushman equation.
2. At $T_A < T < T_B$ the fraction of the tungsten surface covered
   by Th is less than one. The rate limiting step is diffusion
   of Th to the surface.
3. At $T > T_B$, surface diffusion is rate limiting. Evaporation
   of Th is faster than both surface and grain boundary diffusion. Surface coverage decreases continuously and asymptotically approaches zero with increasing temperature.

Point A can be shifted to a higher temperature (A') by either
increasing the rate of grain boundary diffusion or by decreasing
the rate of evaporating Th-atoms.

Point A is defined by the fact that the rate of diffusion
equals the rate of evaporation of Th atoms from a Th-W monolayer.
Since the temperature dependence of the diffusion of thorium
through W (grain boundaries), the temperature dependence of the
diffusion gradient at the surface and the evaporation rate of
thorium from a Th-W-monolayer are known, it is possible to compute
the observed diffusion-enhancement which is caused by the platinum.

The following assumptions were made:
1. The gradient for diffusion of thorium through tungsten is
   not affected by platinum.
2. The evaporation of Th from the tungsten (carbide) surface
   is also independent of platinum.

Assumption one is equivalent to assuming that the rate of
production of metallic thorium is independent of the presence of
platinum. Under this assumption the concentration gradient of
thorium at the surface can be computed for steady-state condition,
i.e. for saturation emission current densities. For a given temperature the Laplace-equation in cylindrical coordinates can be

written as

$$\frac{1}{r} \frac{d}{dr} \left(r \frac{dc}{dr} \right) = B \tag{2}$$

where c is the thorium concentration at a distance r from the wire
axis and B is a constant proportional to the rate of thorium production. This equation can be solved for the 3 boundary conditions

    a. $c = 0$ at $r = R$ (R = wire radius)

    b. $c = c_p$ at $r = 0$

    c. $\frac{dc}{dr} = 0$ at $r = 0$

The solution for the thorium concentration is a parabolic function.

$$c = c_p \left(1 - \left(\frac{r}{R}\right)^2 \right). \tag{3}$$

The gradient of thorium at the surface is

$$\frac{dc}{dr} \Big|_{r=R} = -\frac{2c_p}{R} . \tag{4}$$

It is only dependent on c_p, the thorium concentration at the wire
axis, which is a temperature dependent function of the rate of
production of thorium. The activation energy for the concentration
gradient in thoriated tungsten wires has been determined by Langmuir[11]
who found a value of 44 kcal/g-atom. He likewise determined the
activation energy for diffusion of thorium through tungsten with
94 kcal/g-atom. The resultant activation energy for the diffusional
masstransport (diffusivity x gradient) of thorium to the wire surface is 138 kcal/g-atom. The heat of evaporation of thorium from
a monolayer of thorium on a tungsten surface is 166.5 kcal/g-atom,
using vapor pressure results of Danforth[12] in the equation for rate
of evaporation

$$W_A = 2 \cdot 10^{18} \, p \, \sqrt{MT} \tag{5},$$

where p is the vapor pressure (torr), M is the molecular weight of
the metal and T is the temperature ($^\circ$K).

We can now proceed to compute the diffusion enhancement for
the equation defining point A in Fig. 3.

$$W_A = D \cdot \frac{dc}{dr} \tag{6}.$$

Point A is measured to be 1950° K, point A' is at 2150° K. Both
W_A and $D_{Th \to W}$ at 1950° K are known ($W_A = 4 \times 10^9$ at/cm^2.sec, D_{Th-W}
$= 3.24 \times 10^{-11}$ cm^{2}sec^{-1}), which gives for

$$\frac{dc}{dr}\Big|_{T=1950^{\circ}K} = 1.23 \times 10^{20} \ cm^{-4}.$$

Increasing the temperature from $1950^{\circ}K$ (point A) to $2150^{\circ}K$ (point A')
gives an increase for the gradient to

$$\frac{dc}{dr}\Big|_{T=2150^{\circ}K} = 1.96 \times 10^{20} \ cm^{-4}.$$

From $W_A = 1.5 \times 10^{11}$ atoms/cm^2.sec at $2150^{\circ}K$ the diffusivity of Th
through the platinum-enriched tungsten grain boundary is computed
to be

$$D_{Th \rightarrow W, Pt} = \frac{W_A}{\frac{dc}{dr}} = 7.65 \times 10^{-10} \ cm^2 \ sec^{-1}.$$

If no platinum were present, a diffusivity of $D_{Th \rightarrow W} = 3.09 \times 10^{-10}$
$cm^2 \ sec^{-1}$ would be expected at $2150^{\circ}K$.

The enhancement factor for grain boundary diffusion of thorium
through Pt-enriched grain boundaries is thus

$$\frac{D_{Th \rightarrow W, Pt}}{D_{Th \rightarrow W}} = 2.47.$$

This estimated 2.5-fold increase in the grain boundary diffusivity of Th leads to a similar increase of the maximum emission to
7.5 amp.cm^{-2} as compared to 3 amp.cm^{-2} for the Pt-free thoriated
tungsten cathode. It also implies that at lower temperature ($\leq 2100^{\circ}K$)
more efficient use is made of the available supply of Th, which will
lead to an increase in the life of the cathode.

DISCUSSION

The method used to measure the diffusion enhancement is obviously dependent on a series of assumptions which could not all be
clarified. The fact, however, that a temperature interval of $200^{\circ}K$
is needed to show an enhancement factor of 2.47, demonstrates the
large sensitivity of this method to fairly small chemical effects.

The major unexplained effect is the shift to a smaller value
of A in the Richardson equation above $1900^{\circ}K$ in the Pt-coated thoriated W-wire. We interpret this effect as one which effects the
emission rather than diffusion or evaporation. It could conceivably
be caused by a phase transformation in the surface layer due to the
presence of Pt.

Enhancement of grain boundary diffusion of impurity atoms by the presence of small amounts of solute atoms has been found in a number of experiments before (as reviewed by Gleiter[13]). Evidence from our experiment suggests in addition, that grain boundaries (containing Pt) which have shown a carrier function for the matrix atoms (W), do show this effect for impurities (Th) as well.

CONCLUSIONS

It is possible to raise the thermionic emission of thoriated, carburized tungsten electrodes to 7,5 amp.cm^{-2} at 2200°K, if small amounts of Pt are added to the cathode. Pt serves as a diffusion activator for thorium. The diffusion enhancement factor at 2150°K is 2.47. This experiment demonstrates the fact that group VIII-transition elements can exert a carrier function for both tungsten -as known from activated sintering experiments- and impurities in the tungsten grain boundaries.

REFERENCES

1. J. Vacek, Planseeber. Pulvermet. 7 (1959) 6.
2. H.W. Hayden and J.H. Brophy, J. Electrochem. Soc. 110, 7 (1963) 805.
3. H.W. Hayden and J.H. Brophy, J. Less-Common Metals 6 (1964) 214.
4. J.H. Brophy, L.A. Shepard, J. Wulff, in 'Powder Metallurgy', ed. W. Leszynski, Interscience, New York (1961), p. 113.
5. I.J. Toth and N.A. Lockington, J. Less-Common Metals 12 (1967) 353.
6. V.V. Panichkina, V.V. Skorokhod and A.F. Khrienko, Sov. Powder Met. 7 (55) (1967) 558.
7. G.V. Samsonov and V.I. Yakovlev, Sov. Powder Met. 8 (56) (1967) 606.
8. W. Schintlmeister and K. Richter, Planseeber. Pulvermet. 18 (1970) 3.
9. G.H. Gessinger and H.F. Fischmeister, J.Less-Common Metals 27 (1972) 129.
10. I. Langmuir, Phys. Rev. 22 (1923) 357.
11. M. Hansen, Constitution of Binary Alloys, McGraw-Hill, N.Y. (1958), p. 1235.
12. W.E. Danforth, in 'Electron Tube Techniques', ed. D. Slater, Pergamon Press (1961), p. 143.
13. H. Gleiter and B. Chalmers, Progr. Mat. Science 16 (1972), p.94.

DENSIFICATION KINETICS OF ELECTROLYTIC SILVER POWDER COMPACT

K. Akechi and Z. Hara

Institute of Industrial Science, University of Tokyo

7-22-1, Roppongi, Minato-ku, Tokto, Japan 106

INTRODUCTION

Sintering model studies of two spherical particles in order to determine the mechanisms operative during sintering have been conducted since 1950's, by which, today it is generally recognized that sintering is a diffusion process driven by surface tension. Even the analysis of these models is in dispute as to the principal sintering mechanism being diffusion or plastic deformation by dislocation motion especially at the early stage of sintering[1]. To describe the densification kinetics of powder compacts during sintering, a great number of experimental and theoretical equations have been proposed. They seem to be inconsistent with each other, perhaps because different characteristics of powder compacts, such as geometric transformation, rearrangement, agglomeration are not taken into consideration. It has been shown that the initial, non-isothermal densification kinetic of the very fine powder compact is not compatible with a diffusion controlled process[2].

The present study consists of analysis on densification kinetics of electrolytic silver powder compacts according to Ivensen's theory[3], the equations of which seem to be fairly well consistent with the experimental results on sintering of various metal powder compacts.

EXPERIMENTAL

Commercial electrolytic silver powder (99.7% Ag) was used.
Powder (4g) of various size, i.e., -100+200, -200+250, -250
+350, -350 mesh, was compressed at $0.2-6t/cm^2$ into compacts of 10.3
mm diameter and 4-8mm height. Particle size of -350 mesh powder
was determined as 1-0.5μ with aid of a scanning electron microscope.

A silica tube, in which the compacts have been held between
silica disks was put into a furnace kept at the given sintering
temperature, and the compact was sintered in air. The shrinkage
during sintering was measured by a dilatometer (accuracy 4μ).
The heating-up time required for getting isotherm was a few minutes. The difference between dilatometric results and those obtained by direct measurement with a micrometer after sintering,
was approximately ±1.7%. Microstructures of surface and fractured
surface of the compacts sintered were observed by SEM.

Densification data measured during isothermal sintering were
analyzed according to Ivensen's empirical equations. In order to
assertain the relation representing densification as a process of
reduction of relative pore volume, expressed by the ratio vs/vp
(vp and vs are the pore volumes before and after sintering, respectively). vs/vp is found from the densities before (dp) and
after (ds) sintering and from the density of the bulk metal (db):

$$\frac{vs}{vp} = \frac{dp\ (db-ds)}{ds\ (db-dp)} \qquad\qquad (1)$$

Ivensen assumed that the volume ratio vs/vp is independent
(or nearly independent) of the green density (dp). Within the
limits of initial density for which vs/vp is constant, the variation of this ratio as a function of time under given sintering
condition is determined solely by the properties of the powder and
is not affected by pressing conditions. Under these conditions an
empirical equation describing the variation of vs/vp as a function
of time can be deduced:

$$vs/vp = (vs/vp)_{\tau=0}(qm\tau+1)^{-1/m} \qquad\qquad (2)$$

where vs/vp is the relative pore volume after isothermal sintering
for a time τ, $(vs/vp)_{\tau=0}$ is the relative pore volume at the start
of isothermal sintering, and q and m are constants.

The constant q has the meaning of the relative rate of pore
volume reduction $(d(vs/vp)/d\tau\cdot1/(vs/vp))$ at τ=0; m is a dimensionless constant characterizing the intensity of the decrease in
densification rate. m and q of equation (2) can be obtained from
the experimental densification data.

In this study, m and q were obtained making use of an elec-
tronic-computer, using the least squares method from 30-70 vs/vp
data per each isothermal densification curve.

RESULTS

Dependence of Relative Pore Volume vs/vp on Green Density dp

The retios vs/vp obtained on -350 mesh powder compacts of
green densities (5.0-9.8g/cm^3)and sintered at temperatures (600,
700, 800 and 900°C) for various times τ are shown in Fig.1.
These graphs indicate that, at 800 and 900°C in certain range of
dp, vs/vp is constant for all times and thus independent of dp.

In this study, the sintering process with vs/vp independent
of dp is called "normal" sintering process and the range of dp
corresponding to the normal sintering is called "normal" dp range.
In the range of higher dp than the normal, vs/vp increases steeply
with dp, and in lower dp range, it decreases with decreasing dp.
At the temperatures lower than 800°C, such as 700°C and 600°C,
the abnormal lower dp range expands and the normal dp range is
reduced.

Fig.2 represents the relation between vs/vp and dp in -200
+250 mesh (Fig.2-a) and -100+200 mesh (Fig.2-b) powder compacts at
800°C. In both cases, the constancy of vs/vp independent of dp
can be found, and the normal dp range is expanded up to higher dp
than the normal range in -350 mesh powder.

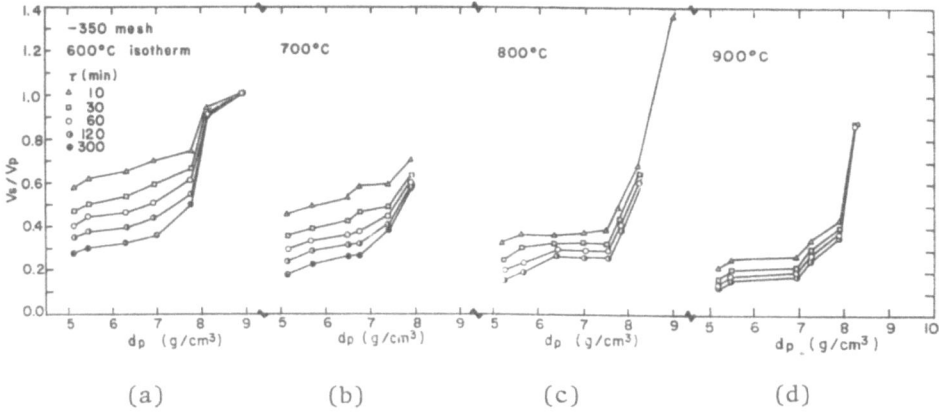

(a)   (b)   (c)   (d)

Fig.1 Relation between vs/vp and dp of -350 mesh powder compacts
       sintered at : (a) 600, (b) 700, (c) 800 and (d) 900°C

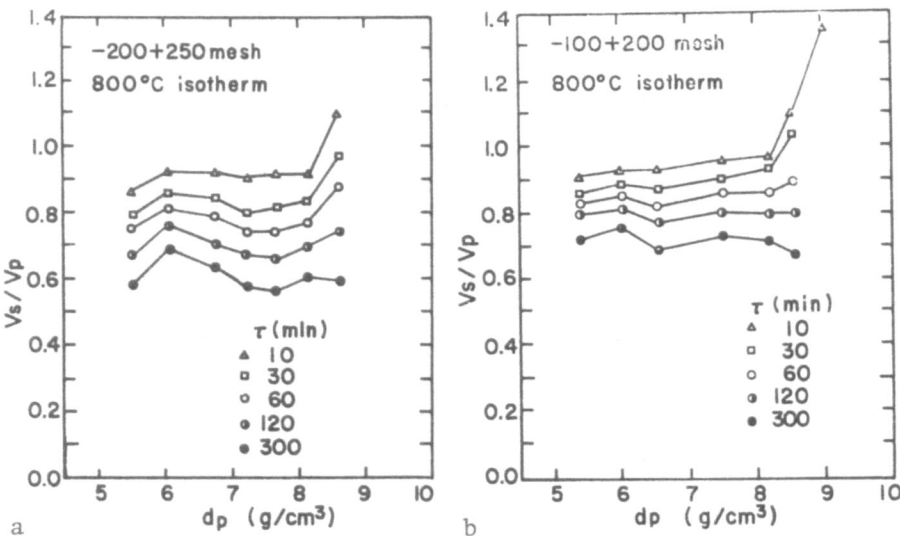

a   b

Fig.2 Relation between vs/vp and dp at 800°C on (a) -200+250 mesh
and (b) -100+200 mesh powder compacts

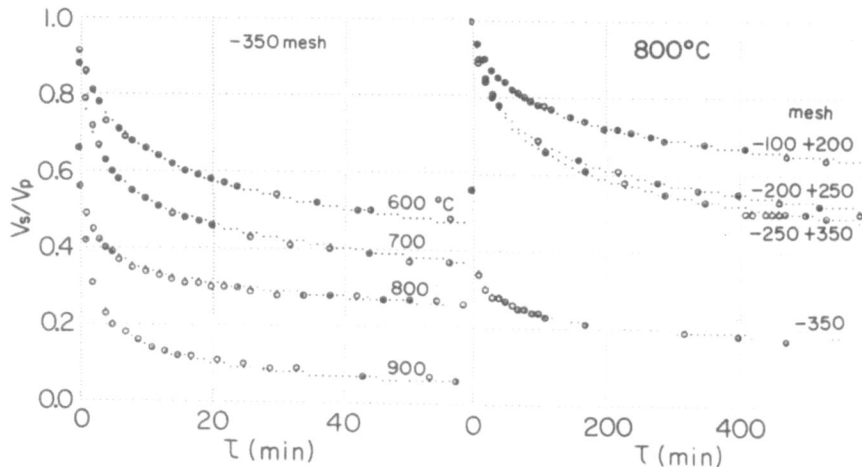

Fig.3 Comparison of the experimental results with the calculated
by computer on the basis of Ivensen's equation (2).

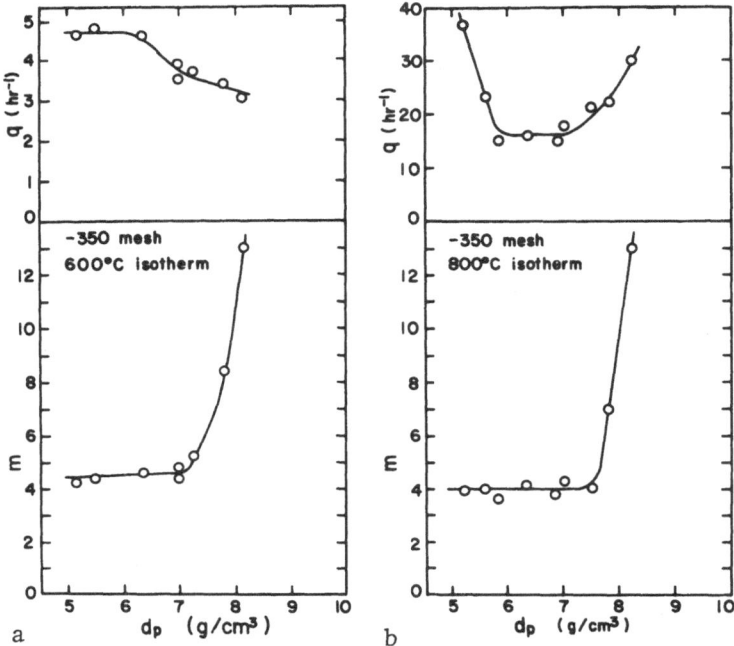

Fig.4   Effect of dp on m and q at (a) 600°C and (b) 800°C (-350 mesh power compacts).

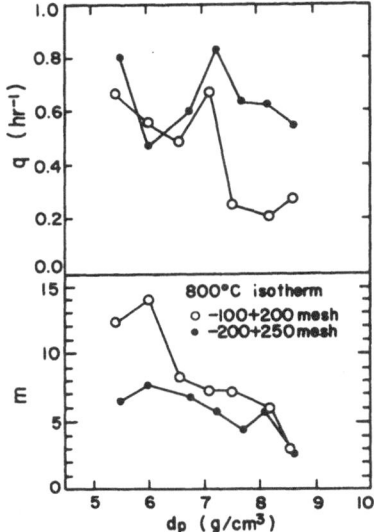

Fig.5   Influence of dp on m and q at 800°C (-200+250 mesh and -100+200 mesh powder compacts).

Application of Ivensen's Equation to Densification Kinetics

From experimental data, constants m and q in eq. (2) were
calculated on the basis of the least squares method, and then substituting the obtained m and q in eq. (2). Densification curves
were plotted by computer (Fig.3). The experimental data agree
well with the calculated curves.

Effect of dp on constants m and q in case of -350 mesh powder
is shown in Fig.4. It is noted that m is independent of dp even
in the abnormal lower dp range. In the abnormal higher dp range,
m increases with dp.

q is nearly constant or independent of dp within the normal

dp range. At the abnormal lower dp range, q increases with decrease of dp, and in the abnormal higher dp range, it is larger or
smaller than q in the normal dp range, depending on temperature
being higher or lower.

Effect of dp on m and q, in case of -200+250 mesh and -100
+200 mesh powder compacts (at 800°C) is shown in Fig.5. In this
case, m is not so constant in some range of dp and in higher dp
range, m decreases with increase of dp. This tendency is opposite
to the case of -350 mesh powder.

Influence of temperature on m and q is shown in Fig.6. q increases usually exponentially with temperature. m increases or
decreases with temperature, depending on the experimental condition (particle size, green density, and temperature).

Micrographs of Sintered Compacts Observed by SEM

Photographs in Fig.7, 8 illustrate the changes in the surface and fracture surface micro-structure of sintered -350 mesh
powder compacts during sintering at 800°C. It is evident from
Fig.7 that during the earliest stage of heating, that is a few
minutes after putting a green compact into the furnace, the smallest particles among -350 mesh powder agglomerate and grow to secondary single grain particles. Due to this process, particles in
the compacts are rearranged resulting in a geometric transformation.

As soon as secondary particle structure is fairly well established (at 800°C, it corresponds to the very beginning of sintering), faceting begins to occur on the surface of the secondary
particles (Fig.7, 8). The faceting increases with isothermal
sintering time and many pronounced facets can be observed on the

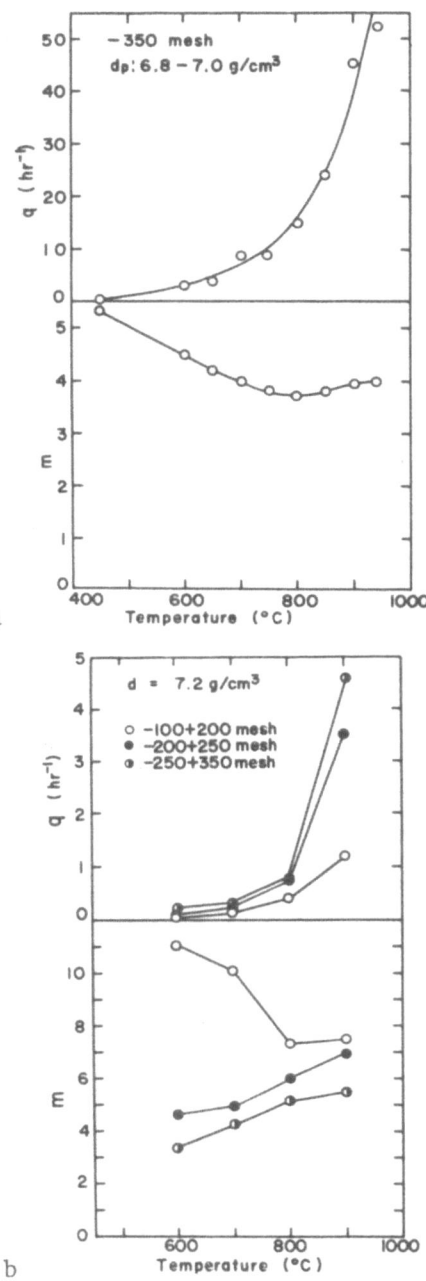

Fig.6 Influence of temperature on m and q.   (a) -350 mesh and (b)
-250+350, -200+250, -100+200 mesh power compacts.

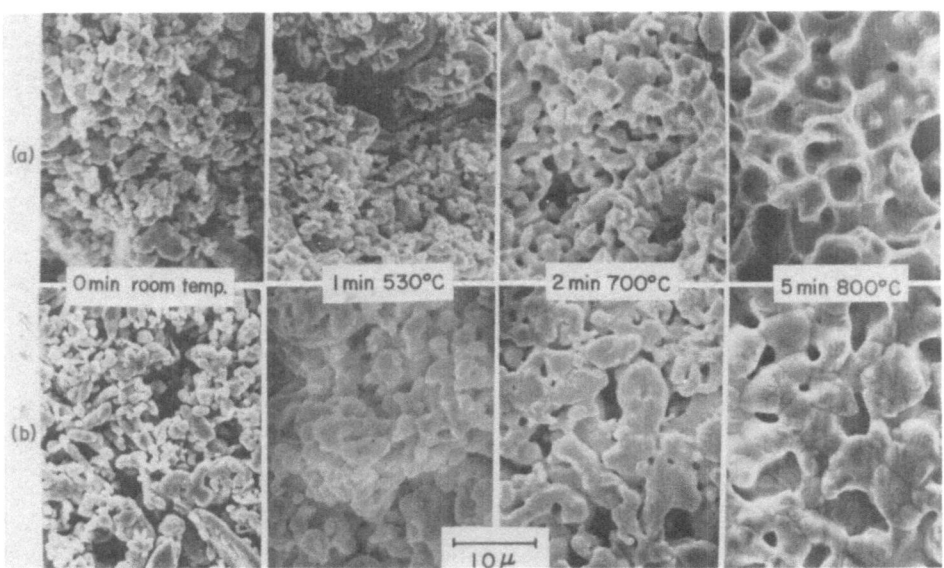

Fig.7 Scanning electron micrograph of (a) fractured surface
       and (b) surface at various heating-up time (-350 mesh powder
       compacts).

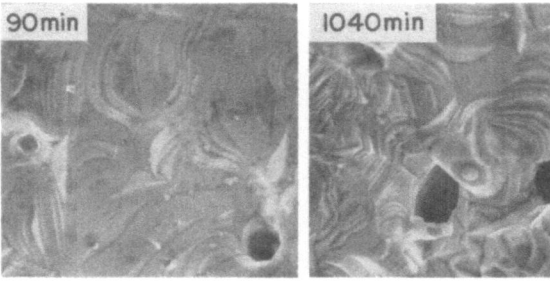

Fig.8 S.E.M. of sur-
faces of compacts at
the later isothermal
sintering stage at
800°C continued from
Fig.7. (the same mag-
nification as in Fig.7)

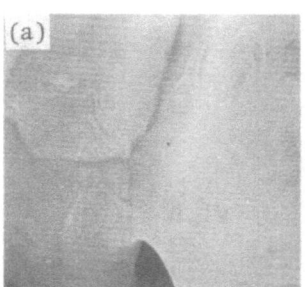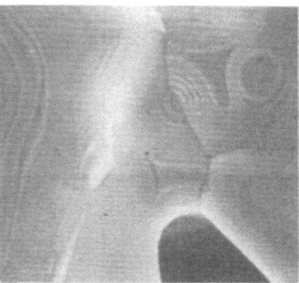

Fig.9 S.E.M. of sur-
faces of (a) -200+250
mesh and (b) -100+200
mesh powder compact
sintered at 800°C for
500min. (the same
magnification as in
Fig.7)

surface of compacts sintered for a long time. It is worth noting
that profiles of the open pores are angular due to the faceting.
Difference between extremely high rate growth of secondary particle structure in the initial sintering stage and slower grain
growth after establishment of secondary particles, is also remarkable.

Fig.9 shows that in the case of -200+250 mesh and -100+200
mesh powders, the grooving at the grain boundaries occurs, but the
particles are not divided into secondary particles like in the
case of -350 mesh compacts.

All of the large particles of electrolytic silver powder have
complicated structure and each of them consists of a lump of fine
spongy particles. So, at the earliest stage of sintering, the
agglomeration very much like that occuring during sintering of
-350 mesh powder compacts occurs in the agglomerate itself and the
lumps themselves do not agglomerate. On the surfaces of -200+250
mesh and -100+200 mesh powder compacts, some facets can be observed,
but their number is less than that of -350 mesh powder compact.

DISCUSSION

At first, the fact as pointed out by Ivensen, that the pore
volume ratio vs/vp of sintered compacts after sintering under the
same condition is constant regardless of the initial density of
green compact, was confirmed by this experiment in the normal dp
range. The reason of the deviation from the constancy outside of
the normal dp range may be due to the following causes. In the
experiment, expansion of compacts occured when the apparent density
of green compacts was higher than about 9 g/cm^3. Therefore in this
range, vs/vp is higher than the constant vs/vp.

The reason why vs/vp decreases with decrease of temperature
and dp in the lower dp range may be due to the fact that, at lower
sintering temperatures or when the green compact density is low,
the geometric transformation with agglomeration of fine particles
will not be completed in the heating-up period and even after the
beginning of isothermal sintering, the transformation will continue and promote densification. The geometric transformation with
agglomeration of fine silver particles at the heating-up period
was observed in this experiment as shown in Fig.7.

Ivensen's empirical equation agress well with experimental
data as shown in Fig.3, as long as the values of m and q are
selected properly. According to the results of these experiments,
m is constant not only in the normal range but also in the lower
dp range. This agrees with Ivensen theory, which indicates that

m and q are constant and characteristic of the powder. But in our experiments q in the lower dp range is higher than q in the normal dp range.

Next, the dependence of m on sintering temperature will be discussed. According to Ivensen's theory, m is the ratio of the rate of defect number decrease to the rate of the material flow in an imperfect crystal (the defects are considered by him to be the origin of the material flow). If the activation energy of the process of defect number decrease and that of the material flow process is defined respectively as Ea and Eb, the activation energy (ΔE) of the densification rate deduced from log m vs 1/T relation can be expressed as the difference between Eb and Ea (ΔE=Eb-Ea). ΔE obtained experimentally was 2.0 Kcal/mole below 800°C and -1.5 Kcal/mole above 800°C. Eb and Ea can be obtained independently of ΔE by the stepwise sintering. Eb and Ea were obtained by this method using -350 mesh powder compacts (dp=6.8 g/cm^3). The ΔE at 400--600°C range was similar to the ΔE value obtained above. But the values of ΔE at higher temperatures were larger than ΔE value obtained from the log m-1/T relation. This difference may be due to the influence of rearrangement with agglomeration of fine particles at initial stage of sintering. At the higher sintering temperature, the value of m is influenced so much by the initial agglomeration that ΔE may be different from the real value at the later stage of sintering. The larger ΔE values mentioned above were obtained from the stepwise sintering experiments conducted at the later stage of sintering. At this stage, faceting, shown in Fig.8, may affect sintering. The details of this problem should be studied in the future.

REFFERENCES

1) J. E. Sherhan, F. V. Lenel and G. S. Ansell: Sintering and Related Phenomena, Ed. G. C. Kuczynski, Plenum Press, New York 1973, p.29
2) C. S. Morgan: Modern Developments in Powder Metallurgy, Ed. H. H. Hausner, Plenum Press, New York 1971, vol.4, p.23
3) V. A. Ivensen: Kinetika Uplotnėniya Metallicheskikh Poroshkov pri Spekanii, Izdatel'stvo Metallurgiya, Moscow 1971

KINETICS AND MECHANISM OF THE INITIAL STAGE OF SODIUM

FLUORIDE SINTERING UNDER NON-ISOTHERMAL CONDITIONS

D.Uskoković, J.Petković[*] and M.M.Ristić

The Institute of Technical Sciences SASA

Beograd, Yugoslavia

1. INTRODUCTION

The investigation of the kinetics and mechanism of sintering of NaF under non-isothermal condition was the subject of some of our previous papers (1-3), in which the course of this complex process was investigated.

Considering the influence of the heating rate and initial density on densification rate (1), we have established that greater linear shrinkage is obtained at lower heating rates. There is a maximum in the shrinkage rate independent of pressing pressure - for a definite heating rate it occurs in the same temperature range, independently on the initial porosity (Fig.1). This phenomenon, which was also noticed in some other materials (4,5), is interpreted, in a phenomenological way, as the result of some considerable change in the system activity (6), due to the rapid recovery of structure macrodefects.

The unsolved problem is the rapid decrease of the shrinkage rate. In previous report (3) the sintering kinetics of NaF was followed by the classical dilatometric measurements as well as by the method of the electronic paramagnetic resonance(EPR). It appeared that EPR spectrum was changing greatly with temperature and time of sintering under isothermal conditions. It is interesting to note that EPR signal (60 G,g=2.005) has a maximum at about 600°C. The increase in temperature above 600°C leads to very quick decrease of the numbers of paramagnetic centres which may be taken either as

---

[*] Now in EI-Ferrites Factory, Beograd.

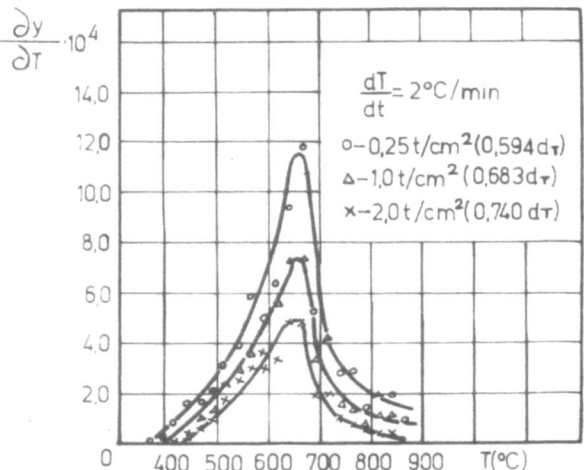

Fig.1 Dependence of Relative Linear Shrinkage Rate on
Temperature and Initial Porosity

electrons localized in the area of anion vacancies, positively
charged pores or charged subsurface layer. It is hard to believe
that the identity of the maximums of the relative linear shrinkage
rate and the intensity of EPR signal is only a coincidence, but
rather that it is a phenomenon the explanation of which would certainly help further comprehension of the sintering mechanism of NaF.

2. EXPERIMENTS

The characteristics of NaF powder and the experimental methods
which were used were given previously (2,3) in some detail. The powder consisted of particles of different shapes, with a little portion of spherical particles. The results of the dilatometric measurements were taken during the constant heating rates of 2,5,10,20 and
50°/min of samples pressed at 0.25 and 2 t/cm^2. The green densities
of the compacts were 0.59 and 0.74 TD.

3. DISCUSSION

Studies of kinetics and mechanism of real systems sintering,
in majority of cases, consist of direct use of the laws and concep-
tions obtained from model studies. The characteristic parameters

(time or temperature exponent "n", experimental activation energies or diffusion coefficients) are used as indicators of mass transfer mechanism.

Johnson's equation (7) derived for the initial stages of iso-thermal sintering:

$$y^{2.06} \frac{dy}{dt} = \frac{2.63 \gamma \Omega D_v y^{1.03}}{k \, Ta^3} + \frac{0.7 \gamma \Omega b D_b}{k \, Ta^4} \tag{1}$$

may be, by proper transformation , brought into the form which holds under non-isothermal conditions of sintering proceeding by the simultaneous effects of the volume diffusion and diffusion along grain boundaries (8):

$$y^n = (K/C) \exp(-\frac{Q + Q'}{RT}) \tag{2}$$

In equations (1) and (2) y is the relative linear shrinkage, γ - surface energy, Ω - volume of vacancies, b - grain boundary width, D_v - volume diffusion coefficient, D_b - diffusion along grain boundary coeficient, t - time, T - absolute temperature, k - the Boltzmann constant, a - particle radius, C - heating rate, Q - activation energy of the densification mechanism, η_o - pre-exponential factor in viscosity term, R-gas constant. The values of constant n, K and Q', which appear in equation 2, for different mass transfer mechanisms, are given in Table 1.

Table 1. The values of the constants from equation 2

Mechanism	n	K	Q' (cal/mol)
Viscous flow	1	$\dfrac{4064 \, \gamma \, R}{a \, \eta_o \, Q}$	5800
Volume diffusion	2	$\dfrac{21500 \gamma \Omega D_{oV} R}{ka^3 Q}$	2907
Grain boundary diffusion	3	$\dfrac{8300 \gamma \Omega D_{oB} Rb}{ka^4 Q}$	2907

In this way the dependence of ln y on 1/T is linear, with a slope of - (Q + Q')/nR. Similarly, the dependence of ln C on 1/T for the constant value of "y" is also linear with slope - (Q+Q')/R. The ratio of these slopes gives the value of the exponent "n",which is dependent on the mass transfer mechanism with n = 1,2 and 3 for

the case of viscose flow, volume diffusion and diffusion along
grain boundaries, respectively. In fig.2 and 3 are shown in this
way analysed results of NaF sintering in non-isothermal conditions,
starting from samples pressed with 0.25 t/cm^2; and in Fig.4 and 5
the results for samples pressed with $2/t/cm^2$ are shown. As can be
seen in Fig. 2 and 4, the dependence of ln y on 1/T represents a
series of curves which may be approximated to straight lines only in
a definite temperature range. The deviation from the linearity above $\sim 650^\circ$C indicates the important changes occuring in the system,
due to the change of the mechanism dominating in the initial stage
of sintering. The non-linearity in the range of low temperatures
is probably the result of the extensive surface diffusion which
leads to smoothing of particle rough surfaces as well as to the
creation of geometric conditions similar to the model system of
spherical particles (7).

The temperature interval corresponding to linear parts in curves given in Fig. 2 and 4, which is in fact the range where it is
possible to apply model equations for the initial sintering stages
to real systems of particles, depends on the heating rates and initial density. In Table 2 and 3 the temperature ranges and shrinkage
intervals in which the maximum changes in the shrinkage rate are
occurring are given as well as the ranges of linearity of ln y =
f(1/T) relation. It can be observed that the linearity range of
the relation ln y = f(1/T), the range within which the laws derived
from models stops with the range in which the maximum rate of the
relative linear shrinkage is registered (Fig.1). The temperature
ranges in which the maximum changes in the shrinkage rate are taking place, as well as the temperature intervals of linearity ln
y = f(1/T) are considerably dependent on the heating rate. In the
case of higher heating rates the interval of linearity is shorter,
when calculated according to the relative linear shrinkage, and
also according to the shrinkage values where the maximum change in
the shrinkage rate is reached. At the end of the temperature interval of linearity the obtained shrinkage, for example, at 50°/min
is relatively small (only about 5%). This effect is probably the
result of the geometric and structural changes occurring in the system, as well as of the competition of the mechanisms during the
initial stage of sintering.

The results obtained on the samples pressed with 2 t/cm^2 agree
with the above explanation. For these samples (Fig.4) the interval
of linearity is observable only at low heating rates and it is very
short (about 3 - 4% of shrinkage). Taking into account the plasticity of NaF, one can expect the radius of the curvature of the
necks between particles to be so much increased during pressing
that the contact of such particles cannot be, considered geometrically identical with the spherical model.

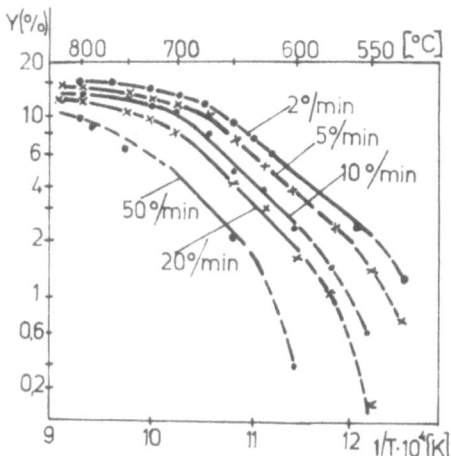

Fig.2 Dependence of Relative Linear Shrinkage on
       Reciprocal Temperature (p = 0.25 t/cm^2)

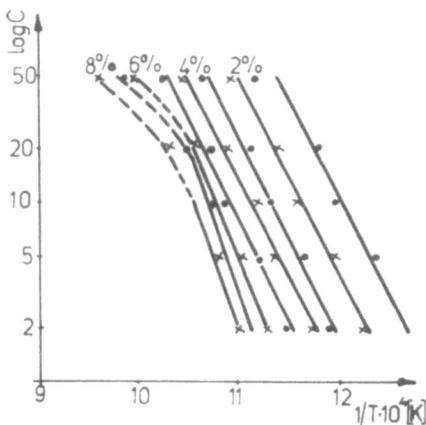

Fig.3 Dependence of Heating Rate at Constant Indicated
       Shrinkage on Reciprocal Temperature (p = 0.25 t/cm^2)

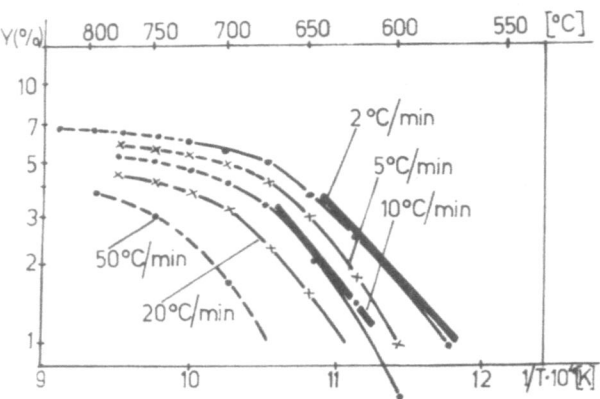

Fig.4 Dependence of Relative Linear Shrinkage on
      Reciprocal Temperature ($p = 2.0$ t/cm^2)

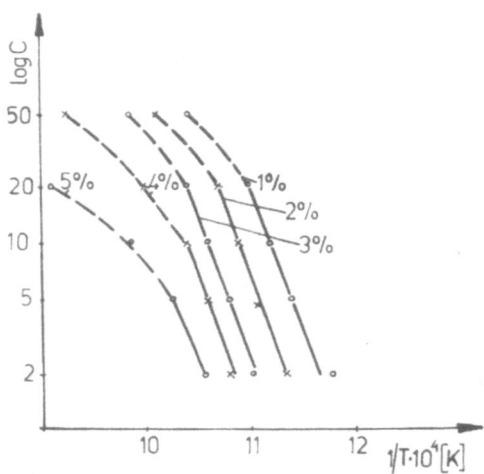

Fig.5 Dependence of Heating Rate at Constant Indicated
      Shrinkage on Reciprocal Temperature ($p = 2.0$ t/cm^2)

Table 2

The temperature ranges and intervals of shrinkage in which the maximum changes in the shrinkage rate $\partial y / \partial T = f(T)$ occur and the areas of linearity $\ln y = f(1/T)$. The pressing pressure of 0.25 t/cm^2, initial density 0.59 TD.

$\frac{dT}{dt}$(°C/min)	A(Δ T)	B(Δ y)	C(Δ T)	D(Δ y)
2	650–675	9,3–12,2	550–675	2–10
5	650–675	7,5–10,7	560–675	2–10
10	650–675	5,4–8,6	580–675	2,5–8
20	650–675	4,3–6,8	600–675	2–7
50	675–725	3,8–6,0	635–700	2–5

Table 3

The temperature ranges and intervals of shrinkage in which the maximum changes in the shrinkage rate occur and the areas of linearity $\ln y = f(1/T)$. The pressing pressure of 2.0 t/cm^2, initial density 0.74 TD.

$\frac{dT}{dt}$(°C/min)	A(Δ T)	B(Δ y)	C(Δ T)	D(Δ y)
2	650–675	3,8–5,0	575–650	1–4
5	650–675	3,1–4,3	600–650	1–3,4
10	650–675	2,2–3,4	625–675	1,5–3,5
20	675–700	2,2–3,5	650–690	1,5–3,2
50	700–725	1,5–2,5	700–725	1,5–2,5

A – temperature range in which the maximum change in the shrinkage rate occurs $\partial y / \partial T$,

B – the interval of the relative linear shrinkage in which the maximum change in the shrinkage rate occurs $\partial y / \partial T$.

C – the temperature range in which the linearity $\ln y = f(1/T)$ exists.

D – the interval of the relative linear shrinkage in which the linearity $\ln y = f(1/T)$ exists.

The value of the slope d ln y/d(1/T) for samples pressed with 0.25 t/cm^2 is ~ 10 600°K, but for samples pressed with 2 t/cm^2 this value is ~ 12 800°K. The corresponding values of the slope d lnC/ /d(1/T) is ~ 26 000°K and 33 900°K respectively. The ratio of the slope for the case p = 0.25 t/cm^2 gives the value n = 2.45, and for p = 2 t/cm^2, n = 2.65. The value of the exponent n, which in both cases are between the values characteristic for the volume diffusion and for diffusion along grain boundaries (Tab.1), indicate that it is impossible to determine the mass transfer mechanism in the initial stage of NaF sintering reliably, and that on the basis of these experiments one cannot speak of the domination of one of the mechanisms. The simultaneous action of the mechanisms of the volume diffusion and diffusion along grain boundaries is in question. One certainly must not neglect the fact that during the "disordered" sintering (9), i.e. during the initial stage, the surface diffusion plays an important role in the smoothing of the rough particle surfaces as well as in the formation of the geometric conditions similar to the model system of spherical particles. It is most likely that for higher heating rates (50°C/min), due to considerable stresses, dislocations affect the mass transfer, so the data obtained under this conditions are displaced in relation to the other heating rates.

A rather frequent approach to the study of mass transfer mechanisms during the sintering process is the evaluation of the apparent activation energy and the comparison of this energy with the activation energy of the elementary processes which may act during sintering. There are several possibilities of the calculation of the apparent activation energy in non-isothermal conditions. We have determined it from the relation (10):

$$\ln \frac{C_j T_i^2}{C_i T_j^2} = \frac{Q}{R} \left(\frac{1}{T_i} - \frac{1}{T_j}\right) \tag{3}$$

where T_i and T_j are the temperatures for which a given amount of linear shrinkage is achieved at two different heating rates C_i and C_j. Activation energies of the processes for all examined heating rates were obtained this way. In all cases maximum shrinkage rate corresponded to the maximum value of the activation energy. The mean value of the activation energy as a function of shrinkage y was computed from all the experiments simultaneously and is given in Fig. 6. The activation energy distribution is analogous to the shrinkage rate distribution, in accordance with a number of our previous observations (4,11,12). The dependence of the activation energy on shrinkage indicates that at lower temperatures, during the initial stage, the mechanism of the diffusion along grain boundaries is dominating, which is characterized by a lower activation energy, but with the increasing shrinkage the volume diffusion be-

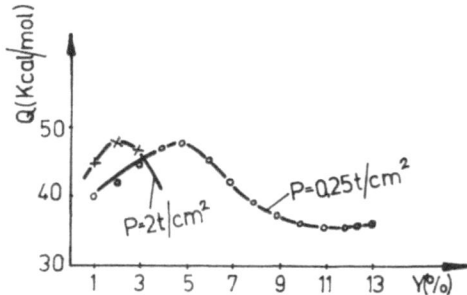

Fig.6 Dependence of Experimental Activation
Energy on Relative Linear Shrinkage

comes dominant until the temperature of the maximum shrinkage rate
is reached. The lack of activation energies of the elementary diffusion processes of sodium fluoride made the exact analysis of mass
transfer during sintering NaF imposible.

4. CONCLUSION

The kinetics and mechanism of the initial stage of sintering of
NaF under non-isothermal conditions, with constant heating rates,
have been studied. It was shown that the equation, derived on the
model system for the initial stages of sintering, may adequately
describe to the range in which the continous increase in the rate
of the relative linear shrinkage is proceeding. The analysis of
the experimental data indicates that the value of the expression
$[d \ln C/d(1/T)]$ / $[d \ln y/d(1/T)]$, which characterizes the responsible mass transfer mechanism is 2,45 for $p = 0.25$ t/cm^2 and
2.65 for $p = 2.0$ t/cm^2, which suggests the simultaneous mechanisms
of volume ang grain boundary diffusion. The non-linearity in the
range of low temperatures is probably the result of the rapid surface diffusion, which leads to smoothing of the rough particle
surfaces and creation of geometric conditions similar to the model
system of spherical particles.

The distribution of activation energies of the sintering process is analogous to the temperature distribution of the relative
linear shrinkage rate, and is continually increasing from 40 100
cal/mol up to 48 500 cal/mol. After reaching the maximum value the
activation energy starts decreasing and becomes $\sim$ 37 000 cal/mol.
The obtained results seem to indicate that the mechanism of diffusion along grain boundaries is dominating at lower temperatures
of sintering, while with the increasing degree of densification
the volume diffusion predominates until the temperature, within
which maximum changes in the shrinkage rate occur, is reached.

After that, due to the intensive grain growth it is impossible to come to adequate conclusions of the mass transfer mechanism in this case.

REFERENCES

1. J.Lazović, Lj.Nikolić, D.Uskoković, J.Phys.Sintering $\underline{3}$(1971)111.

2. Lj.Nikolić, D.Uskoković, M.M.Ristić, Porosh.Metal. No.12(1972) 90.

3. N.G.Kakazei, V.A.Miheev, L.A.Sorin, G.V.Samsonov, D.Uskoković, M.Živković, M.M.Ristić, J.Phys. Sintering $\underline{5}$ (1973) 115.

4. D.Delić, T.Janaćković, D.Uskoković, M.M.Ristić, J. of Japan Soc. Powder and Powder Metal. $\underline{19}$ (1972) 291.

5. Ja. E.Geguzin, Physika Spekaniya, Nauka, M., 1967.

6. Yu.D.Tretiyakov, V.I.Volkov, Neorganicheskie Materiali $\underline{6}$ (1970) 1114.

7. D.L.Johnson, J.Appl. Phys. $\underline{40}$ (1969) 172.

8. D.A.Venkatu, D.L.Johnson, J.Amer.Ceram.Soc.$\underline{54}$ (1971) 641.

9. A.U.Daniels, M.E.Wadsworth, Union Carbide Nuclear Comp.Techn. Rep.XXVI, Utah, 1956.

10. V.V.Skorohod, Reologicheskie Osnovy Teorii Spekaniya, Izd. Naukova dumka, Kiev, 1972.

11. M.Živković, L.Djordjević, D.Uskoković, M.M.Ristić, Porosh, Metal. No. 1 (1972) 197.

12. D.Uskoković, V.Mikijelj, S.Radić, M.M.Ristić, IV Intern.Powder Metal. Conf., Toronto, Canada, July 1973, Eds. H.H.Hausner and W.E.Smith, Publ.Metal.Powder Industries Federation and American Powder Metal.Inst. 1974, 369.

STATISTICAL APPROACH TO THE THEORY OF SINTERING

G. C. Kuczynski

Department of Metallurgical Engineering
and Materials Science
University of Notre Dame
Notre Dame, Indiana 46556

INTRODUCTION

Most of our knowledge of the elementary processes occurring during sintering originated from model experiments. However, the study of systems with well defined and uniform geometry cannot furnish sufficient information to describe in some detail the behavior of the whole particle assembly such as a powder compact. Even the kinetics of the first stage of sintering cannot be simply treated by the rates of formation of interfaces between adjacent particles.[1,2,3] The intermediate stage of sintering which is essentially the shrinkage of a continuous pore phase and the final stage dealing with a discrete pore phase, cannot be represented by simple models, because the Ostwald ripening of the pores is an essential part of the densification process. For this reason the well known Coble[4] theory for the intermediate stage of sintering, most recently refined by Beere[5], is unsatisfactory. Coble replaced the complex geometry of the compact's interior by a system of uniform cylindrical pores situated on the edges of polyhedra of uniform size. The grain growth observed during sintering was interpreted as an increase of the length of the polyhedron's edge with time and the empirical relation for the grain growth was introduced into a differential equation for cylindrical pore shrinkage. In this manner the system of non uniform pores and grains was replaced by one of well defined geometry which reduced the problem to one of the kinetics of shrinkage of an average cylindrical pore in an average geometrical environment. Coble has realized that pore shrinkage and simultaneous increase of their average size, as well as that of the crystalline grains, are mutually interdependent, but within his model he could not formalize these relations.

Another approach to this problem is a statistical one. The intermediate stage of sintering can be modelled by a continuous cylindrical pore of variable diameter meandering through a solid,[11] crossing itself and with time developing dead end branches. The final stage of course can be represented by an ensemble of spherical pores of various diameters.[12]

It has been pointed out above that the phenomenon of pore shrinkage is always accompanied by Ostwald ripening. Therefore our discussion will commence with a quite general statistical description of this phenomenon of growth of a system of discrete particles (pores can be considered as negative particles) immersed in a medium capable of transmitting atoms or, as in the case of pores, vacancies.

OSTWALD RIPENING

Let us consider an assembly of spherical particles randomly distributed in space. We may assume that each particle characterized by its radius r is in a more or less equivalent environment. We may therefore define a distribution function of radius r and time t $f(r,t)$ by stating that $f(r,t)dr$ is the number of particles per unit volume having radius between r and $r + dr$ at time t. It follows, that the total number of particles per unit volume N_v is

$$N_v = \int_0^\infty f(r,t)dr \qquad (1)$$

and the total volume of these particles per unit volume P,

$$P = \int_0^\infty r^3 f dr = N_v \overline{r^3} \qquad (2)$$

where $\overline{r^3}$ is the average particle volume, assuming the form factor is unity. The appropriate continuity equation can be written in the form:

$$\frac{\partial f(r,t)}{\partial t} + \frac{\partial}{\partial r}(vf) = 0 \qquad (3)$$

where $v = \frac{dr}{dt}$ is the radial velocity of particle growth or contraction. From equations (1) and (3) we obtain

$$\frac{dN_v}{dt} = \dot{N}_v = - [vf]_0^\infty \qquad (4)$$

It is obvious that for $r \to \infty$, $f \to 0$ as well as $v \to 0$ because large particles grow slower; hence

$$\dot{N}_v = \lim_{r \to 0}(vf) \qquad (4')$$

Thus the rate of change of the total number of particles depends only on the limiting value of the function vf, when $r \to 0$. As f is always positive or zero and v becomes negative as r approaches zero, $\dot{N}_v$ must always be negative.

In what follows, it will be shown that expressions for N_v and $\bar{r}$, as the functions of time similar to those obtained by Greenwood[6], Lifshitz and Slyozov[7] or Wagner[8] can be derived without the knowledge of the exact form of the distribution function f.

Let us consider a moment:

$$\int_0^\infty r^n f \, dr = N_v \overline{r^n}$$

and its change with time

$$\dot{N}_v \overline{r^n} + N_v \frac{d\overline{r^n}}{dt} = \int_0^\infty r^n \frac{\partial f}{\partial t} \, dr$$

Applying continuity equation (3) we obtain

$$\dot{N}_v \overline{r^n} + N_v \frac{d\overline{r^n}}{dt} = - [r^n vf]_0^\infty + N_v \overline{r^{n-1}v} \qquad (5)$$

But $[r^n vf]_0^\infty = 0$ because $[vf]_0^\infty$ has a finite value. Therefore,

$$\frac{d \ln N_v}{dt} + \frac{d \ln \overline{r^n}}{dt} = \frac{n \, \overline{r^{n-1}v}}{\overline{r^n}} \left(\overline{\frac{\partial r^n}{\partial t}}\right) \qquad (6)$$

An especially important set of equations obtained from (6) is:

$$\frac{d \ln N_v}{dt} = \frac{\bar{v}}{\bar{r}} - \frac{d \ln \bar{r}}{dt} = \frac{2 \, \overline{rv}}{r^2} - \frac{d \ln \overline{r^2}}{dt} = - \frac{d \ln \overline{r^3}}{dt} \qquad (7)$$

because

$$\left(\overline{\frac{\partial r^3}{\partial t}}\right) = \frac{\partial P}{\partial t} = 0 , \qquad (8)$$

and therefore

$$3\overline{r^2 v} = 0$$

From equations (7) it is easy to derive an expression for the change of the total area of the particle system per unit volume A, defined as

$$A = N_v \overline{r^2} \tag{9}$$

$$\frac{d \ln A}{dt} = \frac{2 \overline{rv}}{\overline{r^2}} \tag{10}$$

the form factor of the particles having been again put equal to one.
From (7) it follows immediately that

$$\frac{A \overline{r^3}}{\overline{r^2}} = P \tag{11}$$

The right side of equation (10) is of course a function of
variance σ^2 whose dependence on time is not known unless f is
specified. Therefore, even though the total area of a dispersion
may vary with time as

$$\frac{dA}{dt} = - KA^m \tag{11'}$$

even if v is known, identification of the process responsible for
the exponent m alone is simply not reliable.

Further progress can be made by assuming that the variance
$\sigma^2 = y \overline{r^2}$ where y is a constant. This is equivalent to the assumption of a quasi-stationary distribution. In such a case (11) can
be written in the form:

$$A \overline{r} = A_o \overline{r}_o \ . \tag{12}$$

Using (7) we can also write

$$\frac{d \ln A}{dt} = - \frac{d \ln \overline{r}}{dt} = \frac{\overline{v}}{2\overline{r}} \tag{13}$$

As the total particle area always decreases, $\overline{r}$ must increase because $\overline{v} < 0$. Adopting the mean field approximation[6,8] v can be
written in form

$$v = \frac{\alpha}{r^n} \left(\frac{1}{r^*} - \frac{1}{r} \right) \tag{14}$$

where $\alpha > 0$, a function of temperature, is characteristic for the
controlling process, and r^* is the particle radius which at a
given time t does not grow or shrink. r^* can be determined from
condition (8). The exponent n characterizes the predominating
elementary process. Thus when $n = 0$, the controlling mechanism is
that of interface reaction, $n = 1$ that of diffusion in the external phase and $n = 2$ diffusion of the particle atoms on the surface of a support. With these assumptions, using the expansion of
any function of r in terms of its variance[9], we obtain:

$$\bar{v} = \frac{-2\alpha y}{\bar{r}^{n+1}} \left[\frac{1 - \frac{n^2 - 1}{2}y}{1 + \frac{(n-1)(n-2)y}{2}} \right] \tag{15}$$

It is interesting to note that inasmuch as $\bar{v}$ must be always negative and y always positive, the expression in brackets must always
be positive. This condition is fulfilled for any positive value of
y, when n is 0 or 1. For $n > 2$, in order for the bracketed quantity to be positive, $y < 2/n^2 - 1$, which is a severe restriction on
the magnitude of variance σ^2 or width of the distribution function.
However, this seems to reflect the approximation used.[9] Equations
(13) and (15) and definitions (2) and (9) for quasi-stationary distribution yield

$$\bar{r} = \frac{P}{A} \frac{1+y}{1+3y} \quad .$$

Therefore one obtains

$$\frac{dA}{dt} = - \frac{\alpha y}{P^{n+2}} \left(\frac{1+3y}{1+y} \right)^{n+2} \left[\frac{1 - \frac{(n-1)(n+1)}{2}y}{1 + \frac{(n-1)(n-2)}{2}y} \right] A^{n+3} \tag{16}$$

Thus m in empirical equation (10') is equal n+3. Equation (16)
predicts correct values of m, 3 and 5 for processes controlled by
interface reaction and surface diffusion on the support, respectively. However, it also predicts that the rate of area decrease is
inversely proportional to P, the total volume of particles, which
is in variance with the existing theories[10] and some observations.

Experiments on coalescence of supported catalysts yield m
values ranging from 1 to numbers greater than 10. However, it
should be noted that equation (16) is valid only for a system of
particles of spherical symmetry not interacting with a support,
something like mercury droplets dispersed on a smooth surface of
glass. Such a system of course is far from the actual catalytic
system.

Another very popular method of mechanism identification stems
from the relations first derived by Greenwood.[6]

$$\bar{r}^p - \bar{r}_o^p = B t \tag{17}$$

where B is a function of temperature and the exponent p identifies the predominant mechanism. For $p = 2$, the controlling

mechanism of particle growth is that of interface reaction and for p = 3 that of diffusion in the surrounding medium. Finally p = 4 characterizes surface diffusion on the support surface. Relations (17) can easily be derived for quasi-stationary particle distributions. Indeed using equations (7) we can write:

$$\frac{d\bar{r}}{dt} = -\frac{\bar{v}}{2} \tag{18}$$

Inserting expression (15) for $\bar{v}$ into (18) and integrating, we obtain

$$\bar{r}^{n+2} - \bar{r}_o^{n+2} = 2(n+2)\alpha y \left[\frac{1 - \frac{n^2-1}{2} y}{1 + \frac{(n-1)(n-2)}{2} y} \right] t \tag{19}$$

an equation essentially equivalent to (17).

Finally an equation for the total number of particles per unit volume N_v can be derived as a function of time. It is

$$N_v/N_v^o = \left[1 + \frac{2\alpha y}{\bar{r}_o^{n+2}}(n+2) \frac{1 - \frac{n^2-1}{2} y}{1 + \frac{(n-1)(n-2)}{2} y} t \right]^{-\frac{3}{n+2}} \tag{20}$$

In conclusion it can be said that the general relations between A, $\bar{r}$, N_v and time can be obtained for quasi-stationary particle distributions, without solving the continuity equation (3). Exact values of the constants in these equations contain a parameter $y = \sigma^2/\bar{r}^2$ which can only be determined when the distribution function f is known. However, it should be pointed out that experimental measurements of the rates of Ostwald ripening are seldom accurate enough to warrant this extra effort.

PORE SHRINKAGE

It has been mentioned above that pores can be regarded as a second phase undergoing Ostwald ripening; however, in this case dP/dt ≠ 0 because pores shrink. The process of pore shrinkage is usually divided into two stages. Stage one or the intermediate stage is one with continuous pore phase. This stage ends when the permeability of a compact becomes zero. It is followed by the final stage during which the pore phase is discrete.

The statistical approach to the problem of pore shrinkage was previously proposed by the author of this paper[11,12]. Also a statistical theory of the final stage of sintering was attempted by Aigeltinger and Drolet.[13]

The model of the intermediate stage[11] as briefly described in the introduction is essentially a body perforated by a continuous cylindrical pore of length L_v per unit volume and of variable radius r. Of course this spaghetti-like pore crosses itself in numerous nodes and due to instability of cylindrical surfaces, discrete spherical pores finally result. However, during the whole stage, the pore structure connects one end of the body with another. The final stage of sintering will be modelled by a body with N_v discrete spherical pores of variable radius r, similar to the one adopted in discussions of the Ostwald ripening process. We may define radial distribution functions f_1 and f_2 such that

$$\int_0^\infty f_1(r,t)dr = L_v \quad \text{and} \quad \int_0^\infty f_2(r,t)dr = N_v \qquad (21)$$

The volume of the pore phase per unit volume called porosity P is defined assuming the shape factor equal one. Thus

$$P = L_v \overline{r^2} \quad \text{for the intermediate stage} \qquad (22)$$

and

$$P = N_v \overline{r^3} \quad \text{for the final stage} \qquad (22')$$

As the pores shrink slowly and only while located in the vicinity of grain boundaries[14], Zener's relation should describe adequately their mutual interaction. Therefore the average grain size $\bar{a}$ in equilibrium with the pores is:

$$\bar{a} = K \bar{r}/P \qquad (23)$$

where K is a constant and $\bar{r}$ the average pore radius. For the intermediate stage this expression takes the form

$$\bar{a} = K'(L_v \bar{r})^{-1} \qquad (24)$$

Equations (23) and (24) can be regarded as grain growth equations. From (22) and (22') we obtain

$$\frac{d \ln P}{dt} = \frac{d \ln L_v}{dt} + \frac{d \ln \overline{r^2}}{dt} \quad \text{for the intermediate stage} \qquad (25)$$

$$\frac{d \ln P}{dt} = \frac{d \ln N_v}{dt} + \frac{d \ln \overline{r^3}}{dt} \quad \text{for the final stage.} \qquad (25')$$

For quasi-stationary distributions, the last two equations can be rewritten in the form

$$\frac{d \ln P}{dt} = \frac{d \ln L_v}{dt} + 2 \frac{d \ln \bar{r}}{dt} \quad \text{for the intermediate stage} \quad (26)$$

and

$$\frac{d \ln P}{dt} = \frac{d \ln N_v}{dt} + 3 \frac{d \ln \bar{r}}{dt} \quad \text{for the final stage.} \quad (21')$$

The total pore area per unit volume

$$A = L_v \bar{r} \quad \text{for the second stage} \quad (27)$$

and

$$A = N_v \overline{r^2} \quad \text{for the third stage.} \quad (27')$$

Combining (27), (27') and (23) we obtain

$$A \cdot \bar{a} = \text{const.} \quad (28)$$

Further modification of equations (26) and (26') yields

$$\frac{d \ln L_v}{d \ln P} + 2 \frac{d \ln \bar{r}}{d \ln P} = 1 \quad \text{for the intermediate stage} \quad (29)$$

and

$$\frac{d \ln N_v}{d \ln P} + 3 \frac{d \ln \bar{r}}{d \ln P} = 1 \quad \text{for the final stage.} \quad (29')$$

The simplest but not unique solutions of equations (29) and (29') are

$$L_v = L_v^o \left(\frac{P}{P_o}\right)^{x_1} \quad (30)$$

$$N_v = N_v^o \left(\frac{P}{P_o}\right)^{x_2} \quad (30')$$

where x_1 and x_2 are constants greater than 1 and P_o is the initial porosity. From Zener's relation we obtain

$$\bar{a} \, P^{\frac{1+x_1}{2}} = \text{const. for the intermediate stage} \quad (31)$$

and

$$\bar{a} \, P^{\frac{2+x_2}{3}} = \text{const. for the final stage.} \quad (31')$$

These relations hold fairly well for a variety of materials, metals as well as oxides[11],[12] , and this fact can be considered as a justification of the adoption of solutions (30) and (30') as well as Zener's relation as grain growth equation.

Not specifying the distribution function f, we can find the condition for the continuous pore phase to become discontinuous. The tubular pores are unstable when their axial length is greater than $2\pi\bar{r}$. Therefore if a segment of a pore between two intersections becomes larger than $2\pi\bar{r}$, it should break up into a row of spherical pores. As the average distance between the pore segments is $L_v^{\frac{1}{2}}$, our condition becomes

$$2\pi\bar{r} \geq L_v^{-\frac{1}{2}} \qquad (32)$$

or

$$4\pi^2 L_v \bar{r}^2 \geq 1 \quad \text{which yields}$$

$$4\pi P \geq 1 \qquad (32')$$

(assuming $P = \pi L_v \bar{r}^2$). Thus the critical porosity P_c at which the pore phase becomes discrete is

$$P_c = \frac{1}{4\pi} \cong 0.08 \qquad (33)$$

It is interesting to note that P_c is independent of material and initial particle size. The only condition apparent in this derivation is that the dihedral angle should be $90°$. For more thorough discussion of this problem, the reader is referred to an excellent paper by Beere[5]. The value of P_c given by equation (33) agrees very well with Fischmeister's measurements of permeability in α-Fe compacts[15].

In order to obtain expressions for porosity P as a function of time t, a more detailed model of the interaction of the pores with the grain boundary is necessary. From equation (6) of the previous section we easily obtain

$$\frac{d \ln P}{dt} = \frac{1}{\bar{r}^2}\left(\frac{\partial \bar{r^2}}{\partial t}\right) = \frac{2\overline{rv}}{\bar{r}^2} \quad \text{for the intermediate stage} \qquad (34)$$

and

$$\frac{d \ln P}{dt} = \frac{1}{\bar{r}^3}\left(\frac{\partial \bar{r^3}}{\partial t}\right) = \frac{3\overline{r^2v}}{\bar{r}^3} \quad \text{for the final stage.} \qquad (34')$$

or equivalent equations

$$\frac{dP}{dt} = 2L_v \overline{rv} \quad \text{for the intermediate stage} \qquad (34'')$$

and $\quad \dfrac{dP}{dt} = 3N_v \; \overline{r^2 v}\;$ for the final stage.   (34''')

At high temperatures pore shrinkage is due to the diffusion fluxes through the crystal volume with sinks and sources in the grain boundaries. Thus only the pores in the vicinity of grain boundaries can shrink or expand, as indeed has been observed[14]. In such a case, v given by an expression similar to (14) cannot be directly applied. A pore or its segment shrinks only during some interval of time τ during which the pore remains in the vicinity of a grain boundary. After detachment of the boundary from the pore, the pore ceases to change its volume until another grain boundary approaches it. Therefore the rate v is

$$v = p_{gb} v_o \, \tau \, \nu \tag{35}$$

where p_{gb} is the probability that a given pore or in the case of a tube, its segment, is in the vicinity of a grain boundary; v_o is the rate of radius change during the interval of time τ while pore remains in the grain boundary given by an expression (14); ν is the frequency of pore-grain boundary collisions. In our case that is when the process is diffusion controlled.

$$v_o = \frac{\alpha}{r} \left(\frac{1}{r*} - \frac{1}{r} \right) \tag{36}$$

where $\alpha = \dfrac{C\gamma\Omega D}{kT}$, γ being isotropic surface tension, Ω the atomic volume, D diffusion coefficient, k Boltzmann constant, and T absolute temperature. The constant $C = 3$ in case of a tubular pore[16] and 6 in case of a spherical pore[17].

Probability p_{gb} that a pore segment or a discrete spherical pore of average radius $\bar{r}$ is located in a grain boundary of average size $\bar{a}$ can be taken as proportional to the ratio of the number of pores in the grain boundary layer $2\bar{r}$ thick to the total number of pores or total length of their segments, contained in a grain's interior, i.e., $\bar{r}/\bar{a}$. The frequency of pore-grain boundary collisions should be roughly proportional to $\ell/\bar{a}\tau$ where ℓ is the average distance between adjacent pores or between their segments, $N_v^{-1/3}$ or $L_v^{-\frac{1}{2}}$ respectively. Therefore we can write

$$v = K_o \, \frac{\bar{r}}{\bar{a}^2} \, \bar{\ell} \, \frac{\alpha}{r} \left(\frac{1}{r*} - \frac{1}{r} \right) \tag{37}$$

where K_o is some proportionality constant. Equations (34'') and (34''') become

$$\frac{dP}{dt} = 2K_o L_v^{\frac{1}{2}} \, \frac{\bar{r}}{\bar{a}^2} \, \alpha \left(\overline{\frac{1}{r*} - \frac{1}{r}} \right) \text{ for the intermediate} \atop \text{stage} \tag{38}$$

$$\frac{dP}{dt} = 3K_o'N_v^{2/3} \frac{\bar{r}}{\bar{a}^2} \alpha \left(\frac{\bar{r}}{r^*} -1\right) \quad \text{for the final stage(38')}$$

r^*, the radius defined in the previous section, cannot be determined by the method used previously because relation (8) is no longer valid. However, experiments[15] seem to indicate that the pore distributions are quasi-stationary; hence $r^* = \beta\bar{r}$, where $\beta > 1$. Moreover as the porosity decreases, the term in parenthesis in equation (38) should slowly tend to zero, hence tentatively we may assume that this is proportional to $-P^{s_1}/\bar{r}$ and that in (38') to P^{s_2} where s_1 and s_2 are positive numbers less than one. With these assumptions one can show that the right sides of equations (38) and (38') are functions of porosity P and, through α, of temperature. They integrate into expressions of the form

$$\frac{1}{P^n} - \frac{1}{P_o^n} = K_P\alpha t \tag{39}$$

where

$$n = \begin{cases} \frac{3}{2} x_1 + s_1 & \text{for the intermediate stage} \\ x_2 + \frac{2}{3} + s_2 & \text{for the final stage} \end{cases} \tag{40}$$

and $\quad K_P$ is a proper proportionality constant.

Equations (39) are different from Coble's expression. All available data in the literature[11],[12] can be fitted to equations (39). This, however, is not the ultimate proof of the validity of the present theory. Furthermore, although during the intermediate stage of sintering the network of pores can probably be regarded as rigid and immobile, this cannot be said about the final stage of sintering. There is some evidence[18] indicating that the pores move under the pull of the curved grain boundaries. If this is the case equation (38') does not represent properly the rate of pore shrinkage. However, even in this case it can be shown that the right side of equation (38') is a function of P and T only and therefore after integration should yield a relation similar to that given by equation (39).

As the porosity disappearance is intimately connected with grain growth, through Zener's relation, we can derive an equation for grain growth from equations (39), (31) and (31'). It is of the form

$$\bar{a}^{-m} - \bar{a}_o^{-m} = K_{gb} \alpha t \tag{41}$$

where K_{gb} is a function of temperature only and

$$
m = \begin{cases} \dfrac{3x+2s_1}{1+x} & \text{for the intermediate stage} \\[2ex] \dfrac{3x+3s_2}{2+x_2} & \text{for the final stage.} \end{cases} \tag{42}
$$

If the exponent in the equation (31) or (31') is denoted by r then we have the relation derived before[11],[12]

$$
n = r \cdot m \tag{43}
$$

This relation is rather well obeyed[11],[12].

The above sketch has no pretention to a theory of the advanced stages of sintering. It is an attempt to show that very modest statistical tools can offer insight into the mechanisms involved in the Ostwald ripening process and the phenomenon closely related to it, pore shrinkage.

REFERENCES

1. H.E. Exner, G. Petzov and P. Wellner, "Sintering and Related Phenomena." Materials Science Research, Vol. 6, 351, Plenum Press, New York 1973.

2. C.B. Shumaker and R.M. Fulrath, ibid, p. 191.

3. H.E. Exner and G. Petzov, this volume.

4. R.L. Coble, J. Appl. Phys. 32, 787 and 793 (1961).

5. W. Beere, Acta Met 23, 139 (1975).

6. G.W. Greenwood, Acta Met 4, 243 (1956).

7. I.M. Lifshitz and V.V. Slyozov, J. Phys. Chem. of Solids, 19, 35 (1961).

8. C. Wagner, Z. Elektrochemie, 65, 581 (1961).

9. D. Luss, J. Catalysis, 23, 119 (1971).

10. P. Wynblatt and T.M. Ahn, this volume.

11. G. C. Kuczynski, Phys. Sintering, 5, 41 (1973).

12. G.C. Kuczynski, "Sintering and Related Phenomena." Materials Sc. Research $\underline{6}$, 217, Plenum Press, New York (1973).

13. E. Aigeltinger and J.P. Drolet, "Modern Developments in Powder Met." $\underline{6}$, 323 (1974).

14. B.H. Alexander and R.W. Baluffi, Acta Met, $\underline{5}$, 666 (1957).

15. H. Fischmeister, "Perspectives in Powder Met." $\underline{3}$, (Iron Powder Met) 267, Plenum Press, New York (1968).

16. G.C. Kuczynski, Acta Met. $\underline{4}$, 58 (1956).

17. J. Hornstra, Physica, $\underline{27}$, 342 (1961).

18. W.D. Kingery and B. Francois, Proc. 2nd Int. Conference on "Sintering and Related Phenomena", Gordon and Breach, 471 (1967).

19. B. Francois and W.D. Kingery, ibid, 499.

THE CONTRIBUTION OF DIFFUSIONAL FLOW MECHANISMS TO MICROSEGREGATION IN SILVER-GOLD ALLOYS

A. Mishra,* F.V. Lenel[+] and G.S. Ansell[+]

*Lamp Filament Research Lab., General Electric Co.,
 Cleveland, Ohio 44112
+Materials Division, R.P.I., Troy, N.Y. 12181

ABSTRACT

Homogeneous 12 wt. % gold and 88 wt. % silver alloy wires were sintered at different temperatures for different lengths of time. The cross-sections of the necks between the wires were examined with the help of an electron microprobe to determine the concentration at different parts in and adjacent to the neck. From the concentration profile it was concluded that the initial neck growth as well as segregation takes place by surface diffusion. Peak concentration at the apex of the neck is obtained after sintering 10 minutes at 802°C. At longer times, back diffusion occurs by a combination of surface and volume diffusion and after one hour at 802°C, the samples are homogeneous once again.

INTRODUCTION

Like single component systems when two wires of a homogeneous, two component alloy are sintered, a neck is formed between them to reduce the free energy. Due to the surface tension forces an unbalanced stress system exists at the neck. This causes stress gradients between the apex of the neck where high tensile stresses exist and flat regions adjacent to the neck having zero stresses and between the apex of the neck and the grain boundary between the wires where the stresses are compressive. Due to the stress gradient a chemical potential gradient is produced and the atoms in the wires move generally by diffusion into the neck to lower the chemical potential gradient. The diffusion coefficient D_A and D_B of the two components in the alloy are different and if $D_A > D_B$ then more A atoms move into the neck than B atoms under the same chemical

339

potential gradient and this results in an enrichment or segregation of A atoms in the neck. Segregation of A atoms in the neck gives rise to a chemical potential gradient due to a concentration gradient between the neck and regions adjacent to the neck and acts in a direction opposite to the chemical potential gradient due to the stress gradient. As the atoms move into the neck, the neck grows. As a result the radius of curvature of the neck increases and the stress as well as the stress gradient decreases. With preferencial movement of A atoms into the neck, the concentration gradient also increases and the maximum segregation is obtained when the two individual chemical potential gradients are equal and opposite. After this, the chemical potential gradient due to the concentration gradient predominates and back diffusion from the neck to other regions takes place and subsequently complete homogenization is obtained. In Fig. 1 the variation of the chemical potential gradients with time of sintering have been shown schematically.

Kuczynski, Matsumura and Cullity[1] were the first to observe and explain segregation in solid solutions. They sintered homogeneous, single phase alloy wires having compositions close to the limit of solid solubility. Due to segregation the composition in the neck changed locally to the two phase region and the second phase was then observed under the microscope. Another variation of this method was to sinter the alloy wires at temperatures close to the liquidous temperature for that alloy. Due to segregation the liquid phase appeared at the neck which on quenching was observed as an α rich phase under the microscope. For this study, Cu-In, Cu-Ag and Ag-Cu alloy systems were used.

Attempts were made to measure the amount of segregation in Cu + 3.5 at. % gold powder compacts[1] by measuring the shift in the position of a back reflection line and calculating the amounts of the two constituents.

The aim of this investigation was to study microsegregation more extensively and quantitatively. The extent and distribution of segregation in an adjacent to the neck was measured with an electron microprobe. Silved gold system was used because of its simplicity - phase diagram is isomorphous and the alloy system is suitable for microprobe analysis. From the concentration profiles thus obtained, conclusions have been drawn about the relative importance of the different diffusional flow mechanisms.

Experimental Method

An alloy of 88 wt. % silver and 12 wt. % gold was prepared by mixing pellets of silver and gold of high purity (99.99%) and melting in vycor tubes under a positive pressure of argon in an induction

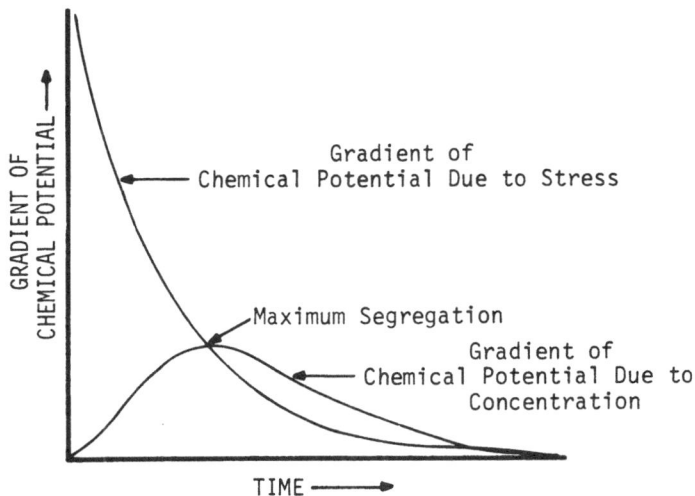

Figure 1 Schematic diagram of the variation of the chemical potential gradient due to a stress gradient and a chemical concentration
gradient with time at the neck.

furnace. To eliminate segregation during solidification, the alloy
was remelted and recast. Ingot was swaged and then drawn into wire
0.5 mm in diameter. Two lengths of wire were then twisted under
constant tension on a lathe. They were presintered by self resistance heating to prevent any unwinding of the wires. Samples of
the twisted wire 18 mm long were sintered in air at the sintering
temperature which was held constant within $\pm 2^{\circ}C$. Samples were
then mounted vertically in epoxy and a scratch free cross section
of the wires containing the neck was obtained. The distribution
of both silver and gold concentrations in the region of the neck
was obtained by using a MAC 400 wavelength dispersive microprobe
using a 90 wt. % silver and 10 wt. % gold standard. Instead of the
usual method of keeping the electron beam fixed and moving the
specimen, the specimen was fixed and the beam was deflected so that
it impinged on the desired spots. A back scattered electron image
of the region of the specimen containing the neck was obtained on
the oscilloscope screen of the microprobe. The screen has a pattern
of gridmarks inscribed on it. The location of a point was determined
by the position of the beam (which could be observed on the oscilloscope screen) with respect to the gridmarks. The raw intensity
data was converted to weight % using the MAGIC IV computer program.
The concentrations obtained were then normallized and the enrichment

(expressed as the concentration of silver at that point minus the original concentration of silver) was plotted on a map containing the contour of the neck and the gridmarks on the oscilloscope screen. This map was obtained from the photograph of the back scattered electron image of the specimen.

Results and Discussion

The following samples of twisted wire pairs of composition 88 wt. % Ag - 12 wt. % Au were sintered.

Sample 1   sintered 10 minutes at 740°C
             corresponding to 2 minutes at 802°C.
Sample 2   sintered 10 minutes at 802°C
Sample 3   sintered 15 minutes at 802°C
Sample 4   sintered 30 minutes at 802°C
Sample 5   sintered 60 minutes at 802°C
Sample 6   sintered 15 minutes at 849°C
             corresponding to 40 minutes at 802°C.
Sample 7   sintered 30 minutes at 849°C
Sample 8   sintered 60 minutes at 849°C

Segregation was observed in Samples 1,2,3,4, and 6.

Results of the microprobe analysis for samples 1 and 2 are shown in Figs. 2 and 3. No segregation was observed further away from the neck. The concentration gradient in both cases was along the surface of the neck and the maximum concentration of silver in each case was at the apex of the neck where chemical potential due to stress is the largest. However, in the case of sample 1 (Fig. 2) the maximum enrichment of silver was 2% whereas in sample 2 (Fig. 3) a peak value of 4% silver was obtained. Sample 1 (Fig. 2) is an intermediate stage of enrichment build up and sample 2 (Fig. 3) represents an advanced stage at or close to the maximum segregation. Unlike sample 2, sample 1 shows a region of depletion of silver near the neck surface. These observations are consistent with Huntington's[2] model for microsegregation. This model assumes a cylindrical geometry similar to the ones used in the experiment. Huntington also assumes that during the segregation build up atoms are transported primarily by surface diffusion. The analysis predicts that during the early stages of segregation there will be two regions namely a) deposition region and b) erosion region. The deposition region is the region of the neck into which the faster moving atoms arrive by surface diffusion. The erosion region supplies the material moving into the deposition region. In the erosion region, due to the depletion of silver atoms, a concentration gradient is set up between the surface and the interior. Therefore atoms move by volume diffusion to eliminate the concentration gradient. Since silver has a higher surface diffusivity than gold,

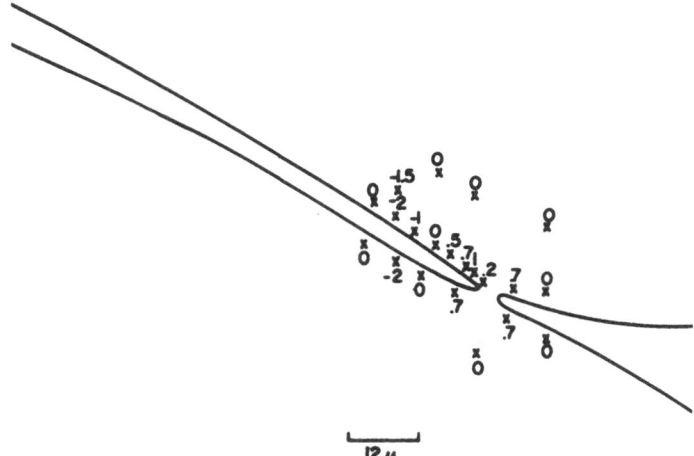

Figure 2   Cross section of the neck between the 88 wt. % Ag –
12 wt. % Au alloy wires after sintering for 10 minutes at 740°C.
Enrichment and depletion of silver observed near the neck surface.

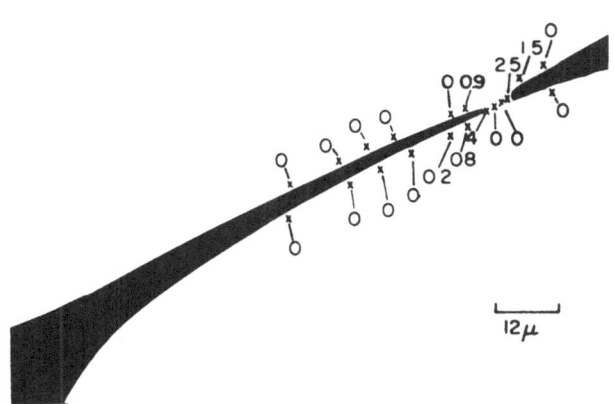

Figure 3   Cross section of the neck between the 88 wt. % Ag –
12 wt. % Au alloy wires after sintering 10 minutes at 802°C.   Enrichment of silver observed near the neck surface.   No silver depleted regions observed.

the surface of the erosion region will be depleted in silver while
in the region of the neck apex, there will be an enrichment of silver atoms. In sample 1 (Fig. 2) the silver depletion region is observed because volume diffusion of silver from the interior to the
surface of the wires was not sufficient to overcome the depletion
caused by the surface diffusion of silver atoms into the neck due
to the short time available for diffusion (10 minutes at 740°C
corresponds to approximately 2 minutes at 802°C). In the more advanced stage of segregation build up (sample 2, Fig. 3) no silver
depletion regions are observed because volume diffusion from the
interior to the surface has decreased the concentration gradient
to within the limits of detectability of the electron microprobe.
In sample 2 (Fig. 3) the size of the neck was large enought so that
two points along the diameter could be measured. No depletion or
enrichment was observed at these two points. This shows that the
grain boundary between the wires is not an important vacancy sink
or material source for either grain boundary or volume diffusion
during segregation build up. Therefore Anthony's[3] analysis which
is based on diffusional flow with grain boundaries as vacancy sink
is not applicable to the results of this investigation. In Fig. 4
(sample 3) and Fig. 5 (sample 4) contour maps for silver enrichment
have been plotted for the regions surrounding the neck. After sintering for 15 minutes at 802°C, the maximum segregation at the apex
of the neck has decreased to 2-3 % showing back diffusion has set
in. Segregation in this sample is observed over a wider area. The
total amount of excess silver as obtained by integrating over the
enriched regions seems to be greater in sample 3 (Fig. 4) than sample 2 (Fig. 3). This is due to the difference in spatial distribution of the enrichment and also because the total flux of silver
due to the stress gradient is larger than that due to the concentration gradient , even sometime after the point of maximum segregation
is reached. Comparison of samples 3 and 4 (Figs. 4 & 5) gives the
effect of sintering 15 more minutes at 802°C. From the two figures
it is evident that the total enrichment is lower in sample 4 (Fig.5)
than sample 3 (Fig. 4). The distribution of enrichment in sample
4 (Fig. 5) is more circular and also extends much less along the
flanks of the neck than that for sample 3 (Fig. 4). This effect is
due to the homogenizing effect brought about primarily by volume
diffusion. Sample 6 (Fig. 6) represents an advanced stage of homogenization or back diffusion. In this case a region of silver segregation surrounding the neck was obtained and the maximum segregation
of silver was less than 1 %. When the time was increased to 60 minutes at 802°C (sample 5) complete homogenization was obtained. Similarly no segregation was observed in samples 7 and 8.

This experiment on the silver-gold system shows that diffusional
flow mechanisms are not the same for segregation and subsequent
homogenization. Segregation takes place primarily by surface diffusion from the flanks of the neck between the wires. Volume diffusion with surface or grain boundary as vacancy sink or grain boundary

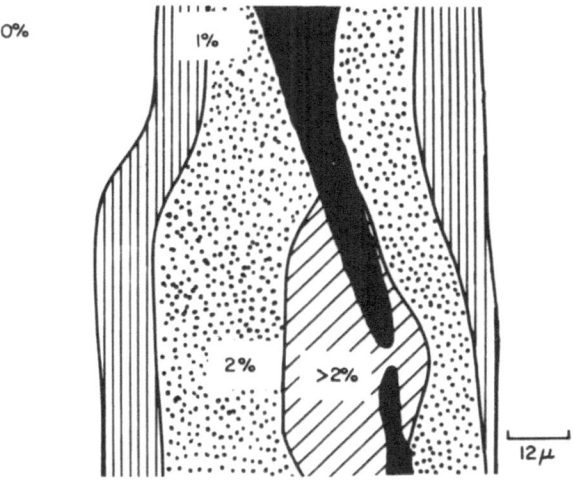

Figure 4 Cross section of the neck between the 88 wt. % Ag –
12 wt. % Au alloy wires after sintering 15 minutes at 802°C. Contours of zones enriched in silver surrounding the neck observed.

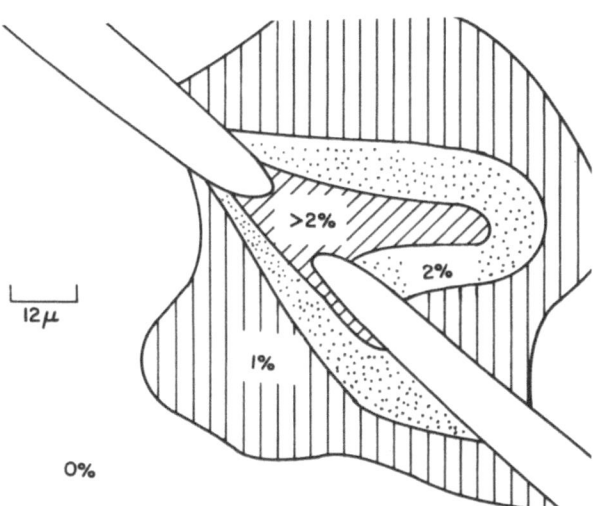

Figure 5 Cross section of the neck between the 88 wt. % Ag –
12 wt. % Au alloy wires after sintering 30 minutes at 802°C. Contours of zones enriched in silver surrounding the neck observed.

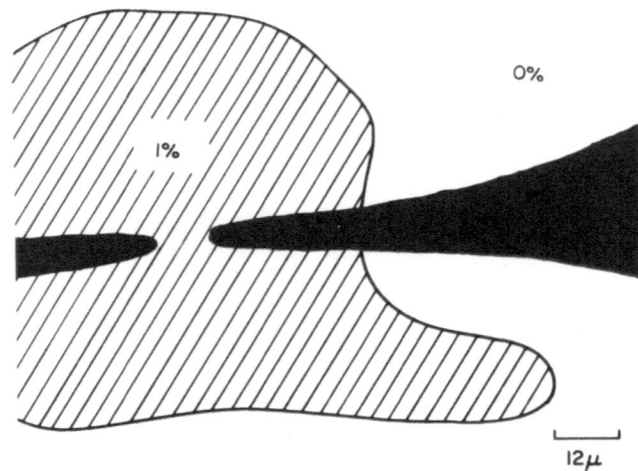

Figure 6   Cross section of the neck between the 88 wt. % Ag –
12 wt. % Au alloy wires after sintering 15 minutes at 849°C.   Only
one enrichment contour with small amount of enrichment observed.

diffusion are not important for segregation. The rehomogenizaion
process, however, takes place primarily by volume diffusion and to
a small extent by surface diffusion. Therefore the analysis of
Kuczynski, Cullity & Matsumura[1] are not valid for this investigation as they assume the same mechanism for both segregation and re-
homogenization.

SUMMARY

1. When 20 mils (0.5 mm) diameter wires of an alloy of 88 wt. %
silver and 12 wt. % gold are sintered at 802°C then build up of
segregation takes place during the first 10 minutes. The maximum
enrichment of silver at the apex of the neck is of the order of 4%.
After 15 minutes of sintering the maximum enrichment decreases to
about 2.7%. This is due to back diffusion of silver. Nevertheless,
when segregation is measured by integrating the silver enrichment
over the entire area perpendicular to the wire axis where it occurs,
segregation continues to increase at least during the first 15 minutes of sintering. After 1 hour at 802°C complete homogenization
is achieved.

2. The growth of neck between the wires and segregation take place
primarily by surface diffusion of material from the flanks of the
neck. Since the surface diffusivity of silver in the alloy is

greater than that of gold, material at the apex of the neck becomes enriched in silver. The stages of build up in silver enrichment are in qualitative agreement with Huntington's analysis for microsegregation.

3. Homogenization takes place primarily by volume diffusion with surface diffusion assisting the process.

REFERENCES

1. G.C. Kuczynski, G. Matsumura and B.D. Cullity, Acta Met. $\underline{8}$, 209 (1960).

2. H.B. Huntington, Rensselaer Polytechnic Institute, p. 349 this volume.

3. T.R. Anthony Acta Met. $\underline{17}$, 603 (1969).

ACKNOWLEDGEMENT

This research was supported by a grant of the Research Division of the U.S. Atomic Energy Commission at N.A.S.A. Materials Research Center of Rensselaer Polytechnic Institute.

SOME ANALYSIS OF MICROSEGREGATION BY SURFACE TRANSPORT DURING

SINTERING

Hillard B. Huntington

Rensselaer Polytechnic Institute

Troy, New York

Microsegregation in an homogeneous alloy during sintering was first demonstrated by Kuczynski et al. (1), for a bulk transport process. The effectiveness of bulk transport in causing segregation at internal voids has been explored by Anthony (2), and his analysis (3) of the transport process in the sintering geometry working from a grain boundary source is satisfactorily complete. That microsegregation can also be obtained by surface transport has recently been demonstrated most satisfactorily by Mishra et al., (4) in the preceding paper. Here the situation is somewhat more complex since the segregation that is developed by surface transport in the early stages of sintering tends to be obliterated by the bulk diffusion that takes place during the later stages of sintering. Moreover the deficit of the faster diffusing component in the erosion region disappears early while in the region of the sintering neck the excess of the faster component persists much longer.

This paper is an attempt to analyze this experiment in a simple manner without resorting to detailed machine solutions for particular conditions. Because of the complexity of the operation, it has not been possible to model the whole process in a single solution but an approximate piecemeal approach does show the main features of the experimental results for both the early and late stages.

1. Basic Equations for Segregation by Surface Diffusion

For simplicity we assume that sintering takes place between
two rod-like specimens which are in uniform contact along their
length. The translational symmetry along the axis of the rods
makes the problem effectively two dimensional. Let the specimens
be alloys of the constituents A and B, and let the surface diffusivities be respectively D_A and D_B where it will be assumed that
$D_A > D_B$. Also, N_A and N_B stand respectively for the numbers of
atoms per unit surface of each constituent. The basic diffusion
equations for surface transport are then

$$\frac{\partial}{\partial s} \{D_A \frac{\partial N_A}{\partial s} - \frac{F(s)}{kT} D_A N_A\} = k_A(s) \tag{1}$$

$$\frac{\partial}{\partial s} \{D_B \frac{\partial N_B}{\partial s} - \frac{F(s)}{kT} D_B N_B\} = k_B(s) \tag{2}$$

Here s is the length of arc drawn on the surface of specimens perpendicular to their axes. The origin for s is taken at the median
plane between the two rods (or wires). The k_A and k_B are the
numbers of atoms respectively of A and B that are deposited on the
surface per unit time and unit surface. When the k is < 0, there
is erosion. The driving force $F(s)$ comes from the action of the
surface tension,

$$F(s) = - \gamma\Omega \frac{\partial}{\partial s} (\frac{1}{\rho(s)}), \tag{3}$$

where γ is the surface tension, Ω is the atomic volume, and ρ is
the radius of curvature of the surface in the plane perpendicular
to the wire axes, It will be convenient to write

$$\frac{F(s)}{kT} = - f \frac{\partial}{\partial s} (\rho_0/\rho(s)) \tag{4}$$

where f is a dimensionalless constant, $f = \gamma\Omega/kT\rho_0$. The complete
behavior of the surface curvature as a function of the time and s
has been developed in considerable detail by Nichols and Mullins
(5). In a general way, one can take ρ to have its minimum value,
ρ_0, at the symmetry joining plane, to have an inflection point a
particular $s = s_1$ and to become essentially constant at $s = s_0$,
where the constant value is R, the radius of the specimens. Of
course shape of the profile changes with time as s_1 and s_0 increase.
This consideration considerably complicates the problem of describing the progress of the alloy segregation.

We treat the diffusion constants, D_A and D_B as true constants independent of composition and hence of position s. In general this is not a serious approximation since the extent of alloy segregation is usually limited to only a few percent. We can then divide equations (1) and (2) by D_A and D_B respectively and add to obtain

$$Nf \frac{\partial^2(\rho_0/\rho)}{\partial s^2} = \frac{k_A}{D_A} + \frac{k_B}{D_B} \,, \tag{5}$$

since $N_A + N_B = N$, the number of total atoms per unit area.   Let the amount of segregation be given by δN_A and δN_B, where $N_A - N_{Ao} = \delta N_A$, etc. The equation for N_A^A then becomes

$$\frac{\partial^2 \delta N_A}{\partial s^2} + f \frac{\partial(\rho_0/\rho)}{\partial s} \frac{\partial \delta N_A}{\partial s} + f \frac{\partial^2(\rho_0/\rho)}{\partial s^2} \delta N_A$$

$$= - f \frac{\partial^2(\rho_0/\rho)}{\partial s^2} N_{Ao} + k_A/D_A \,. \tag{1a}$$

To proceed further with the solution for the N's one must develop relations between them and the k's to use along with Eq. (5) to form the solution of Eq. (1,2).

2.   Relation Between the Variations in the Composition and the k's

This section requires a careful consideration of the influence of volume diffusion in the region of erosion ($s_1 < s < s_o$), but in the region of deposition ($s < s_1$) the treatment is simple.
   a) Deposition region. Here the newly deposited material has the composition directly determined by the ratio of the k's.

$$N_A/N_B = k_A/k_B \tag{6}$$

While volume diffusion certainly occurs in this region, it has no opportunity to influence the composition of the most recently deposited layer.
   b) The erosion region. The problem here is that of volume diffusion to a surface which is being eaten away at the same time that it is being leached preferentially of one of its constituents. One needs to explore the relation between the diffusion flux to the surface and the variation of the surface composition from the uniform composition of the bulk. The relationship is, however, not a direct unique one but depends on the surrounding conditions

and on prior history. We attempt here only piecewise solutions subject to quite rough approximations.

First we assume that the surface is being eroded back at a velocity v_e so that

$$k_A + k_B = - v_e N/d, \tag{7}$$

where d is the thickness of the atomic layer parallel to the surface. The number of A-atoms that must be supplied per unit time-per unit area by diffusion is $-k_A - v_e N_A/d$. Substituting for v_e/d one obtains

$$- (k_A + v_e N_A/d) = C_A k_B - C_B k_A = D_v \frac{\partial n_A}{\partial z} \tag{8}$$

where $C_A = N_A/N$ and $C_B = N_B/N$. The quantity D_V is the volume coefficient for chemical diffusion for A and B, and n_A is the volume concentration of constituent A. The final quantity is evaluated at the moving surface. The z coordinate is measured perpendicular to the surface. Next comes the standard equation for bulk diffusion,

$$\frac{\partial}{\partial z} \{D_V \frac{\partial n_A}{\partial z}\} = \frac{\partial n_A}{\partial t} . \tag{9}$$

Throughout, we shall neglect diffusion along the s-direction except for surface diffusion, although the volume diffusion might become quite important in the latter stages of the sintering.

We shall attempt to synthesize suitable solutions for $n_A(z,t)$ from superposition of error function solutions to the one-dimensional problem we have posed.

$$n_{Ao} - n_A (z,t) = \int_0^t K(t') [1 - \mathrm{erf} \{(z - v_s t')/2(D_V(t - t'))^{\frac{1}{2}}\}] dt' \tag{10}$$

where

$$\mathrm{erf} (y) = 2(\pi)^{-\frac{1}{2}} \int_0^y e^{-x^2} dx . \tag{11}$$

The instantaneous position of the surface is given by $v_e t'$ and the function K(t') is at our disposal. The corresponding physical situation is that of inserting an infinites step in composition continually at a moving boundary in the material so that the normal gradient at the boundary (i.e. free surface) satisfies the boundary condition Eq. (8).

For a first try we set $K(t') = K_o$ a constant. The resulting diffusion flow out of the surface is then evaluated at $z_S = v_e t$

$$D_V \frac{\partial n}{\partial z}]_{z_S} = D_V K_o \pi^{-\frac{1}{2}} \int_0^t \exp(- v_e^2(t - t')/2D^{\frac{1}{2}})(D_V(t - t'))^{-\frac{1}{2}} dt'$$

$$= (D_V K_o/v_e) \, 2 \, erf(v_e^2 t/4D)^{\frac{1}{2}} \tag{12}$$

Consistent with the general roughness of the treatment we shall approximate the error function as follows:

$$erf \, z = z \qquad for \, z < 1$$

$$erf \, z = 1 \qquad for \, z > 1 \tag{13}$$

Of course a more careful fitting or even numerical evaluation could be employed if it should appear that the increased precision were justified. For short time $(t < t_1 = 4D_V/v^2_e)$

$$D_V \nabla n]_{z_S} = K_o \, D_V^{\frac{1}{2}} \, t^{\frac{1}{2}} \tag{14}$$

where the normal derivative is approximated by the gradient. For long times $t > t_1$

$$D_V \nabla n]_{z_S} = 2K_o \, D_V/v_s \tag{15}$$

The significance of the time t_1 is that it denotes that interval after which surface segregation is limited by bulk diffusion. The D_V is known, about $3 \times 10^{-10} cm^2/sec$ for the system and temperature of the experiment (4), but it is difficult to estimate the erosion velocity from the experimental results.

One might protest that this solution $(K(t') = K_o)$ does not fit the natural boundary condition of the problem, namely, that the diffusion flux starts with finite slope at zero time and levels off after t_1. Instead, Eq. (14) shows it starts as $t^{\frac{1}{2}}$ with infinite slope at zero time. To meet this criticism we chose $K(t') = K_1(t')^{\frac{1}{2}}$. Accordingly, the equation for the diffusion flux becomes

$$D_V \nabla n_A]_{z_S} = D_V K_1 \int_0^t (t')^{\frac{1}{2}} \frac{\partial}{\partial z}(erf\{(z - v_e t')/2D_V^{\frac{1}{2}}(t - t')^{\frac{1}{2}}\} \, dt' \tag{18}$$

In the interest of consistency and simplicity, we again invoke the approximation Eq. (13) so that the derivative of the erf function is a constant for $t < t_1$ and zero at longer times. We obtain then for short time

$$D_V \; \nabla n_A]_{z_S} = \tfrac{1}{2} D_V^{\frac{1}{2}} K_1 \int_0^t (t - t')^{-\frac{1}{2}} (t')^{\frac{1}{2}} \, dt'$$

$$= \frac{\pi}{4} \, D_V^{\frac{1}{2}} K_1 t \tag{19a}$$

and at longer times, $t > t_1$,

$$D_V \; \nabla n_A]_{z_S} = \frac{\pi}{4} \, D_V^{\frac{1}{2}} K_1 t_1 \tag{19b}$$

This behavior for the diffusion flux is more in accord with what one expects since it starts with a constant slope and saturates later as steady state conditions are established. It is reasonable to associate this steady limit with the constant value we have rather simplistically assumed for v_e, the rate of erosion. On this basis one finds that $K_1 \sim 2N/d(t_1)^{3/2}$.

On the smme basis the deviation of the concentration at the surface becomes for short times

$$- \delta n_A = \frac{2}{3} K_1 t^{\frac{2}{3}} - K_1 \int_0^t (t')^{\frac{1}{2}} v_e (t - t')^{\frac{1}{2}} / 2D_V^{\frac{1}{2}} \, dt' \tag{20a}$$

$$= \frac{2}{3} K_1 t^{\frac{3}{2}} \left[1 - \frac{3\pi}{16} \left(\frac{t}{t'} \right)^{\frac{1}{2}} \right];$$

and for $t > t_1$, $-\delta n_A = 0.27 \, K_1 t_1^{\frac{3}{2}}$ \hfill (20b)

This time the values for the ratio of the deviation in composition to the diffusion flux is for $t < t_1$,

$$- \frac{\partial n_A}{D_V \; \nabla n_A} \Big]_{z_S} = \frac{8}{3\pi} \frac{t^{\frac{1}{2}} \left(1 - \frac{3\pi}{16} \left(\frac{t}{t'} \right)^{\frac{1}{2}} \right)}{D_V^{\frac{1}{2}}} \tag{21a}$$

and for $t > t_1$, $- \dfrac{\partial n_A}{D \; \nabla n_A} \Big]_{z_S} = .69 \, v_e^{-1}$ \hfill (21b)

By comparison with Eqs. (17 a,b) it can be seen that the change of the functional form of $K(t')$ has had little effect on the ratio.

Also one can generalize the calculation to take into account that during the period of linear increase in the diffusion flux (Eq. 19a) the rate of erosion is more accurately $v_s(t) = a\,t$ than a constant. This refinement does not affect Eqs. 19 a,b, but the equation for $-\delta n_A$ becomes

$$- \delta n_A = K_1 [\frac{2}{3} t^{\frac{3}{2}} - \int (t')^{\frac{1}{2}} \{ \int_{t'}^{t} v_e(t'') dt'' / 2 D_V^{\frac{1}{2}} (t - t')^{\frac{1}{2}} \} dt'$$

(22)

For the period $t \leq t_1$ this reduces to

$$- \delta n_A = \frac{2}{3} K_1 t^{\frac{3}{2}} [1 - (9\pi/64) (t/t_1)^{\frac{1}{2}}]$$

(23)

which shows a somewhat slower saturation than Eq. (20a).

The results of this section show that a change in the erosion condition at the surface is reflected by a corresponding change in the concentration at the surface that increases at a rate $t^{\frac{1}{2}}$ times the erosion change and that saturates approximates at a time t_1 thereafter.

3.   Solution to the Surface Transport Equations

a)   The deposition region.   From the Eqs. (5) and (6) one finds

$$k_A = N_A f (\rho_0/\rho)'' \frac{D_A D_B}{D_{sc}} ,$$

(24)

where $D_{sc} = C_A D_B + C_B D_A$, the coefficient for chemical diffusion on the surface. (Differentiation with respect to s is hereafter shown by primes.) When this result has been substituted back into Eq. (1a), one has for the diffusion equation in this region determining segregation

$$(\delta N_A)'' + f (\rho_0/\rho)' (\delta N_A)' + f (\rho_0/\rho)'' \delta N_A = N_A C_B f (\rho_0/\rho)'' \frac{(D_B - D_A)}{D_{sc}}$$

(25)

In general, the dimensional number f is small compared to one so that the second and third terms on the left hand side of Eq. (25) are small compared with the first term. If they are discarded, then a simplified Eq. 25 results.

$$(\delta N_A)'' = N_{Ao} C_B \frac{D_B - D_A}{D_{sc}} f (\rho_0/\rho)''$$

(25a)

If one disregards the spatial dependence of the C's and D's one can write as a sort of steady state solution for δN_A

$$\delta N_A = N_A C_B \{ (D_B - D_A)/D_{sc} \} f (\rho_0/\rho) + M + P,$$

(26)

where the constants of integration, M and P are to be determined by the appropriate boundary conditions at the inflection point $s = s_1$ which marks the boundary between the deposition and erosion regions. The solution is steady state only in a piecewise sense since the parameters which determine the shape of $\rho(s)$, such as s_1 and s_o are changing with time, reasonably slowly perhaps, as the sintering proceeds.

b) The erosion region. We now return to the erosion region and its more complex behavior. The end result of the treatment of section 2 for this region was a functional relation between the diffusional flux of Eq. (8) and the alloy segregation at the surface as evaluated first in Eqs. (16a, b). Examples of this relation appear in Eqs. (17a, b) and (21a, b). Their general form is

$$-\frac{D_V \nabla n_A}{\partial n_A}]_{z_S} = (\frac{D_V}{t})^{\frac{1}{2}} g(t,t_1) \tag{27}$$

where the function $g(t)$ is initially constant, but increases as $t^{\frac{1}{2}}$ at long times. From Eq. (21), for example, the initial constant would be $3\pi/8$ and for t large $g(t) \sim 2.9 (t/t_1)^{\frac{1}{2}}$. One now combines Eqs. (7) and (8) to solve for k_A

$$k_A = (v_e/d) N_A + \delta N_A (D_V/t)^{\frac{1}{2}} g(t,t_1) \tag{28}$$

A second equation for k_A can be obtained from Eqs. (5) and (7)

$$k_A (\frac{1}{D_A} - \frac{1}{D_B}) = Nf (\rho_0/\rho)'' + \frac{v_s N}{d D_B} \tag{29}$$

Between these two equations one eliminates v_e which is a function of s that duplicates to some extent the role of the second derivative of the curvature.

$$\frac{k_A}{D_A} = \frac{D_B C_A}{D_{sc}} Nf (\rho_0/\rho)'' + \frac{C_A}{C_B} \frac{(D_V/t)^{\frac{1}{2}} g(t,t_1)}{D_{sc}} \frac{\delta N_A}{d} \tag{30}$$

We substitute into Eq. (1a), and as in the deposition region, drop the last two terms on the left hand side.

$$(\delta N_A)'' = \frac{(D_B - D_A)C_B N_A}{D_{sc}} f (\rho_0/\rho)'' + \frac{C_A}{C_B} \frac{(D_V/t)^{\frac{1}{2}}}{D_{sc}} g(t,t_1) \frac{\partial N_A}{d}$$

$$\tag{31}$$

The short time situation is easily accounted for. Clearly the last term in Eq. (31) will dominate the derivative on the left hand side and one can solve the right side for δN_A to find

$$(\delta N_A)'' = c_1 (D_A - D_B) c^2_B N_A df (\rho_0/\rho)'' (t/D_V)^{\frac{1}{2}}, \tag{32}$$

where c_1 is a constant of the order of unity. Since $D_A > D_B$ by hypothesis and the second derivative of the curvature is negative, we see that δN_A is negative as expected in the erosion region.

The long time regime is somewhat more complex. At times long compared with t_1, $(D_V/t)^{\frac{1}{2}} g(t,t_1)$ becomes constant in time with a spatial dependence given approximately by $3v_e$. This means that the last term on the right of Eq. (28) will be small under these circumstances compared with the first since $\delta N_A/N_A \ll 1$. From this and Eq. (24) one finds the following for v_e,

$$v_e \simeq df \; (\rho_0/\rho)'' \; \frac{D_A \; D_B}{D_{sc}} \tag{33}$$

and, when this result is put back in Eq. (31), one finds

$$(\delta N_A)'' = (f \; (\rho_0/\rho)'' \; d/D_{sc}) \; \{(D_B - D_A) \; C_B \; N_A -$$
$$c_2(C_A/C_B)(D_A \; D_B/D_{sc})\delta N_A\} \tag{34}$$

where c_2 is a numerical constant of the order of 3. This equation has the trivial, particular solution of $\delta N_A =$ a constant

$$\delta N_A = \frac{C_B^{\;2}}{c_2 \; C_A} \; \frac{(D_B - D_A) \; D_{sc}}{D_A \; D_B} \; N_A, \tag{35}$$

but a constant can of course never fit the boundary conditions at s where δN_A goes to zero. Boundary fitting can be accomplished by adding general solutions of the homogeneous equation of a step-wise character such as the error function in the transition region to constant curvature. This procedure takes into account the inward flow of constituent A from the parts of the specimen which have not yet been otherwise affected. The fact that δN_A is essentially independent of s in the region is indicative that difference in diffusion rates is no longer a driving force toward segregation.

4. Summary and Discussion

The problem of treating the segregation problem mathematically is complicated because it occurs against a changing geometry evolving from the sintering process and because the surface conditions are quite different in the erosion region from those in the region of deposition. Consequently, the solutions are quite approximate in character, being piecemeal treatments in a pseudo-steady state geometry.

The results obtained in section 3 show that in the deposition region one can expect an enrichment of the faster diffusing constituent starting with the first sintering and continuing as this

region is enlarged. The strength of the segregation appears from
the mathematical result to be proportional to the negative reciprocal curvature, at least far from the inflection point s_1. One
can expect that the boundary matching near the erosion region may
alter all this somewhat. In the erosion region itself, initially
the surface is depleted of the faster diffusing constituent. Here
$-\delta N_A$ turns out to be proportional to the second derivative of the
curvature and hence has the opposite sign from the deposition region where the proportionality was to the curvature. The curvature is still negative in most of the erosion region, but its
second derivative is not. As time proceeds, the bulk diffusion
zone widens and slows down the segregation. The equation for long
times shows a saturation in δN_A at a constant value quite independent of spatial variation of the erosion current. Actually it is
quite easy to see that the segregation must actually decrease with
time because of two effects which are difficult to incorporate
formally into the mathematics. (1) The surface sintering fluxes
decrease appreciably with time because the sintering tends to
blunt the driving forces arising from the variation of the curvature of the specimens as the compaction continues. This decreases
$- D_A \nabla C_A$ and immediately the effect is transmitted to δN_A.
(2) A decrease of δN_A induces a diffusional flux from the $s > s_0$
region - up to the moment unaltered by the sintering - and this
is most important in bringing the N_A level back to its bulk value.
Of course, if the heating is continued long enough the equilibrium
value of the composition will be established everywhere.

The above partial analysis appears then to be in reasonable
qualitative accord with the results observed by Mishra et al. (4).
At this time it does not seem feasible to attempt to obtain more
quantitive checks since some of the physical parameters such as
the surface diffusivities D_A and D_B are essentially unknown.

The author wishes to thank Professor Lenel and Dr. A. Mishra
for stimulating his interest in this problem, for supplying physical insight, and for showing confidence in the significance of
the results.

REFERENCES

1. G.C.KUCZYNSKI, G. MATSUMURA, and B.D. CULLITY: Acta Met. 8, 209
   (1960).
2. T.R. ANTHONY: Diffusion in Solids: Recent Development, Chap 7,
   (Eds. Nowick and Burton) Academic Press,Inc., New York, 1975.
3. T.R. ANTHONY: Acta Met. 17, 603 (1969).
4. A. MISHRA, F.V. LENEL and G.S. ANSELL, preceding paper.
5. F.A. NICHOLS, and W.W. MULLINS: J. Appl. Phys. 36, 1826, (1965).

SINTERING IN THE PRESENCE OF LIQUID PHASE

W. J. Huppmann

Max-Planck-Institut für Metallforschung
Institut für Werkstoffwissenschaften
7 Stuttgart-80, Büsnauerstr. 175, Germany

INTRODUCTION

In recent years considerable progress in materials development has been achieved by applying the methods of liquid phase sintering well known from heavy alloy and hard metal production. Sintering of cobalt rare earth magnets (1-4), hot pressing of silicon nitride structural ceramics in the presence of a liquid phase (5-8),supersolidus sintering of superalloys (9-12), accelerated diffusion alloying for powder forging (13) or the development of hardenable cemented carbides (14-16) may be listed here as the most prominent examples. These technological achievements, however, were for a long time hardly matched by the scientific work which is necessary not only for a basic understanding of liquid phase sintering but also for an improvement of the materials made by this method.

Therefore the present review(for earlier reviews see (17-19))intends to put less emphasis on materials than on the mechanisms of liquid phase sintering, particularly on the basic developments of the last few years. For this purpose mainly the densification behavior will be used here to discern and discuss the different sintering stages and mechanisms although other aspects, e.g. microstructural development can also provide valuable information.

THE STAGES OF LIQUID PHASE SINTERING

Based on the fundamental observations made by
Lenel et al.(20,21) and the work of Gurland and Norton
(22), in his theoretical treatment Kingery discerns
three stages of liquid phase sintering (23-25)(Fig.1):

1. Immediately after melting during particle rearrangement rapid densification of the powder mass is brought
about by the solid particles sliding over one another
and by the collapse of particle bridges under the action
of capillary pressure.

2. After rearrangement densification has been completed
the solution-reprecipitation process will start if the
solid particles have some solubility in the liquid
phase. The solubility at the contact points between
particles is larger than the solubility of other solid
surfaces and this results in the transfer of material
away from the contact points allowing the center-to-
center distance between particles to be decreased and
densification to take place.

3. If in the course of sintering solid particles come
into direct contact without the interposition of melt
(coalescence) only material transfer within the solid
phase can lead to further densification. This will
happen at a comparatively slow rate as normally encountered in solid state sintering.

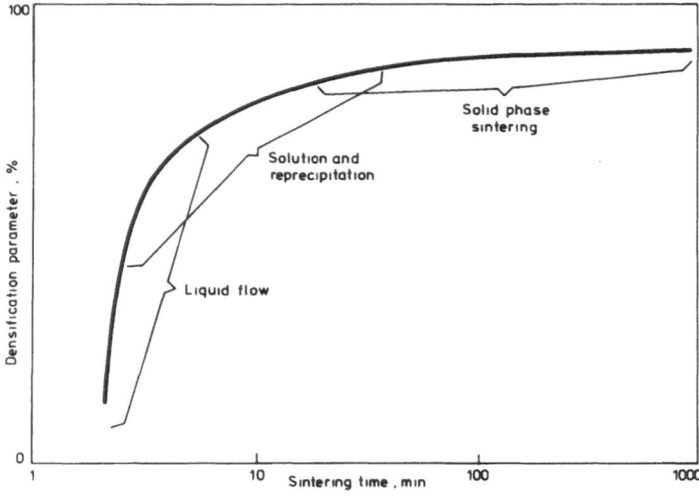

Fig. 1.   Densification during liquid phase sintering,
          schematic (21).

Although the theory of Kingery has vastly increased
our knowledge of liquid phase sintering and has
stimulated a considerable amount of experimental work
all too often the experimental data were interpreted
on the basis that the three mechanisms occur completely
separately. It must, however, be assumed that in most
systems considerable interaction between the different
sintering mechanisms will occur and this has been taken
into account in the schematic representation of Fig. 1
by overlapping of the brackets which indicate the re-
spective time intervals of the three basic mechanisms.

PARTICLE REARRANGEMENT

The Driving Force

In order to derive the kinetics of rearrangement
shrinkage

$$\Delta L/L_o \sim t^{1+y} \qquad\qquad [1]$$

Kingery replaces the individual interparticle forces by
an external pressure acting on the powder mass as a
whole (23). On the basis of recent calculations (19,26-31)
a different approach treats rearrangement densification
as a direct function of the force existing between two
solid particles which are connected by a melt bridge
(31). This method offers the advantage that the influence
of such parameters as amount of liquid phase and wetting angle can be treated adequately. On the other hand
it requires the exact knowledge of the melt bridge configuration because of the strong dependence of the interparticle force on contact geometry (19,26-28).

The interparticle force F between two spheres connected by a liquid bridge

$$F = 2\pi r \, \gamma \cos\Phi - \pi r^2 \Delta P \qquad [2]$$

consists of a surface tension term acting at the wetting perimeter $2\pi r$ and a term arising from the capillary pressure ΔP in the liquid (19,26,28,29). Here γ
is the surface free energy of the liquid-vapor interface and r and Φ are defined in Fig. 2. Heady and Cahn
(26) have solved eq. [2] by numerical methods and have
also shown that for most situations the so called circle approximation (illustrated in Fig. 2) is a very
satisfactory approximation.

In order to prove the applicability of eq. [2] model
experiments (31) were performed in which the shrinkage
and swelling behavior respectively of planar arrays of

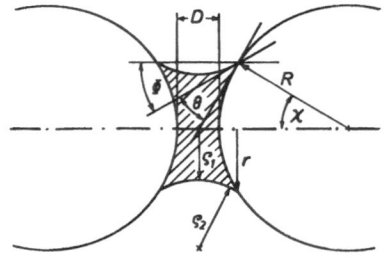

Fig.2. Geometry of a particle
        contact region

spherical tungsten particles which were uniformly coated with copper was investigated. Figures 3 to 6 and
Table 1 summarize the results. In Fig. 3 and 4 the interparticle force F as a function of interparticle distance D is plotted for different amounts of melt V in
the contact region for $\Theta = 8°$ and $85°$ respectively. For $\Theta = 8°$
(Fig. 3) a strong increase of the attractive (positive)

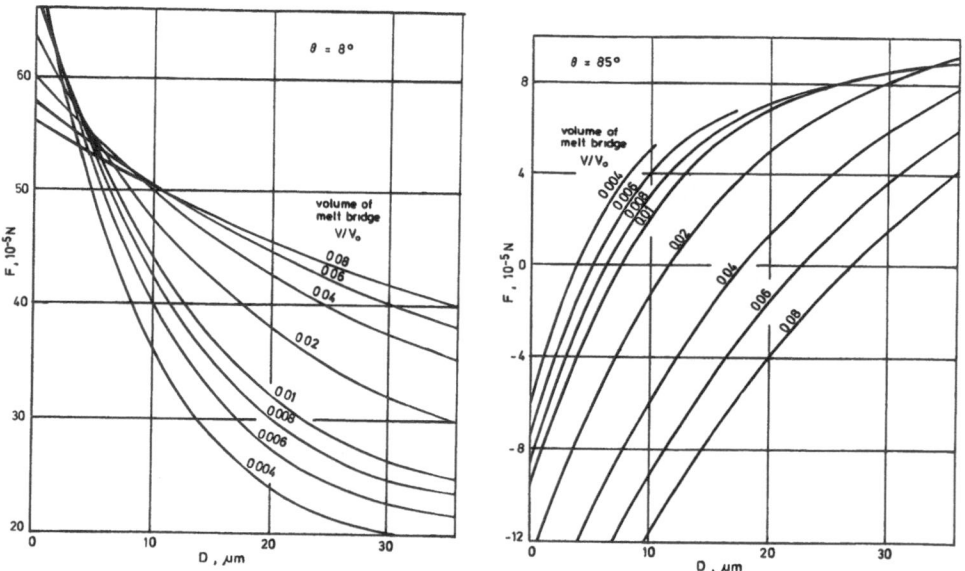

Fig. 3 and 4. The force acting between two tungsten
              spheres connected by a copper bridge as
              a function of the interparticle distance
              for $\Theta = 8°$ and $\Theta = 85°$. $\gamma = 1.28$ N/m (32).
              Tungsten sphere diameter 214 μm. V volume
              of melt bridge, V_0 volume of tungsten sphere.

force F is observed as D decreases and therefore attraction of the tungsten particles until they come into contact is expected upon melting of the copper layer.

This is in agreement with observations on metallographic sections through the contact regions of the sintered (4 min, 1100°C) arrays. For $\Theta = 85°$ (Fig.4) F is
attractive for large D's, passes through zero as the
interparticle distance decreases and becomes repulsive
(negative) for small D values. The experimentally determined values of D after melting of the copper are
in good agreement with those for which the calculated
F becomes zero (designated $D_{F=0}$ in Table 1).Fig.5
and 6 finally show scanning electron micrographs of the
different neck formation in the case of $\Theta = 8°$ and 85°.

Table 1. Dimensional changes upon liquid phase sinter-
         ing in densely packed planar arrays of tungsen
         spheres uniformly coated with copper. Tungsten
         sphere diameter 214 /um. Wetting angle $\Theta = 85°$

Amount of liquid phase (Vol.%)	$\Delta L/L_0$* (%)	D exp. (/um)	$D_{F=0}$ calc. (/um)
5.4	−1.2±0.1	6.6±0.2	6.2
7.7	−0.9±0.1	7.9±0.2	7.8
12.4	0.3±0.1	9.0±0.2	11.1

* positive values: shrinkage, negative values: swelling

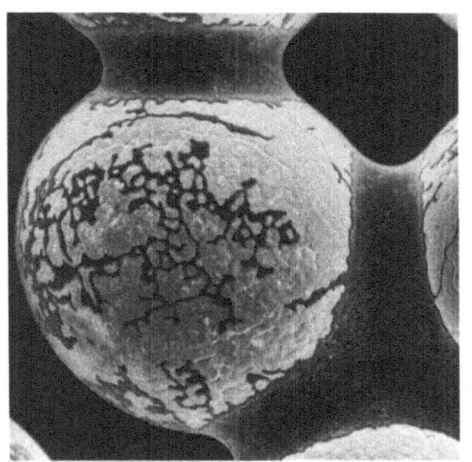 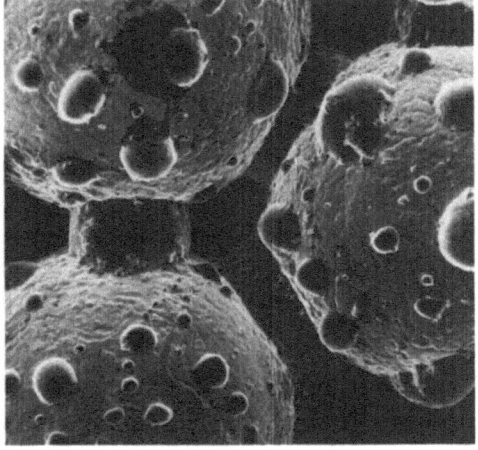

Fig.5. $\Theta = 8°$ Fig.6. $\Theta = 85°$
Fig. 5 and 6. Scanning electron micrographs of neck
              region in liquid phase sintered arrays
              of copper coated tungsten spheres.

These experiments not only confirm that the calculations
(19,26,28-31) yield satisfactory values for the interparticle force F but also suggest that considering
these interparticle forces as the driving force for
particle rearrangement in random arrays might give
meaningful results.

Rearrangement in Randomly Packed Arrays of Particles

Although generally a rigorous separation of rearrangement and solution-reprecipitation densification
is impossible, in a few selected systems exhibiting
mutual insolubility (e.g. W-Cu (33)) particle rearrangement can be isolated. Therefore experiments were carried
out (31,34) whereby the planar array method illustrated
in Fig. 7 was used to study the influence of wetting
angle, amount of liquid and packing density on the dimensional and microstructural changes during rearrangement of tungsten spheres uniformly coated with copper.

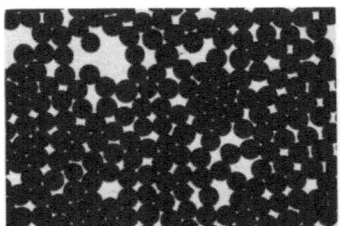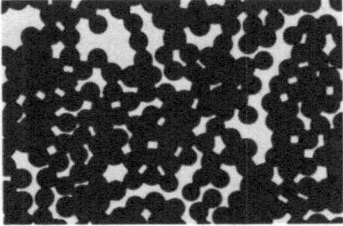

Fig. 7. Planar arrays of tungsten spheres (diam.214/um)
        coated with copper in the presintered condition
        (left) and after liquid flow (4 min, 1100°C)
        (right).

Considerable local variations in shrinkage were observed; a clear picture evolved when the total linear
shrinkage of an array (300 to 500 particles) was measured as a function of the mean interparticle force F.
This force could be calculated as outlined in the previous section by inserting the quotient of the total
amount of copper divided by the number of contacts per
array as melt bridge volume. According to

$$\Delta L/L_0 = (\Delta L/L_0)_{isotr} + (\Delta L/L_0)_{rearr} \qquad [3]$$

the linear shrinkage is split into two terms. The isotropic term is the shrinkage value measured in densely
packed areas and simply arises from the two-particle

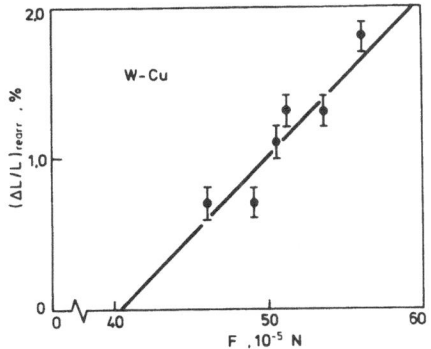

Fig.8. Shrinkage due to particle
rearrangement in random-
ly packed planar arrays
(mean coordination num-
ber $\bar{n}=4$) of W-Cu as a
function of mean inter-
particle force F.

approach when the copper coating melts whereas the rearrangement term arises from gross particle movement.
Fig. 8 shows the rearrangement shrinkage as a function
of the mean force F whereby it must be stressed that
the individual data points in this diagram correspond
to different values of wetting angle ($\Theta = 8^{\circ}$ and 25°)
and amounts of liquid phase (5.4, 7.7 and 12.4 vol.%).
Although more work is needed to establish the true relation between interparticle force and rearrangement
shrinkage the relation

$$(\Delta L/L_O)_{rearr} = k(F - F_O) \qquad [4]$$

may serve as a first approximation. A qualitative interpretation of eq. [4] would be that in the array certain configurations, e.g. particle bridges, of different
strength exist. Only when F exceeds the weakest configurations characterized by the intrinsic force
$F_O \approx 40 \cdot 10^{-5}N$ can particle rearrangement take place.
Since ever stronger bridges will collapse as F increases,
rearrangement shrinkage will be the more pronounced the
more F exceeds F_O. There is direct experimental evidence
for this interpretation, (see Fig.9.) In this figure the
pore size distribution in planar arrays before and after
liquid flow is shown for $\Theta = 8^{\circ}$ and 25°. Whereas for
$\Theta = 8^{\circ}$ the area (frequency) of all pore sizes decreases
during sintering, for $\Theta = 25^{\circ}$, i.e. smaller F, the area
of large pores increases upon liquid flow. This effect
can directly be seen in Fig. 7 and is explained by some
contacts being broken during sintering which in turn requires the existence of certain strong particle bridges
in the array. Similar results are obtained if F is decreased either by increasing the amount of liquid phase
or by lowering the packing density (31,35) or by in-
troducing mixing inhomogeneities (36,37).

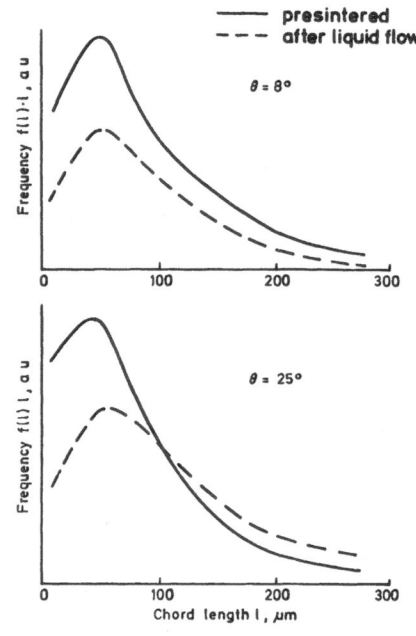

Fig.9. Pore size distributions in randomly packed arrays before and after liquid flow for two wetting angles. Mean coordination number $\bar{n} = 4.5$; 7.7 vol.% liquid.

It is worth mentioning that the work on the influence of the degree of mixing on densification (36,37) from which Fig.10 is taken was done with threedimensional arrays of spherical particles and yields results consistent with the measurements on planar arrays. This

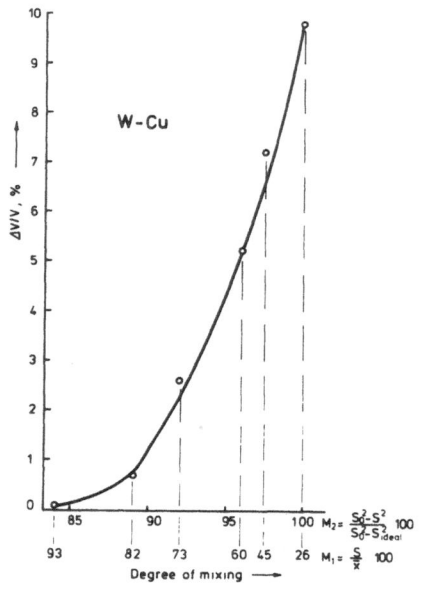

Fig.10. Volume shrinkage as a function of the degree of mixing in threedimensional arrays of spherical tungsten (60 μm diameter) with 15 vol.% Cu (36,37).

suggests that quite generally bridge formation which
has been mentioned before by Kingery (23) and Masuda
(38) plays an important role during particle rearrangement and therefore completely regular packing is hardly
ever achieved during the earliest sintering stage.The
scanning electron micrograph of Fig.11 shows the uneven densification due to bridging in a real powder
compact.

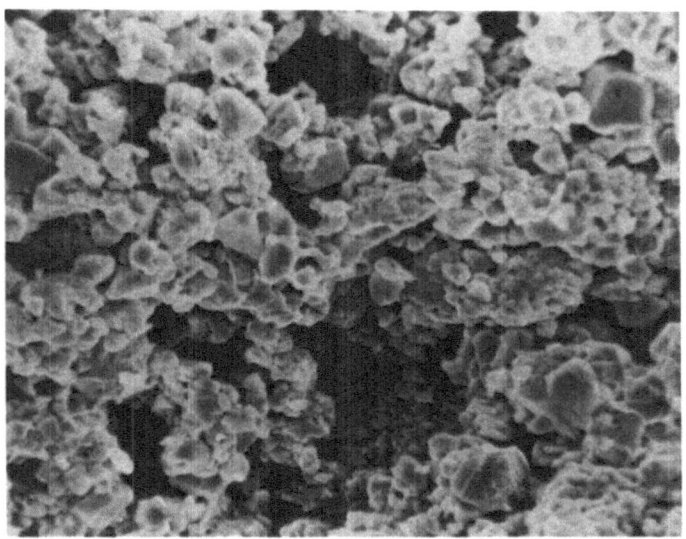

Fig.11. Uneven densification in a WC-12 vol.%
         (50Cu+50Ni) compact sintered 8 min at 1330°C.
         Scanning electron micrograph of fracture
         surface. 2 000 x.

     While eq. [4] and the described microstructural
changes during sintering provide some insight into the
mechanism of particle rearrangement,the only kinetic
treatment (eq. [1]) of the process is by Kingery (23,24).
Numerous investigations (19,39-46) have tried to verify
eq. [1] with variable success. One of the reasons for
the observed discrepancies might lie in the assumption
of a constant frictional force (viscosity) during the
whole rearrangement process.In the light of the above findings it appears likely that initially the viscosity of the
liquid determines the frictional force (proportionality
constant) in eq. [1] while later the mechanism of bridge
collapsing may become rate determining. This is also
expressed in a recent study on the densification behavior of diamond-metal systems in which Naidich et al.

(47) come to the conclusion that the major contribution
to the frictional force in eq. [1] arises from interparticle friction, interlocking and bridging. The apparent discrepancy mentioned by Nelson and Milner (48)
that the rearrangement process usually lasts considerably longer than the spreading of the melt (compare
(49)) may also be explained by the fact that two different processes occur during the first sintering stage,
namely liquid flow and gross regrouping when bridges
of particles collapse.

Interaction Between Solution Processes and
Particle Rearrangement

An observation often made in experiments on liquid
phase sintering is that densification during the early
sintering stage is more rapid and more complete in systems where the solid phase has some solubility in the
liquid than in systems exhibiting negligible intersolubility. The data of Fig.12 and 13 (19,39,48,50) are but
examples for this behavior. The data collection by
Eremenko et al. (19) can be considered as the most important evidence for acceleration of rearrangement
shrinkage if some intersolubility between the phases in
a powder compact exists, since it shows that the time
exponent of densification (y in eq. [1]) increases as
the solubility of the solid component in the liquid
phase increases. Finally the observation that in certain

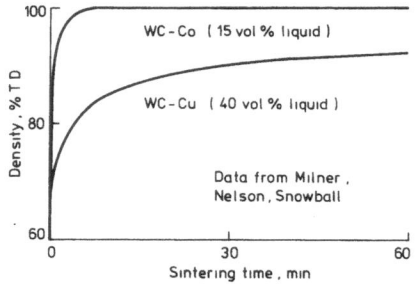

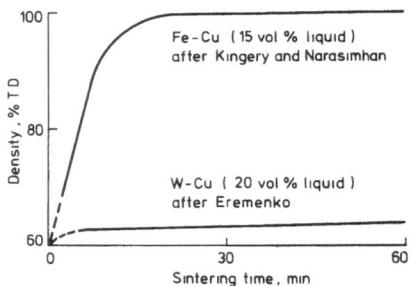

Fig.12. Comparison of the
densification be-
havior of WC-Co
(solubility) and
WC-Cu (insolubi-
lity) (48,50)

Fig.13. Comparison of the
densification behav-
ior of Fe-Cu (solu-
bility) and W-Cu
(insolubility) (19,39)

hard metals substantial shrinkage occurs below the melting point of the liquid phase (25,40,50-52) is an indication that particle rearrangement takes place in the

solid state and adds a further argument to the concept
that rearrangement densification is influenced by intersolubility of the two phases in a powder compact.

Although at present little is known about the mechanism by which solution processes and particle rearrangement interact this sintering stage must be clearly separated from densification due to contact flattening
by solution-reprecipitation which leads to much lower
densification rates and will be described in the
following chapter. In light of the findings on
bridge formation (compare previous section) Kingery's
view (23) that particle bridges will collapse by the
solution of small amounts of material at the contact
points appears very reasonable.

In the work of Stephenson and White (53,54) on
ceramic materials it was found that early shrinkage is
the more pronounced the lower the dihedral angle (the
angle at which the two solid-liquid interfaces intersect at a solid-solid-liquid junction) and thus the
area of particle contact, i.e. contiguity (55) which
develops when the solid and liquid phases react. This
was interpreted (53,54) as an indication of low resistance against rearrangement and thus as favouring
shrinkage. Warren and Waldron (56) point out that even
a value of contiguity typical for a continuous skeleton
of solid phase needs not to be inconsistent with densification by particle rearrangement.

Yet another mechanism for accelerated shrinkage due to the interaction of rearrangement and solution processes was proposed by Lund and Bala (57) who
consider that particle rearrangement may be facilitated
by the shape change of particles according to Kingery's
second stage model of sintering (see following chapter).

Clearly more basic and quantitative experiments
are needed in this area. The method which might prove
very useful in such experiments would be to determine
particle contiguity (55) and continuity (58) as a function of sintering time for relatively short sintering
periods and to compare these microstructural changes
with the measured shrinkage values. Activation energy
measurements might also allow some conclusions as for
example in the work of Exner et al. (40,52) who calculated the activation energy for densification in WC-Co
and found an identical value (typical for a dissolution
process) both below and above the melting point of the
liquid phase.

SOLUTION-REPRECIPITATION AND
COALESCENCE (PARTICLE GROWTH)

The Driving Force

After rearrangement has been completed the contact points between particles are still under compressive stress (see Fig.3 for D = 0). According to Kingery (23) a thin film of liquid present in the contact regions carries the major part of this stress and therefore the melt in the contact region is under substantially larger pressure than the melt in the surrounding area. This pressure differential ΔP causes an increase in chemical potential u, activity a and thus solubility C of the solid in the liquid phase according to

$$u - u_0 = RT \ln \frac{a}{a_0} = RT \ln \frac{C}{C_0} = \Delta P V_0 \qquad [5]$$

where V_0 designates the molecular volume. This increase in activity provides a driving force for transferring material from the contact zones to areas in the melt with lower solubility where solid phase is precipitated. It is this process which eventually leads to contact flattening and thus shrinkage. While Kingery's treatment was for the case of complete wetting, Gessinger et al. (59,60) have shown that the case of $\Theta > 0$ can be treated in an analogous way as long as the solid particles are separated by a liquid film. Thus quite generally the same driving forces which cause particle rearrangement are responsible for densification by particle flattening during the solution-reprecipitation stage. However, these forces do not directly cause densification of the powder compact but via increased material dissolution at the contact points and transfer through the liquid phase. While the driving force is unchanged if a solid boundary exists between particles (dihedral angle >0) the solution-reprecipitation mechanism can no longer act in this case and densification will proceed at a much slower rate determined mainly by grain boundary diffusivity (59,60).

The great achievement of Kingery's analysis (23) is that it can explain contact flattening (compare Fig.14) which is not possible on the basis of Ostwald ripening as it is usually understood, i.e.the growth of particles while their shape is maintained. For the W-Ni-Cu-system the result of an Ostwald ripening process would be a microstructure of spherical particles in the liquid matrix while the experiment (Fig.14)(65) shows almost perfect particle shape accomodation due to contact flattening and this also explains that complete densification is often observed even with small volume fractions of liquid phase.

The driving force for Ostwald ripening which plays a role in the final stage of liquid phase sintering must be sought in the activity and solubility differences of interfaces showing different curvature. Because during Ostwald ripening little if any densification occurs it will be treated rather shortly here. The interested reader is referred to the recent review literature on this subject (61-64).

Densification

Based on eq. [5] and assuming spherical particles of radius r Kingery (23) derived equations for the kinetics of shrinkage which may be rewritten to contain the length of a powder compact L_O at the time t_O of the beginning of the solution-reprecipitation process

$$\Delta L/L_O = \text{const} \cdot r^{-4/3} \cdot (t - t_O)^{1/3} \qquad [6]$$

$$\Delta L/L_O = \text{const} \cdot r^{-1} \cdot (t - t_O)^{1/2} \qquad [7]$$

whereby eq. [6] applies to conditions where diffusion in the liquid phase is rate determining while eq. [7] aplies if the phase boundary reaction leading to solution is rate controlling. If instead of spherical particles a prismatic particle shape is assumed similar relations but with the time exponents 1/5 and 1/3 for diffusion and phase boundary control respectively are obtained (23).

While contact flattening had been known to occur during the solution-reprecipitaion stage since the classical work of Price, Smithells and Williams (65) (see Fig.14) only the establishment of the kinetic laws [6] and [7] stimulated considerable experimental effort (19,39-46,67) to prove the validity of the underlying principles quantitatively. However, all these experiments were done on powder compacts with highly complex and changing geometry and therefore the results are only of limited value to quantitatively verify the validity of eqs. [6] and [7].

As an example of the problems which can arise in interpreting measurements on powder compacts the work of Prill et al. (42,43) and Exner and Fischmeister (17,66) may be quoted. These authors have shown that the experimentally determined time exponent for solution-reprecipitation sintering depends critically on the point of time which is chosen as t_O in eqs. [6] or [7]. Since in sintering a powder compact usually rearrangement and solution-reprecipitation shrinkage overlap choosing

t_o is somewhat arbitrary and therefore an exact determination of the time exponent becomes difficult. As a consequence of this Kingery and Narasimhan (39) considered the sintering of Fe-Cu as diffusion controlled while Prill et al. (42) came to the conclusion that a phase boundary reaction is rate controlling.

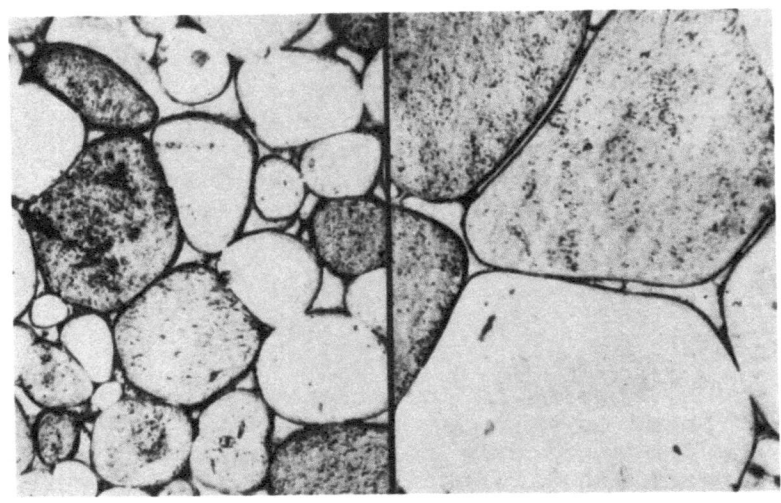

t = 1h   t = 6h

Fig.14. Contact flattening and particle growth during sintering (1400°C) of 93%W – 5%Ni – 2%Cu heavy metal (65).

The positive assertion of Kingery's theory stems from accurate model experiments of well defined geometry (68,69). Unambiguous conclusions about the rate controlling processes can be drawn from these experiments and depending on the system studied both diffusion (68) and phase boundary reaction control (68,69) have been observed.

Although additional work of this kind is desirable to gain more quantitative data on solution-reprecipitation sintering, particularly on the influence of particle radius (compare eqs. [6] and [7]), the real remaining problem in this area is to make the known changes in contact geometry applicable to the sintering of real powder compacts.

Coalescence and Particle Growth

Little if any densification occurs during the terminal stage of liquid phase sintering. In this review the emphasis has been on shrinkage during sintering and therefore the microstructural changes in the last sintering stage will be discussed here rather briefly.

For a wide variety of systems (53,54,67,70-77) the cubic rate law for particle growth

$$\bar{r}^3 - \bar{r}_o^3 = K \cdot t \qquad [8]$$

($\bar{r}$ mean particle radius) as postulated by diffusion controlled Ostwald ripening (78,79, compare also the reviews 61-64) was found. In spite of the observation of the rate law [8] it is in many cases difficult to decide definitely whether diffusion through the liquid phase or a reaction at the solid-liquid interface is the rate controlling mechanism. The difficulties in finding out the true mechanism of the observed coarsening in microstructure are partly due to the so far not completely understood mechanism of particle coalescence which for isometric grains seems to be a direct function of the dihedral angle (53, 54,56) while for anisometric grains no such relation exists (54). Partly the difficulties also arise from the unknown effect of contiguity on the rate constant for particle growth (77).

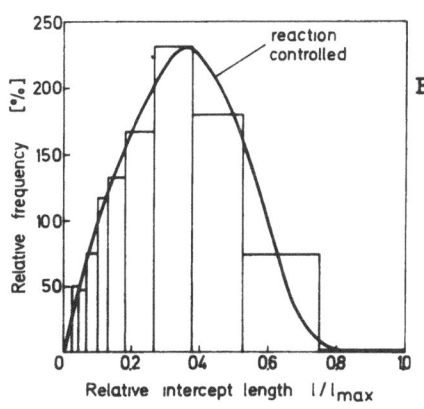

Fig.15. Intercept distribution of VC particles in a Ni matrix (16h, 1500°C) and comparison with theoretical distribution assuming a reaction controlled growth mechanism (80).

Sophisticated quantitative microstructural analysis as depicted in Fig.15 (80) must be used to ascertain the predictions of the theory of Ostwald ripening for each system of interest. Even then it can be difficult to obtain unambiguous conclusions about the active growth mechanism (81).

Thus there is considerable evidence that particle growth is by a solution-reprecipitation process although a particle coalescence mechanism may also play a role in certain systems, as emphasized by Humenik and Parikh (82,83). Final clarification of several details of the process of microstructural development during the terminal stage of liquid phase sintering will require very refined measuring techniques and theoretical analysis.

CONCLUSIONS

I hope that this review has made clear the progress achieved during recent years by studying carefully selected model systems:

1. We have succeeded in modelling particle rearrangement in systems of negligible intersolubility. The measurements show that basically two processes, liquid flow and particle bridge destruction, occur.

2. Extension of the model experiments to systems with soluble components is required to better understand the later stages of particle rearrangement during which densification is thought to occur by solution effects at particle contacts in the skeleton of bridges formed during the early stages of rearrangement.

3. The predictions of Kingery's solution-reprecipitation theory were proven repeatedly in model experiments.

4. The first results of experiments using recently developed methods of quantitative microstructural analysis seem to indicate that quantitative theories on particle growth could be used in the not too distant future for developing ceramics or cemented carbides with specified microstructure and properties.

5. The most important problem of materials scientists concerned with liquid phase sintering will be to derive from the results on model systems concepts sufficiently general so that they can be applied to real systems.

ACKNOWLEDGEMENTS

I would like to thank Prof.G.Petzow for his continued support of the work on liquid phase sintering in this laboratory and Prof.F.V.Lenel and Dr.H.E.Exner for the frequent discussions during preparation of this manuscript. The stimulating comments of Prof.G.C. Kuczynski during the initial period of this work are also appreciated.

REFERENCES

1. M.G.Benz and D.L.Martin, Appl.Phys.Lett.17 (1970)176

2. M.G.Benz and D.L.Martin, J.Appl.Phys.43 (1972)3165

3. J.B.Y.Tsui, K.J.Strnat, and J.Schweizer, Appl.Phys. Lett.21 (1972)446

4. T.Shibata and T.Katayama, Japan.J.Appl.Phys.12 (1973)1020

5. S.Wild, P.Grievson, K.H.Jack, and M.J.Latimer, Special Ceramics, ed.P.Popper 5 (1972)377

6. I.Coloquhoun, D.P.Thompson, W.I.Wilson, P.Grievson, and K.H.Jack, Proc.Brit.Ceram.Soc.22 (1973)181

7. R.I.Weston and T.G.Carruthers, Proc.Brit.Ceram.Soc. 22 (1973)197

8. S.Hofmann and L.J.Gauckler, Powder Met.Int.6 (1974)90

9. E.J.Westerman, Trans.AIME 224 (1962)159

10. J.F.Strachan and A.J.R.Soler-Gomez, Proc.6th Plansee-Seminar, ed.F.Benesovsky, Metallwerk Plansee, Reutte, Austria (1969)539

11. A.J.R.Soler-Gomez and A.G.Todd, Proc.7th Plansee-Seminar, ed.F.Benesovsky, Metallwerk Plansee, Reutte, Austria (1971) vol.2, paper 36

12. R.Kieffer, G.Jangg, and P.Ettmayer, Proc.4th Int. Powder Met.Conf.CSSR, Czech.Scientific and Technical Soc., Zilina (1974) vol.2, p.19

13. H.F.Fischmeister and L.E.Larsson, Powder Met.17 (1974)227

14. M.Epner and E.Gregory, Planseeber.Pulvermet.7 (1959) 120

15. M.Epner and E.Gregory, Trans.AIME 218 (1960)117

16. J.L.Ellis, E.Gregory, and M.Epner, Progress in Powder Met., MPIF, New York 16 (1960)76

17. H.E.Exner and H.Fischmeister, Metall 19 (1965)941

18. F.Thümmler and W.Thomma, Met.Reviews 12 (1967)115

19. V.N.Eremenko, Yu.V.Naidich, and I.A.Lavrinenko, Liquid Phase Sintering, Consultants Bureau, New York - London (1970)

20. F.V.Lenel, Trans.AIME 175 (1948)878

21. H.S.Cannon and F.V.Lenel, Proc.1st Plansee-Seminar, ed.F.Benesovsky, Metallwerk Plansee, Reutte, Austria (1953)106

22. J.Gurland and J.T.Norton, Trans.AIME 194 (1952)1051

23. W.D.Kingery, J.Appl.Phys.30 (1959)301

24. W.D.Kingery, in Kinetics of High Temperature Processes, ed.W.D.Kingery, M.I.T., Cambridge, Mass. (1959) p.187

25. W.D.Kingery, E.Niki, and M.D.Narasimhan, J.Am.Ceram. Soc.44 (1961)29

26. R.B.Heady and J.W.Cahn, Met.Trans.1 (1970)185

27. J.W.Cahn and R.B.Heady, J.Am.Ceram.Soc.53 (1970)406

28. Yu.V.Naidich, I.A.Lavrinenko, and V.N.Eremenko, Poroshkovaya Metallurgiya, Jan.Feb.(1964)5; Int.J. Powder Met.1 (1965)41

29. W.Pietsch and H.Rumpf, Chem.-Ing.-Techn.39 (1967) 885

30. H.Schubert, In Proc.Colloquium: Physics of Adhesion, ed.R.Polke and P.Hohn, Inst.f.Mech.Verfahrenstechnik, Karlsruhe, Germany (1969)

31. W.J.Huppmann and H.Riegger, Acta Met. (in press)

32. W.Gans, F.Pawlek, and A.v.Röpenack, Z.Metallkde.54 (1963)147

33. M.Hansen and K.Anderko, Constitution of Binary Alloys, Mc.Graw-Hill, New York (1958)

34. W.J.Huppmann, H.Riegger, and G.Petzow, Proc.4th Int. Powder Met.Conf.CSSR, Czech.Scientific and Technical Soc., Zilina (1974) vol.1, p.31

35. H.Riegger, Diplomarbeit Universität Stuttgart (1974)

36. W.Bauer, Diplomarbeit Universität Stuttgart (1975)

37. W.J.Huppmann, W.Bauer, and G.Petzow, to be published in Proc.4th Powder Met.Conf.Poland, Zakopane, 7-9 Oct.(1975)

38. Y.Masuda, J.Soc.Mat.Science, Japan 19 (1970)514

39. W.D.Kingery and M.D.Narasimhan, J.Appl.Phys.30 (1959)307

40. H.E.Exner, Thesis, Montanist.Hochschule Leoben, Austria (1964)

41. V.N.Eremenko, Yu.V.Naidich, and I.A.Lavrinenko, Sov. Powder Met.and Met.Ceramics No.4, 10 (1962) 282

42. A.L.Prill, H.W.Hayden,and J.H.Brophy, Trans.AIME
    233 (1965)960

43. J.H.Brophy and A.L.Prill, Trans.AIME 236 (1966)805

44. T.J.Whalen and M.Humenik, In Sintering and Related
    Phenomena, ed.G.C.Kuczynski, N.A.Hooton, C.F.Gibbon,
    Gordon and Breach, New York (1967) p.715

45. P.J.Jorgensen and R.W.Bartlett, J.Appl.Phys.44
    (1973)2876

46. G.H.Gessinger and E.de Lamotte, Z.Metallkde.64
    (1973)771

47. Yu.V.Naidich, I.A.Lavrinenko, and V.A.Evdokimov,
    Poroshkovaya Met. No.9 (1972)36

48. R.J.Nelson and D.R.Milner, Powder Met.14 (1971)39

49. Yu.V.Naidich, V.M.Perevertailo, and G.M.Nevodnik,
    Poroshkovaya Met. No.7 (1972)51

50. R.F.Snowball and D.R.Milner, Powder Met.11 (1968)23

51. H.F.Fischmeister and H.E.Exner, Planseeber.Pulver-
    met.13 (1965)178

52. H.E.Exner, Metall 20 (1966)448

53. I.M.Stephenson and J.White, Trans.Brit.Ceram.Soc.
    66 (1967)443

54. J.White, in Sintering and Related Phenomena,
    ed.G.C.Kuczynski, Plenum Press, New York - London
    (1973) p.81

55. J.Gurland, Trans.AIME 215 (1959)601

56. R. Warren and M.B.Waldron, Powder Met. 15 (1972)166

57. J.A.Lund and S.R.Bala, Modern Dev.in Powder Met.,
    ed.H.H.Hausner, W.E.Smith, MPIF, Princeton, N.J.
    (1974) vol.6, p.409

58. J.Gurland, Proc.4th Plansee-Seminar, ed.F.Benesovsky,
    Metallwerk Plansee, Reutte, Austria (1962) p.507

59. G.H.Gessinger, H.F.Fischmeister,and H.L.Lukas,
    Powder Met.16 (1973)119

60. G.H.Gessinger, H.F.Fischmeister, and H.L.Lukas,
    Acta Met.21 (1973)715

61. H.Fischmeister and G.Grimvall, In Sintering and
    Related Phenomena, ed.G.C.Kuczynski, Plenum Press,
    New York - London (1973) p.119

62. G.W.Greenwood, in The Mechanism of Phase Transformations in Crystalline Solids, Inst.Metals Monograph $\underline{33}$, London (1969) p.103

63. A.J.Ardell, ibid., p.111

64. Che-Yu Li and R.A.Oriani, in Oxide Dispersion Strengthening, ed.G.S.Ansell, T.D.Cooper, F.V.Lenel, Gordon, New York (1968) p.431

65. G.H.S.Price, C.J.Smithells and S.V.Williams, J.Inst. Metals $\underline{62}$ (1938)239

66. H.E.Exner, Jernkont.Ann.$\underline{154}$ (1970)159

67. N.C.Kothari, J.Less-Common Metals $\underline{13}$ (1967)457

68. G.Matsumura, Int.J.Powder Met.$\underline{5}$ (1969)55

69. G.C.Kuczynski and O.P.Gupta, Phys.Sinter.$\underline{5}$ (1973)187

70. K.W.Lay, J.Amer.Ceram.Soc.$\underline{51}$ (1968)373

71. H.F.Fischmeister, A.Kannappan, Lai Ho-Yi, and E.Navara, Physics of Sintering $\underline{1}$ (1969) paper G

72. A.Kannappan, Thesis Chalmers University of Technology, Gothenburg (1971)

73. D.S.Buist, B.Jackson, I.M.Stephenson, W.F.Ford, and J.White, Trans.Brit.Ceram.Soc.$\underline{64}$ (1965)173

74. R.Warren, J.Mat.Sci.$\underline{3}$ (1968)471

75. R.Warren, J.Less-Common Metals $\underline{17}$ (1969)65

76. R.Warren, J.Mat.Sci.$\underline{7}$ (1972)1434

77. R.Warren and M.B.Waldron, Powder Met.$\underline{15}$ (1972)180

78. I.M.Lifshitz and V.V.Slyozov, J.Phys.Chem.Solids $\underline{19}$ (1961)35

79. C.Wagner, Z.Elektrochemie $\underline{65}$ (1961)581

80. H.E.Exner, E.Santa Marta, and G.Petzow, Mod.Dev.in Powder Met., ed.H.H.Hausner, Plenum Press, New York - London (1971) vol.4, p.315

81. E.Santa Marta and H.E.Exner, Z.Metallkde.$\underline{64}$ (1973) 273

82. M.Humenik and N.M.Parikh, J.Amer.Ceram.Soc.$\underline{39}$ (1956)60

83. N.M.Parikh and M.Humenik, J.Amer.Ceram.Soc.$\underline{40}$ (1957)315

SINTERING OF $MgAl_2O_4$ IN THE PRESENCE OF LIQUID PHASE

Emilija M. Kostić

Boris Kidrič Institute of Nuclear Sciences

Lab.170,11001 Beograd, P.O.B.522, Yugoslavia

INTRODUCTION

Highly dense ceramics have ever wider application in production of components of instrumental parts, functioning at high temperatures. One of the materials often used is sintered $MgAl_2O_4$. In this paper the results obtained on the sintering of partially synthesized spinel powder in the temperature range 1500-1850°C with additions of SiO_2 and SiO_2 and CaO mixture (molar ratio = 2:1) are presented. By addition of SiO_2 , the densities up to 92%TD can be achieved, while addition of SiO_2 and CaO mixture can produce 97%TD. Higher densification rate by adding above mentioned materials, in respect to the partially synthesized spinel powder densification, is achieved by liquid phase formation but only in these cases in which liquid phase enables better packing and development of favourable microstructure. Optimal quantity of additives for obtaining dense spinel material, varied with the sintering temperature under our experimental conditions.

EXPERIMENTAL PROCEDURES

The mixture of spinel (80%) and unreacted Al_2O_3 and MgO powders was used. The powder was obtained by isothermal heating of $MgCO_3$ and Al_2O_3 mixture (molar ratio MgO : Al_2O_3 = 1:1) in air, during 120 minutes. Specific surface area of starting powder as determined by BET method was 2,5 m^2/g. Rounded particles of the powder were below 1 micron in size, and were agglomerated. Before sintering, the starting powder, as well as powder with addition of 0,25- 5 wt.% SiO_2, and SiO_2 and CaO (molar ratio = 2:1) mixture, had been pressed to obtain cylindrical pellets. The homogenization of starting powder and additives, in the form of oxides, was obtained by wet mixing. Green pellets of cylindrical shape were pressed up to 55%TD. The sintering was performed in the temperature interval 1500-1850°C in vacuum (10^{-4} Torr). The isothermal heating time ranged from 15-300 minutes.

RESULTS AND DISCUSSION

a) Kinetic Analysis

Sintered density values of powder without additives, as well as powders containing 2 wt.% SiO_2 and 2 wt.% SiO_2 and CaO mixture, are given in Fig. 1. Kinetic analysis of densification of used powders is represented in Fig. 2.

Without analysing in detail the mechanisms determining densification of powder without additives it is obvious that the process is fast under isothermal conditions (Fig. 2). Since in the material without additives the reaction of spinel formation terminates at 1700°C after 300 minutes, different types of particle contacts exist within these samples /1/. The reaction of spinel formation is followed by formation of a new phase having larger volume than that of the sintering components. The stresses, set up during this process, affect the rate of material transport and the densification mechanism. In addition, sintering in vacuum causes partial decomposition of spinel, what also influences the densification. All these factors are of importance, depending on sintering temperature and affect the time exponent values.

Kinetic analysis of densification of the powder with 2 wt.% SiO_2 as well as that with 2 wt.% SiO_2 and CaO mixture (Fig. 2), showed that $\Delta V/V_o$ - t dependence obeyed, within acceptable limits of error, the relations derived previously /2/. Starting from 1400°C heating times of 3; 5 and 20 minutes were needed to achieve the temperatures of 1500°C, 1600°C and 1700°C, respectively. According to obtained data, at higher temperatures both the rearrangement and solution-reprecipitation processes terminate in a shorter time. At the same time higher temperature brings about the initiation of solid state mass

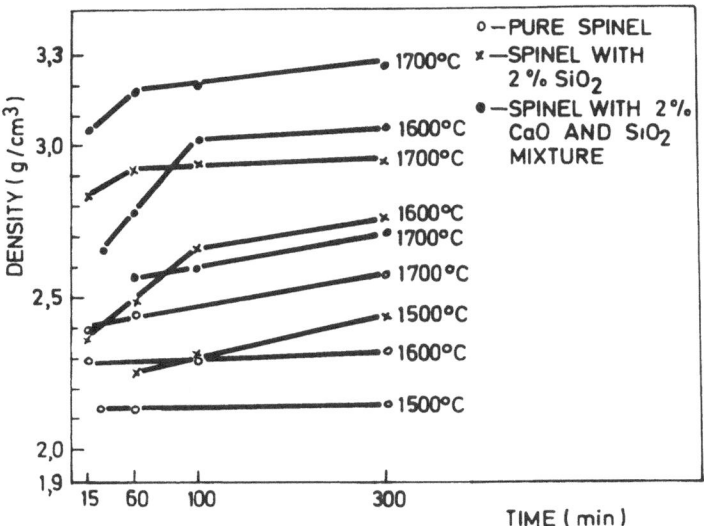

Fig. 1. Density change of spinel samples as a function
of isothermal heating time

transport. Its contribution to overall mass transport,
increases with the presence of the greater number of solid-solid contacts.

The influence of the temperature on the rearrangement and solution-reprecipitation precesses, is connected with the quantity of liquid phase and its composition. During the rearrangement process our samples
achieved higher densification degree in the presence of
SiO_2 and CaO mixture.   If the equilibrium quantities
of liquid phase in $Al_2O_3-MgO-SiO_2$ and $MgAl_2O_4-Ca_2SiO_4$
systems at 1500-1700°C are taken into account, as well
as the fact that during densification the rearrangement
process is affected both by contact angle and by liquid
viscosity /3/, it can be concluded that the conditions
for particle rearrangement were less convenient in the
case of SiO_2 addition. If one accepts the relations
derived from models for liquid phase sintering /2/, on
the basis of the time exponent values given in Fig. 2,

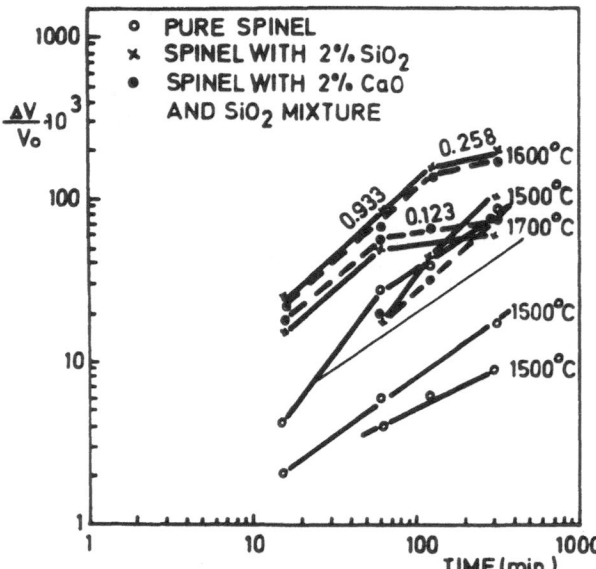

Fig. 2. Isothermal shrinkage of $MgAl_2O_4$

it can be said that the densification takes place by re-arrangement process, within the whole time period at $1500^\circ C$, within 120 minutes at $1600^\circ C$, and within 60 minutes at $1700^\circ C$. The solution-reprecipitation process occurs at $1600^\circ C$ after 120 minutes of isothermal heating, while at $1700^\circ C$, already after 60 minutes, densification takes place according to the relationships valid for solid state sintering.

The studies of the reaction between spinel and the slag containing SiO_2 and CaO, showed that by substitution of CaO by SiO_2, solution of spinel took place faster, while low-melting silicates and aluminates were formed as reaction products /4/. The contact angle values for $MgAl_2O_4$ - liquid with SiO_2, as well as for $MgAl_2O_4$ - liquid with SiO_2 and CaO mixture, used in this work in the range of temperatures from 1500 to $1850^\circ C$, were not found in the literature. They anyway depend on liquid phase composition as well as on temperature. With increasing temperature, the composition of liquid phase in equilibrium /5/ contains ever lower

Table 1.Grain size of spinel samples with additives / microns/

samples	sintering temperature /°C/	time at temperature /min/				
		15	60	120	240	300
spinel with 2 % SiO₂	1600			4		7
	1700	9		14,5		25
	1750		12		15	
spinel with 2 % CaO and SiO₂	1600			3		8,5
	1700	10		12,8	16	17,5
	1750		24		27	

amount of SiO₂ (i.e. SiO₂ and CaO) what affects decreasing of both solid-liquid interface energy and dihedral angle/8/.Partial decomposition of solid phase,due to loss of MgO from spinel at high temperatures,influences contact angle value too /6/.It should be born in mind that the partially synthesized spinel powder was used for experimental work and additives were mechanically mixed with it,but it is likely that equilibrium composition of the liquid has been achieved already after a short time of isothermal heating.In the system where liquid phase appeared,spinel formation reaction had already been over at 1500°C,after 120 minutes, as was proved by X-ray analysis data.

Except influencing densification,liquid phase affects the microstructural characteristics of sintered samples.The samples without additions do not show remarkable grain growth untill 1700°C(after heating at 1700° for 300 minutes, average grain size was 2 microns),and possess great amount of open porosity(26%).Microstructural analysis of sintered samples reveals grain growth in doped samples already at 1500°C.It can be noticed that the grain size of samples containing SiO₂ is greater at 1700°C(Table 1), although the density is lower than density of samples containing SiO₂ and CaO mixture (Fig.1).Better dissolution of spinel in the liquid,in the presence of SiO₂ ,will probably influence the value of dihedral angle.A thin layer of liquid phase surrounding the solid grains can be seen in the samples containing SiO₂ addition.On the polished surface of spinel samples with SiO₂ and CaO mixture,the liquid is gathered

in the "pockets",what is obvious from Figs.4a,b.Accor-
dingly,better penetration of liquid with SiO_2 , into
the framework of spinel grains, and more rapid solution
at the solid- liquid interface, contribute to the faster grain growth. The difference in grain shape of samples with SiO_2 ,and those with SiO_2 and CaO mixture,
is evident from Figs. 5 a,b. It can be explained by
lower solubility of spinel in the presence of SiO_2 and
CaO addition at 1700°C/7/. The grain size of doped
spinel samples sintered at 1750°C(Table 1) indicates
that samples with SiO_2 and CaO addition possess a greater number of solid-solid contacts (in comparison with
SiO_2 doped samples). These ones could enable the grain
growth in spite of density decreasing from 1700-1750°C
(Fig.3).

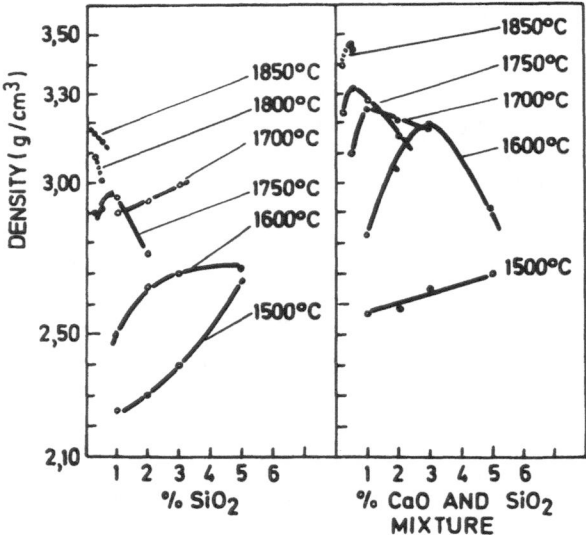

Fig. 3 Sintered densities of spinel,as a function of
        addition amount after 120 minutes of isother-
        mal heating

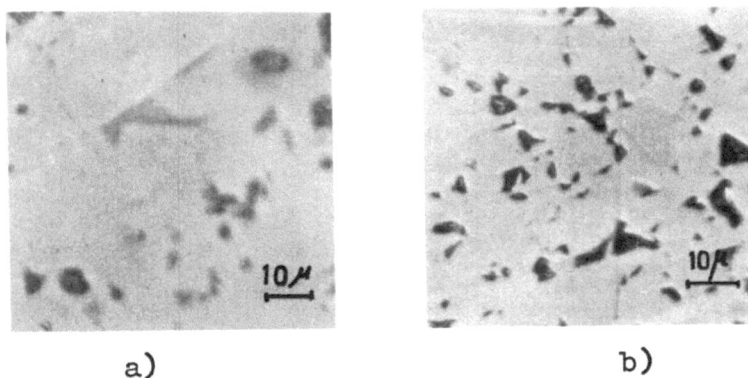

Fig.4 Polished surface of spinel   a) with 2 wt.% SiO_2,
      b) with 2 wt.% CaO and SiO_2 at 1700°C, 120 min.

Fig.5 Etched surface (H_3PO_4) of spinel a) with 2 wt.%
      SiO_2, b) with 2 wt.% CaO and SiO_2, at 1700°, 300min

Table 2. Sintered densities of spinel after isother-
         mal heating at 1850°C/g/cm³/

isothermal time/min/	spinel with-out addition	spinel 0,25% SiO_2	spinel with 0,4% (CaO,SiO_2)
120	3,27	3,16	3,46
240	3,29	3,24	3,47
480	3,29	3,27	3,47

b) The Influence of Quantity of Additives
   on Densification

Theoretical considerations of models, bearing on
the influence of the quantity of liquid on the densification degree, point to the conclusion that the density
of the samples, sintered in the presence of liquid phase depends on its amount /2/. In Fig. 3 it can be seen
that the densities of sintered spinel samples depend on
the quantity of additives. Obtained data indicate that
under our experimental conditions with SiO_2 doped sam-
ples at $1500^{\circ}C$, $1600^{\circ}C$ and $1700^{\circ}C$ (up to 3 wt.%addition)
as well as with SiO_2 and CaO doped samples at $1500^{\circ}C$
densities increase with increasing quantity of additive.
At the temperatures above mentioned there is a certain
quantity of additives which causes a maximum densification. In the presence of greater quantity of additives
(i.e. amount of liquid) above the optimal one, density
decrease   is noticed. This behaviour can be explained
by setting up the repulsion forces between particles in
samples when the amount of liquid is greater than the
optimal quantity for densification /7/.

Microstructural analysis of the samples, sintered in the presence of different quantity of additives,
showed, that the quantity of the liquid within the samples, affected grain growth, as well as the open porosity. For example, with pellets containing SiO_2 and
CaO addition,   heated isothermaly at $1600^{\circ}C$ for 120
minutes, grain growth from 2 to 3 microns can be observed if the additive quantity is changed from 1-5 wt.%,
while at $1700^{\circ}C$ grain size of sintered samples decreases
from 22 to 16 microns when quantity of addition varies
from 1-3 wt.%. The interdependence of densification and
microstructural characteristics indicates that the excess of liquid phase above optimal value at each temperature, makes the process of rearrangement slower. As a
consequence of the decrease of capillary attraction, the
process of solution-reprecipitation will be slowed down.

With the raise of temperature, the liquid phase
composition in equilibrium is closer to that of the solid. For example, in the case of SiO_2 and CaO mixture addition, spinel content in the liquid phase varies
in the temperature range $1500-1700^{\circ}C$ from 40% to 54%
(for 2 wt.% addition). This approach of compositions decreases solid - liquid interfacial energy /8/. Accordingly the contact angle as well as dihedral angle will be
getting lower with increasing temperature. Therefore,the
sintering will be the faster the greater liquid-solid
interface and the thinner liquid layer between particles.

In Fig. 3 it can be seen that maximum values of densities move towards lower quantity of addition, as the temperature increases.

More favourable conditions for development of rearrangement process in the case of SiO_2 and CaO mixture addition, affect the achievement of higher densities in the system even for long periods of heating at 1850°C, in comparison with the SiO_2 doped samples. The greater densification level can also be noticed with spinel samples without additives than with SiO_2 doped ones at the temperature (Table 2). This could indicate that diffusion processes in the solid phase are fast enough at 1850°C. The addition of SiO_2 and CaO mixture produces a intense grain growth too. After 120 minutes at 1850°C grain size in SiO_2 and CaO doped samples is 60 microns; in spinel samples containing 0,25 wt.% SiO_2 and in those without additives the grains of 24 and 35 microns in size wereobtained.

CONCLUSION

In the temperature range 1500-1700°C, where no considerable mass transport through solid phase occured, it has been found that addition of 2 wt.% SiO_2, as well as 2 wt.% SiO_2 and CaO mixture (molar ratio=2:1), enhanced densification of spinel, by mechanisms acting during sintering in the presence of liquid phase. By increasing the temperature to 1850°C, positive effect on sintering rate was noted only with samples containing SiO_2 and CaO, while with those containing SiO_2 densification was slower, even in comparison with spinel samples without additives. From the data obtained on the influence of additives (up to 5 wt.%) on the densification it could be concluded that above 1700°C in samples containing SiO_2, and above 1500°C in those containing SiO_2 and CaO, the additive quantities necessary for achieving the maximum densification level decrease with the raise of heating temperature. At each sintering temperature the grain growth in samples containing additives depended on the composition and quantity of liquid phase, as well as on the contribution of solid state diffusion. At the temperatures where solid state diffusion was relatively slow, grain growth was more rapid in the system with liquid phase having lower contact angle; the increase of additives quantity, above the value necessary for achieving the maximum density brought about the slower grain growth. With increase of the

temperature, grain growth was faster in samples in
which during rearrangement process higher densities
were obtained.

REFERENCES

1. E. Kostić, J. Petković, V. Petrović, V. Marković,
   "Modern inorganic materials", Izd. Institut Jožef
   Stefan, Ljubljana, 1974
2. W.D. Kingery, J. Appl.Phys., 30,301(1959)
3. G. Degre, M. Tenenbaum, "Physical Chemistry of Slag-
   Metal Reaction," AIME, New York, 1951
4. Ya.V. Klyucharov, S.A. Suvorov, Yu.D. Kuznecov,
   Visokotemperaturnaya himiya silikatov i okislov,"
   Izd. "Nauka", Leningrad, 1972
5. E.M. Levin, C. Robbins, H.E. McMurdie,"Phase Diagrams
   for Ceramists," The Amer. Cer.Soc., Columbus, Ohio,
   1964
6. E.S. Foster, F.R. Lorenz, USAEC Report WAPD-121,1955
7. V.N. Eremenko, Yu,V. Naidich, I.A. Lavrienko,
   "Spekanie v prisustvii zhidkoi metallicheskoi fazi,"
   Izd. "Naukova dumka", Kiev, 1968
8. O.K. Riegger, The Role of the Solid-Liquid Interpha-
   se Boundary in Microstructures,"Doctoral Dissertat-
   ion, University of Michigan, 1962.

THE GROWTH OF SOLID PARTICLES IN SOME TWO-PHASE ALLOYS DURING

SINTERING IN THE PRESENCE OF A LIQUID PHASE

R. Watanabe   and   Y. Masuda

Department of Metal Processing

Tohoku University, Sendai, Japan 980

INTRODUCTION

The particle coarsening in many liquid-sintered alloys has been considered to occur mainly by a solution-precipitation process which was first proposed by Price, Smithells and Williams (1). From this point of view the experimental results have been discussed on the basis of the theoretical kinetic equations derived by Greenwood (2), Lifshitz and Slyozov (3), and Wagner (4) for the Ostwald ripening of the second-phase particles in a saturated solution. However, the experimental observations have not satisfactorily been explained by those kinetic theories and in some cases inexplicable inconsistencies exist between the experimental and the theoretical.

On the other hand, there have been some metallographic observations which indicate particle coalescence. The term " coalescence " used here means the process that a number of particles bonded to each other coalesce to form a larger particle. The coalescence mechanism was originally proposed by Humenik and Parikh (5) without any quantitative consideration and often been called " coalescence hypothesis ". Although particle coalescence may actually contribute to the growth of solid particles in liquid-sintered alloys, only a few works have been reported on the kinetics of coalescence process.

The main object of this paper is to clarify how particle coalescence could contribute to the overall growth process. For this purpose, the relative changes in shape and size of solid particles and interparticle welds have been studied on the three alloy systems: Fe-Cu, Cu-Ag and W-Ni-Cu alloys, by means of quantitative microscopic analysis.

EXPERIMENTAL

The following powder materials were used in this investigation: carbonyl iron powder of c-p type (4.0~6.0 μ) and atomized copper powder (-325 mesh) for Fe-Cu system, electrolytic copper powder (-325 mesh) and electrolytic silver powder (-200 mesh) for Cu-Ag system, reduced tungsten powder (<5μ), carbonyl nickel powder (<5μ) and atomized copper powder (-300 mesh) for W-Ni-Cu system. The powders were mixed in various weight ratio and compacted to a bar of 10 mm in height and 10 mm in diameter. The compacts were sintered in the presence of a liquid phase for various time intervals. The sintering temperatures were 1150°C and 1200°C for Fe-Cu alloys, 800°C for Cu-Ag alloys and 1400°C for W-Ni-Cu alloys, respectively. For Fe-Cu and W-Ni-Cu alloys a hydrogen atmosphere was used to maintain good wetting, while Cu-Ag alloys were sintered in vacuum to avoid swelling. The metallographical observations were made on the polished sections and their photomicrographs with respect to solid particles and interparticle welds. The methods of the determination of the structural parameters were the same as described in a previous paper (6).

RESULTS AND DISCUSSION

Kinetic Behaviours of Particle Growth

Some examples of microstructures of sintered alloys are shown in Fig. 1. The well-known characteristic feature of the heavy alloy, in which round particles are bonded to each other by a number of interparticle welds, can be seen in all alloys.

Particle growth isotherms. The increase in the mean intercept for the three alloy systems investigated are shown as log-log plot

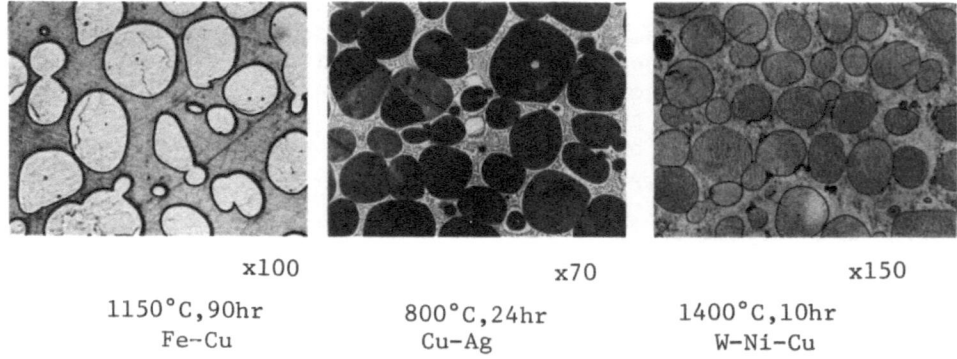

    x100   x70   x150
1150°C,90hr   800°C,24hr   1400°C,10hr
   Fe-Cu   Cu-Ag   W-Ni-Cu
Figure 1. Typical Microstructures of the three alloys.

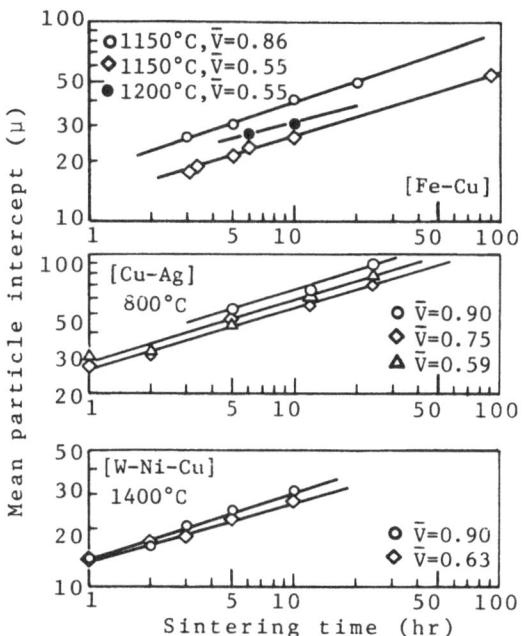

Figure 2. Particle growth in the three alloy systems. $\bar{V}$ is the
average volume fraction of solid particles.

in Fig. 2. All experimental plots are approximated by the straight
lines, the slopes of which are nearly one-third. That is, the experimental growth isotherm will be expressed by the cubic law.
The similar relation has been observed in many alloy systems and
explained by the theory of the diffusion-controlled growth.
However, in the later section, another possibility will be suggested for the growth mechanism which will also result the cubic law.

     The volume fraction effect. Another characteristic aspect of
the particle growth isotherms is their dependency on the volume
fraction of solid particles. In order to reveal this effect more
clearly, the rate constants of the growth isotherms were evaluated
according to the general formula of the cubic growth law: $l^3 = K/T \cdot t$,
where l is the mean particle intercept, K is a rate constant and
T and t have their usual meanings. The results are shown in Fig. 3
together with the data of Okamoto (7) and Klock (8) which were
roughly evaluated from their original diagrams. The theoretical
rate constants calculated from Lifshitz-Slyozov equation are also
presented in the same figure. The theoretical values seem to agree
well with the extrapolated points of the experimental curves. It
is clearly seen in these diagrams that the rate constants of growth

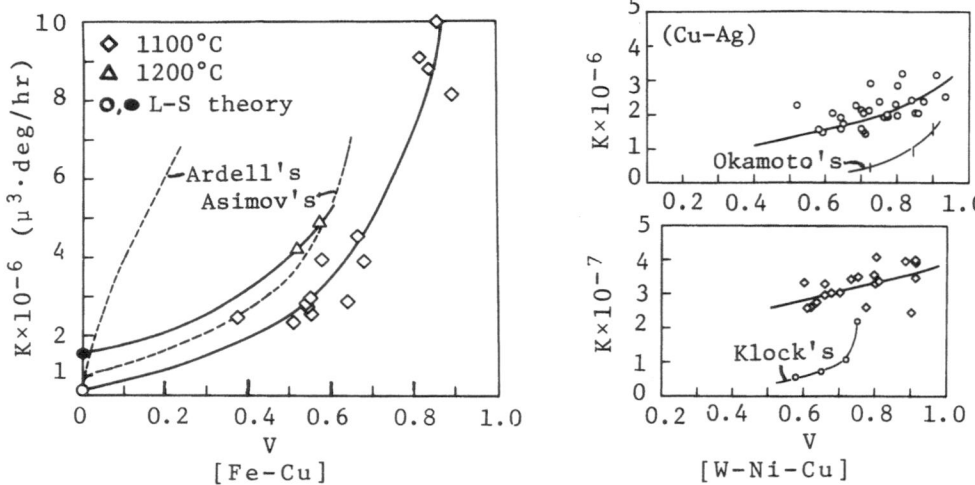

Figure 3. Relation between the rate constants of particle coarsening and the volume fraction of solid particles.

isotherms become greater with increase in the solid content. The volume fraction effect on the particle coarsening especially in polydisperse system of discrete particles has been considered as a problem of the overlapping of diffusion-flux field due to the decrease in the particle separation distance (9)-(12). In Fig. 3 the theoretical curves for Fe-Cu system are also presented. Our results apparently agree with Asimov's theory. However, this agreement is considered to be fortuitous since these theories fail to explain the majority of the experimental results obtained on the complete polydispersed systems in which the main assumptions of the theories are fulfilled better than in our alloy systems.

Particle size distribution. The experimental size distributions were normalized (13) in order to confirm whether they are steady-state distributions or not. The normalized distributions are shown in Fig. 4, where the theoretical size distributions for diffusion-controlled and for reaction-controlled growth are also presented. In all cases the size distributions prove to be fairly stable within the scatter of the experimental points. This may be another evidence that the particle coarsening in these alloys will proceed in steady-state. However, the experimental distributions are somewhat different from the theoretical in that the size positions of the peaks of the curves deviate to the smaller size.

Geometrical similarity of the microstructure during coarsening. In Fig. 5 and 6 the changes in the size and the number of the interparticle welds are shown for the three alloy systems.

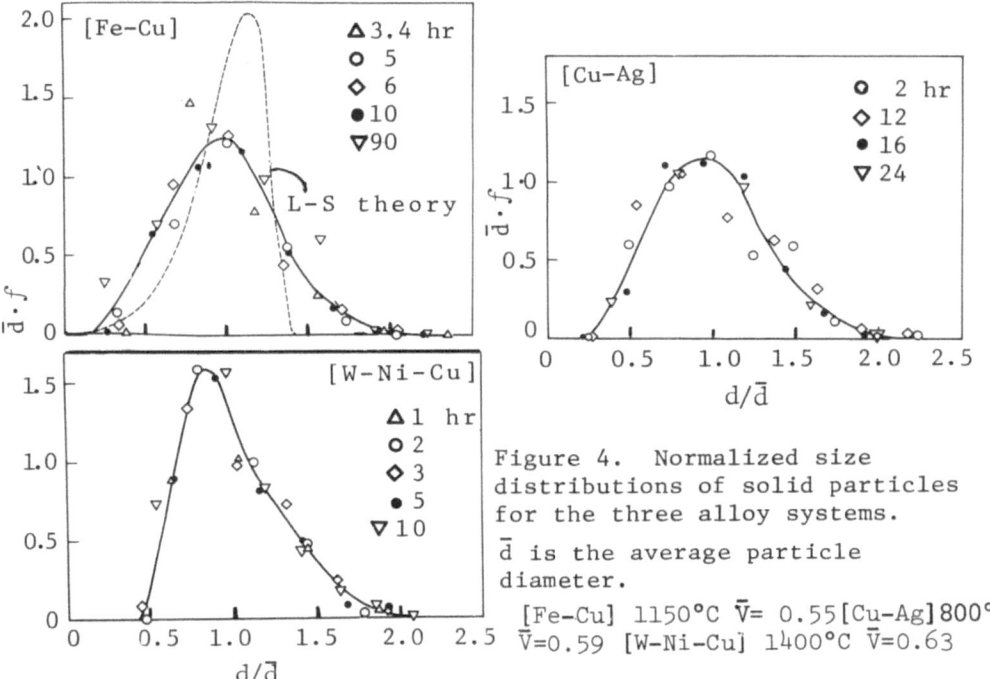

Figure 4. Normalized size distributions of solid particles for the three alloy systems.

$\bar{d}$ is the average particle diameter.

[Fe-Cu] 1150°C $\bar{V}$= 0.55 [Cu-Ag] 800° $\bar{V}$=0.59 [W-Ni-Cu] 1400°C $\bar{V}$=0.63

The size of welds increases as $t^{1/3}$ and the number of them decreases as t^{-1} in quite similar ways to the changes in the size and the number of solid particles, respectively. This indicates that the solid skelton formed in the liquid-sintered alloy retains the geometrical similarity during coarsening.

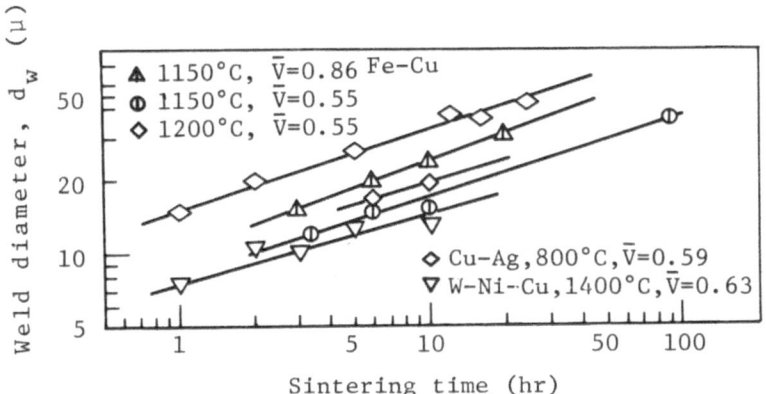

Figure 5. Increase in the size of interparticle welds during particle coarsening.

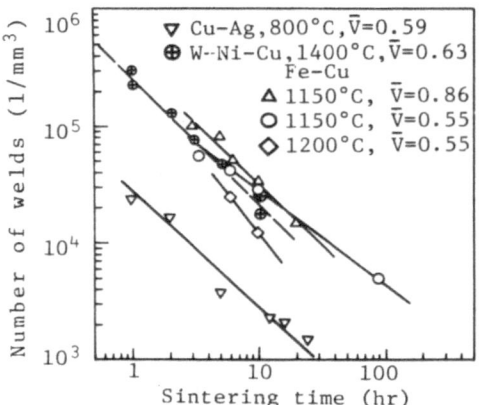

Figure 6.   Decrease in the number of interparticle welds during
coarsening.

As shown in Fig. 7, the ratios of weld diameter to particle diameter
are approximately constant which shows quantitatively that the
geometrical similarity of the microstructure is retained.   This
point will be discussed briefly in the following. Two types of
interparticle weld could be observed in this experiment as shown
in Fig. 8. The first type is of the general form common to the
three alloy systems and is considered to be in a geometrical equilibrium with a certain dihedral angle at the triple point of solid-
liquid interface and crystal grain boundary. The second type, often
seen in Fe-Cu alloys, occasionally in Cu-Ag and W-Ni-Cu alloys, re-
sembles the neck of the solid sintering system of metal particles.
The interparticle weld of the second-type is only formed under the
condition that no grain boundary exists in the contact region.
According to White's analysis of the equilibrium geometry of the
first type weld (14), size ratio of weld to particle for a couple
of particles in equal size should be equivalent to $\sin\phi/2$, where
ϕ is dihedral angle. For a couple of two particles different in

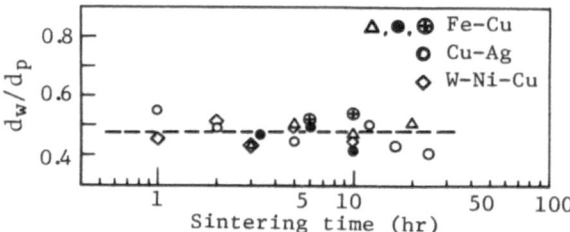

Figure 7.   Size ratio of the interparticle weld to the solid
particle. Particle diameters were evaluated from the three dimensional size distributions.

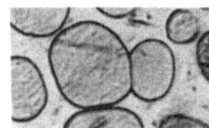 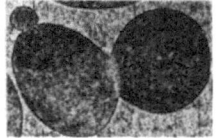

Type 1   Type 2

Figure 8.   Two types of the interparticle weld.

size, the size ratio should be smaller than sinϕ/2. The dihedral
angles for these types of alloys would be of the order of 30° and
the value of sinϕ/2 is about 0.26, whereas the values of the size
ratio observed in this experiment are about 0.5 in average as seen
in Fig. 7. Such a large discrepancy has also been observed by White
(14) for the oxide systems.

Since the retention of the geometrical similarity is not
possible to be interpreted satisfactorily by the White's analysis
it must be explained from a different stand point. In this case,
it is necessary to take into account at least that the size and
the number of the interparticle welds change in close relation to
the growth of solid particles.

Particle Coalescence

Some metallographical evidences. In Fig. 9 are shown some
examples which may be evidences of particle coalescence.   The
boundary shown as B is supposed to have very low energy since the
dihedral angles are seen to be nearly 180°. Another microstructural
evidence of coalescence is the entrapping of the liquid globule
within solid particles, as shown as E in Fig. 9.

The mode of coalescence of solid particles. A detailed
argument was made by Warren and Waldron (15) on the effect of
relative mobility of the weld-boundary to that of the solid-liquid
interface on coarsening. They suggested that in carbide-metal
systems the coarsening of the solid particles could be controlled
by the mobility of the weld boundaries. As seen in Fig. 1, almost

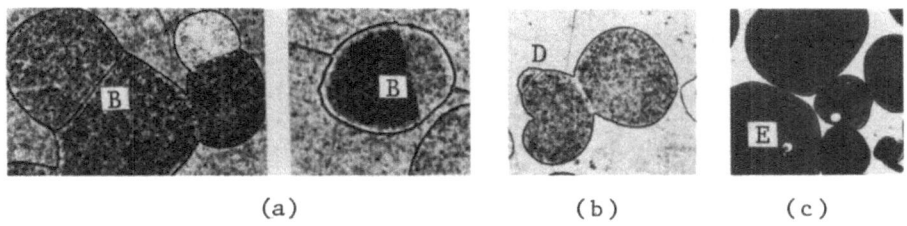

(a)   (b)   (c)

Figure 9.   Some metallographical evidences showing particle
coalescence.   (a), (b): Fe-Cu,   (c): Cu-Ag.

all the boundaries formed at the weld are flat. This implies that
the weld boundaries migrate much faster than the solid-liquid
interface. In this case, two types of morphological change might be
possible. The one is that the size of particle couple or aggregates
bonded to each other by the first-type welds increases in the
manner of Ostwald ripening in keeping the geometrical equilibrium
with a definite dihedral angle inherent to the individual weld-
boundaries. This process should always be accompanied by the
" engulfment " (15) of the smaller particles into the larger particle, which is occasionally observed in the actual structure shown
as D in Fig. 9. Another mode of the morphological change of the
bonded particles, which seems to be the main type of particle coalescence in these alloys, could be characterized by the gradual increase
in size ratio of weld to particle. The morphological change shown
in Fig. 9-a requires the gradual decrease of the boundary energy
which will result an increase in the dihedral angle. With progressing
of this process the bonded particles coalesce to form a larger one.
The continuous decrease in the energy of weld-boundary could be
resulted from the grain boundary rotation for lowering the total
energy of the system (16). The remaining boundaries often seen in
the coalesced particles are considered to be the low-energy coincidence boundary or twin boundary. If the weld has no grain boundary
at the outset of the coarsening process, or the boundary is
eliminated by any reasons, the coalescence of the particles could
occur without limitation. Anyway, the latter type of morphological
change will inevitably be accompanied by the increase in the size
ratio of weld to particle. On the other hand, in concequence of
the complete coalescence of a couple of particles a second-type
weld disappears and the fraction of the second-type weld in the
whole number of welds reduces. Thus, by the counteracting effects
of the coalescence on the size ratio it might be possible that the
mean ratio would remain constant, as observed in our experiment.

   Contibution of the coalescence to the particle growth. If the
time needed to complete the coalescence of a couple of particles
having a radius r is t, the average rate of the coalescence
growth is given by (17):

$$dr/dt = \sqrt[3]{2} \cdot r/t \qquad\qquad (1)$$

It follows from Herring's analysis (18) that the time needed to
complete the particle coalescence should be proportional to r^n,
where n=1 for viscous flow mechanism, n=2 for solution-precipitation
controlled by interfacial reaction, n=3 for volume diffusion, and
n=4 for interfacial diffusion, assuming that the geometrical
similarity is maintained. The rigorous functional form of the time of
coalescnece has been derived by Frenkel (19) for viscous flow
mechanism and by Nichols (20) for surface diffusion mechanism. Here,
only the dependence of the coalescence time on the particle size
will be concerned for the first approximation.

From the argument described above, the rate of coalescence growth as a function of particle size can be given for various transportation mechanisms. Among them, only the volume diffusion mechanism (solution-precipitation controlled by the solute diffusion in the liquid phase) leads to the cubic growth law, which is required from the experimental results. In an actual sintering system, the particle growth in the sense of the Ostwald ripening theory and the coalescence growth may be operative simultaneously. The overall rate of particle growth should be given by the sum of the rates of both the processes; hence

$$(dr/dt)_t = (dr/dt)_0 + (dr/dt)_c \qquad (2)$$

where $(dr/dt)_t$ is the total growth rate, $(dr/dt)_0$ is the rate of Ostwald growth and $(dr/dt)_c$ is the rate of coalescence growth. From Eqs. 1 and 2 the average radius of particles is given by the following formula for diffusion controlled growth.

$$\bar{r} = (k_0 + \alpha k_c)\ t^{1/3} \qquad (3)$$

where k_0 and k_c are the rate constants of Ostwald growth and of the coalescence growth, respectively, and α is a factor of relative contribution of the coalescence to particle growth. The rate constants k_0 and k_c should be identical within a numerical factor for the same elementary process (in our case the elementary process is considered to be the diffusion of solid constituent in the liquid phase). The factor α would depend on the mode of coalescence and the number of the welds through which coalescence will take place. Roughly speaking, the contribution of coalescence growth will become larger with increase in the number of welds. Since the number of welds increases as the liquid content decreases, the factor α would increase with increase in the volume fraction of solid particles. This, together with the kinetic Eq. (3), could be a possible explanation of the experimentally observed dependency of the growth rate on the volume fraction of solid particles already presented in Fig. 3.

In order to make a more quantitative description on the mechanism of the coalescence growth and its contribution to the particle growth in the liquid-sintered alloys, further investigation is to be needed on the nature of the particle coalescence.

CONCLUSION

1. The well-known cubic law was confirmed to hold in the particle coarsening in the three alloy systems studied.

2. The rate constant of particle growth increases with increase in the volume fraction of solid particles.

3. The size distributions of solid particles are able to be normalized to a fixed distribution, indicating that the particle
   coarsening is proceeding in the steady-state.

4. The size of the interparticle weld increases as $t^{1/3}$ and the
   number of them decreases as t^{-1}.

5. The ratio of the weld size to the particle size is approximately
   kept constant and the geometrical similarity is maintained
   during steady-state coarsening.

6. From a primary argument on the nature of particle coalescence,
   a possible mechanism which could explain the contribution of
   particle coalescence to the overall particle growth and the
   volume fraction effect has been suggested.

ACKNOWLEDGEMENT

The authors are grateful to Mr. M. Momose and Mr. K. Takemura
for their experimental assistence.

REFERENCES

1. G. Price, C. J. Smithells and S. V. Williams: J. Inst. Metals,
   62, 239 (1938).
2. G. W. Greenwood: Acta Met., 4, 243 (1956).
3. I. M. Lifshitz and V. V. Slyozov: J. Phys. Chem. Solids,
   19, 35 (1961).
4. C. Wagner: Z. Electrochem., 65, 581 (1961).
5. N. M. Parikh and M. Humenik: J. Am. Cer. Soc., 40, 315 (1957).
6. R. Watanabe and Y. Masuda: Trans. JIM, 14, 320 (1973).
7. Y. Okamoto: J. Japan. Soc. Powder Met., 9, 1 (1962).
8. R. H. Klock: Plansee Proceedings,1964 F. Benesovsky, Ed.,
   (1965) p.256.
9. S. Sarian and H. W. Weart: J. Appl. Phys., 37, 1675 (1966).
10. R. Asimov: Acta Met., 11, 72 (1963).
11. A. J. Ardell: Acta Met., 20, 61 (1972).
12. H. Fischmeister and G. Grimvall: Proc. Third Conf. on Sintering
    and Related Phenomena, Notre Dame 1972, p.119.
13. A. J. Ardell and R. B. Nicholson: J. Phys. Chem. Solids,
    27, 1793 (1966).
14. J. White: Proc. Third Conf. on Sintering and Related Phenomena,
    Notre Dame 1972, p.81.
15. R. Warren and M. B. Waldron: Powder Met., 15, 166 (1972).
16. T. L. Wilson and P. G. Shewmon: Trans. AIME, 48, 236 (1966).
17. C. Greskovich and K. W. Lay: J. Am. Cer. Soc., 55, 42 (1972).
18. C. Herring: J. Appl. Phys., 21, 301 (1950).
19. J. Frenkel: J. Phys., 9, 385 (1945).
20. F. A. Nichols: 37, 2805 (1966).

INVESTIGATION OF MICROSTRUCTURE DEVELOPMENT IN RuO$_2$-LEAD

BOROSILICATE GLASS THICK FILMS

A. N. Prabhu and R. W. Vest
School of Materials Engineering
Purdue University
West Lafayette, Indiana 47907

Introduction

A thick film resistor can be described as a combination of conductives (noble metals and/or oxides) and oxide glasses, several
microns in thickness, and thermally bonded to a ceramic substrate.
A typical thick film resistor is fabricated by screen printing the
formulations consisting of inorganic constituents (small particle
size glass and conductives) and organic screening agents through
masks on to a ceramic substrate and firing in a tunnel kiln with an
appropriate time-temperature profile so as to develop a microstructure that gives the desired electrical characteristics. The maximum
temperature for firing the resistors varies between 700 and 900°C,
depending upon the conductive ingredient in the formulation. The
resistor is at the maximum temperature for about 2-5 minutes and
the total firing time might vary between 10-30 minutes. The resistor system selected for this study was ruthenium dioxide (RuO$_2$)
conductive, lead boro-silicate glass (63 w/o Pbo-25 w/o B$_2$O$_3$-12
w/o SiO$_2$), and 96% Al$_2$O$_3$ substrate. This particular system was
chosen because it is one of the simplest of all possible systems,
and because formulations based upon the conductive RuO$_2$ are being
increasingly used for thick film resistors [1].

One of the most interesting features of any thick film resistor system is the plot of resistivity versus volume fraction of
conductive to glass. Electrical continuity is achieved at 2 volume
percent RuO$_2$ to glass whereas, commonly applied theories for conduction in a random system consisting of a conducting phase and an
insulating phase predict a percolation threshold near 30 volume
percent [2]. Therefore, the electrically conducting phase, which is

399

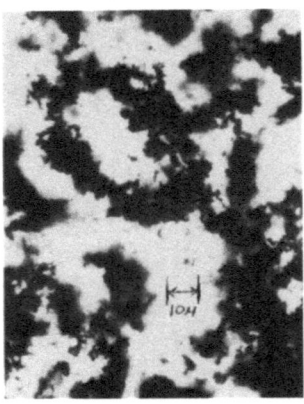

Figure 1.   Resistor in Transmitted Light.

present in the formulation as discrete particles, cannot be randomly
distributed in the resistor but rather, must develop into a physically continuous network along the length of the resistor during
firing. This continuous network in the glassy matrix can be observed in Fig. 1, which represents a typical low value resistor.

For our resistor system a typical firing schedule would be 5
minutes at 800°C, but the same microstructure can be achieved after
longer times at lower temperatures where the resistance change with
firing time can be easily monitored. The observed behavior shows
an initial rapid decrease in the resistance followed by a slower decrease and finally, further increase.

In order to account for the observed behavior a model consisting of the following basic steps for microstructure development in
thick film resistors has been suggested [3].

    (1) Glass sintering and wetting of the conductive particles
    (2) Rearrangement of the conductive particles to form a network
    (3) Sintering of the conductive in the presence of glass
    (4) Growth of larger particles at the expense of smaller ones

Knowledge of the kinetics of each of these stages, which are
also the basic features of a typical liquid phase sintering process,
is essential in order to select optimum materials and to choose a
firing profile for resistor fabrication so that desired electrical
characteristics are achieved. The investigation reported here is
directed towards obtaining quantitative information on the different stages of microstructure development.

For a system undergoing sintering the relationships involving
time dependence of neckgrowth between two spherical particles,
shrinkage of compacts during initial stage sintering, and the growth of particles during ripening have been reported by many workers
and only those of interest to this study are listed.

Glass Sintering: The ratio of neck radius (x) to particle radius
(a) during initial stage sintering by Newtonian viscous flow is
given by [4]:

$$\left(\frac{x}{a}\right)^2 = \left(\frac{3}{2} \quad \frac{\sigma_{SV}}{\eta}\right) a^{-1} t \tag{1}$$

where σ_{SV} is the solid-vapor interfacial energy, η is the viscosity
and t is time.

Conductive Sintering: The shrinkage due to initial stage liquid
phase sintering by solution-precipitation depends on the rate
limiting process and two possibilities are [5]:

a. Diffusion controlled process.

$$\frac{\Delta L}{L_o} = \left[\frac{6K_1 \delta DC_o \sigma_{LV} \Omega}{K_2 RT}\right]^{1/3} a^{-4/3} t^{1/3} \tag{2}$$

b. Phase boundary reaction controlled process.

$$\frac{\Delta L}{L_o} = \left[\frac{2K_T K_1 \sigma_{LV} C_o \Omega}{K_2 RT}\right]^{1/2} a^{-1} t^{1/2} \tag{3}$$

In equations (2) and (3) $\Delta L/L_o$ is the relative shrinkage, K_1 is a
constant of proportionality relating the ratio of the contact area
to the projected particle area, δ is the thickness of the film sepa-
rating the particles, D is the diffusion coefficient of the slowest
moving species, C_o is the equilibrium solubility, σ_{LV} is the liquid-
vapor interfacial energy, Ω is the molar volume, K_2 is the ratio of
pore to particle radii, R is the universal gas constant, T is the
absolute temperature, and K_T is the phase boundary reaction constant.

Ostwald Ripening: The growth of particles by solution-precipitation
also depends on the rate limiting process and two possibilities
are [6]:

a. Diffusion controlled process.

$$\overline{a}(t)^3 - \overline{a}(o)^3 = \left[\frac{8}{9} \quad \frac{\sigma_{SL} C_o D\Omega^2}{RT}\right] t \tag{4}$$

b. Phase boundary reaction controlled process.

$$\overline{a}(t)^2 - \overline{a}(o)^2 = \left[\left(\frac{8}{9}\right)^2 \frac{K_T C_o \sigma_{SL} \Omega^2}{RT}\right] t \tag{5}$$

In equations (4) and (5) $\bar{a}$ (t) and $\bar{a}$ (o) are the average particle radii at time t and time zero respectively.

Preliminary Experiments

As indicated from equations 1-5, the sintering processes are controlled by factors such as high temperature surface tension and viscosity of the glass, solubility of RuO_2 in the glass, wettability of the glass to RuO_2, and changes in the intrinsic properties of the ingredient materials due to interactions among them. Since no information existed in the literature regarding these properties they were determined independently.

High temperature surface tension and viscosity of the glass were determined by the "modified dipping cylinder method" and the "sphere method" respectively, and the results have been reported elsewhere [7]. The surface tension increased with decreasing temperature from about 160 dynes/cm at 800°C to about 200 dynes/cm at 650°C. The viscosity was found to be about 390 poise at 800°C and increased to 7100 poise at 725°C. The equilibrium solubility of RuO_2 in the glass at different temperatures was found [8] to be very low with a value of about 8 ppm at 600°C and about 15 ppm at 1000°C. Contact angle measurements of the glass on pressed and sintered pellets of RuO_2 at temperatures from 650 to 1000°C indicated that the glass completely wets RuO_2. Phase analysis by X-ray diffraction measurements suggested that no new crystalline phases formed either due to reaction between the glass and RuO_2, or RuO_2 and the substrate. Alumina does go into solution because of the reaction between the glass and the substrate, but new crystalline phases were observed only after very extended heating, i.e., more than 50 hours at 900°C.

Initial Stage Sintering Studies

A hot stage metallograph was modified so that neckgrowth data could be recorded continuously. The regular camera system of the metallograph was replaced by a video camera and a second video camera was used to monitor a digital voltmeter that measured the sample thermocouple emf and a digital clock. A special effects generator was used to mix the two video signals so that the thermocouple emf and time were positioned in a corner of the image from the viewing camera. All information thus obtained was recorded on a video recorder with stop frame capability and observed on a television monitor. The whole system was capable of giving a magnification of about 1000X. This method of recording data is useful because it creates a virtually continuous and complete record of the sintering process. Even very rapid coalescence at high temperatures can be observed and quantitative data obtained because 60 frames, each representing one datum point, are obtained each second.

When the neckgrowth data for spherical glass particles were plotted as $(x/a)^2$ versus t, (Fig. 2), straight lines were obtained, indicating that the sintering of lead borosilicate glass occurred by Newtonian viscous flow.

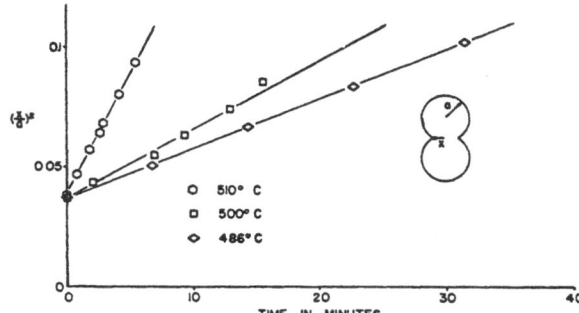

Figure 2. Initial State Sintering Kinetics for Glass Spheres.

Preliminary experiments conducted to observe neckgrowth between RuO_2 particles in the glass using the hot stage with transmitted light were not successful in that neckgrowth between 100 μm particles was not observed even after 60 hours at 800°C. Smaller particles could not be used due to the resolving power of the optical system. An attempt was then made to follow initial stage conductive sintering by observing the shrinkage during isothermal heating of RuO_2-glass compacts. After preheating at 450°C for 1 hour in order to allow for the initial contraction due to glass sintering to be completed, the length changes were recorded, using a horizontal dilatometer whose sensitivity was adjusted so that length changes as small as 10^{-5} cm could be easily detected. The total shrinkage was observed to be less than 1/2% even after heating at 800°C for 10 hours. Microstructure investigation of the fired compacts with SEM revealed considerable closed porosity. As the firing time increased the pores increased in size, but the total pore volume remained almost the same.

Ripening Studies

Due to the difficulties encountered in obtaining any quantitative information on the conductive sintering by initial stage sintering studies, it was decided to determine the ripening kinetics by studying the growth of RuO_2 particles in the glass and then to use these results to predict the initial stage sintering kinetics for RuO_2 particles. As discussed in the introductory section, the final stage of microstructure development during firing of the resistors has been characterized as Ostwald ripening where the small particles dissolve and precipitate on the larger ones. The rate controlling mechanism for such a process can be determined by following the average particle size with time at any temperature.

Samples for the ripening study were prepared by heating well dispersed mixtures of RuO_2 and glass in a platinum boat at 800, 850,

900, 950 and 1000°C. Since the aim was to observe the growth of RuO_2 particles only, the glass was completely leached out after heat treatment by treating the samples with HCl, Hf and hot H_2O, a procedure which was proven to leave RuO_2 completely unaffected. The RuO_2 powder thus obtained was analyzed for particle size by microscopy (SEM and TEM), X-ray diffraction line broadening and BET surface area measurements.

Fig. 3 shows the sintering and growth of RuO_2 particles at 1000°C with increasing time. A sintered network of RuO_2 particles in addition to large crystals of RuO_2 began to form with increasing time. Although the growth of RuO_2 particles could be observed using SEM, quantitative microscopy was not feasible to study the growth kinetics because: (1) at shorter periods of heating the particles are essentially of uniform size but it was not possible to separate them out individually, due to agglomeration; (2) longer periods of heating resulted in a wide distribution of particle sizes ranging from 1000Å to 1 μm, a large number of smaller crystals being sintered to a few big crystals; and (3) still longer periods of heating at temperatures higher than 1000°C were not feasible for this system because of the volatalization of RuO_2 and glass.

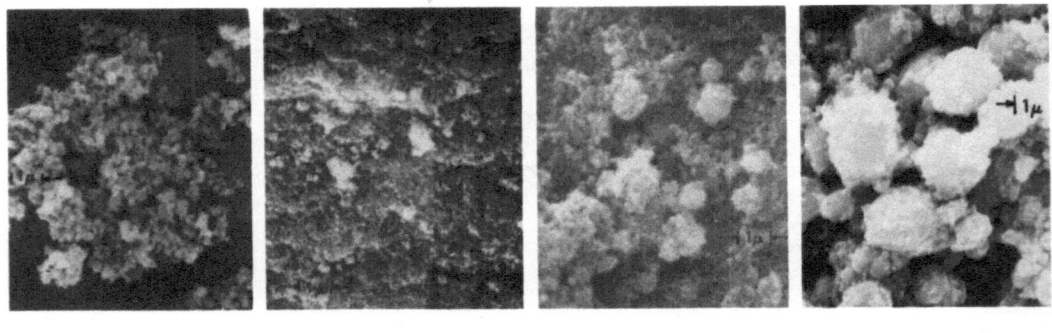

a. 0 Minutes   b. 10 Minutes   c. 30 Minutes   d. 6 Hours

Figure 3. Sintering and Growth of RuO_2 Particles at 1000°C (10,000X).

For the X-ray diffraction line broadening experiments three different peaks (110, 101 and 211) were scanned at a speed of 1/10 degree/minute. After the integral line breadths were corrected for instrumental and K-α doublet broadening, the pure diffraction breadths were obtained by graphical methods [9] and the average crystallite sizes were computed, using the Scherrer equation [9]. Surface area of the powder was determined, using an Aminco Sor BET surface area meter and the average particle sizes were computed, assuming all the individual particles to be uniform spheres.

The average particle size of the RuO_2 powder (Fig. 3a) used for making all samples was found to be 120Å from the surface area measurements whereas, the X-ray results gave the average crystallite size to be 60Å. A BET surface area measurement is expected to give a greater average particle size because of the agglomeration of the smaller particles, thus decreasing the surface area available for nitrogen adsorption. The average X-ray crystallite size was computed, neglecting the contribution to broadening due to strain and other defects; hence, the average X-ray crystallite sizes obtained are smaller than the actual sizes. In the light of these considerations it seems reasonable to assume that the crystallites in the starting powder are individual particles with an average diameter between 60 and 120Å. However, the SEM investigations indicated the average particle size of the starting powder to be approximately 1000Å. This order of magnitude difference is due to the agglomeration of the ultrafine RuO_2 particles. Agglomeration was a problem in TEM study also, although a few particles as small as 50Å could be clearly seen.

The average particle sizes obtained from X-ray and surface area measurements as a function of time at various temperatures are shown in Fig. 4. The discrepancy between X-ray and surface area results after long hours of heating arises from the wide distribution of particle sizes generated during the experiments. For a powder sample composed of crystallites with a wide distribution of sizes the

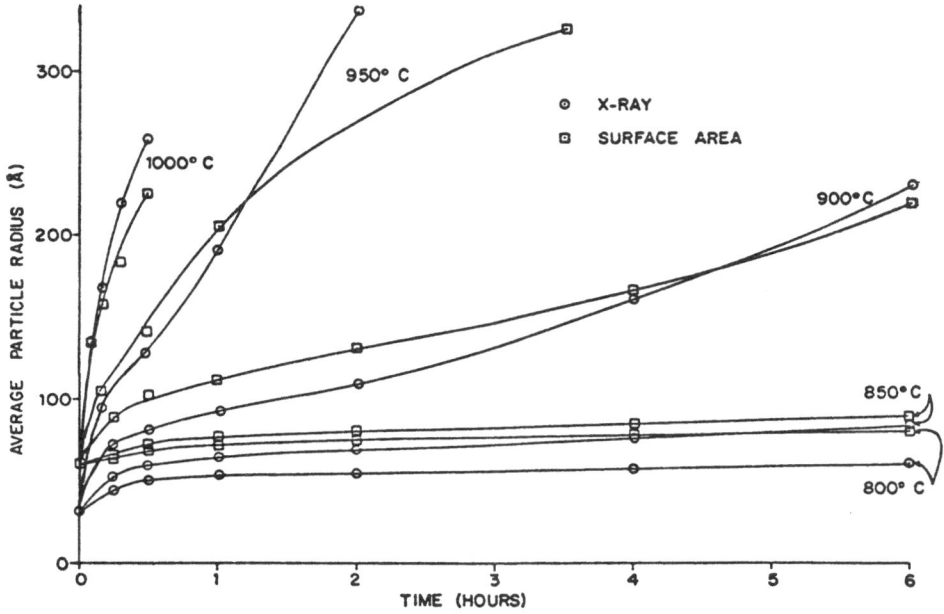

Figure 4. Particle Size of the Samples From X-Ray & Surface Area.

average size calculated from the line breadth is weighted towards
larger sizes [10]. The difference between the observed and true mean
sizes increases if a few large crystallites and a large number of
small crystallites are present, which is exactly what is happening
in this system. Therefore, in order to analyze the surface area and
X-ray results needed to obtain ripening data, only those data points
were used for which the differences in the two values were small.

From the least square fit obtained on the $\bar{a}\,(t)^2 - \bar{a}\,(o)^2$ versus
time plot (Fig. 5) the phase boundary reaction is considered to be

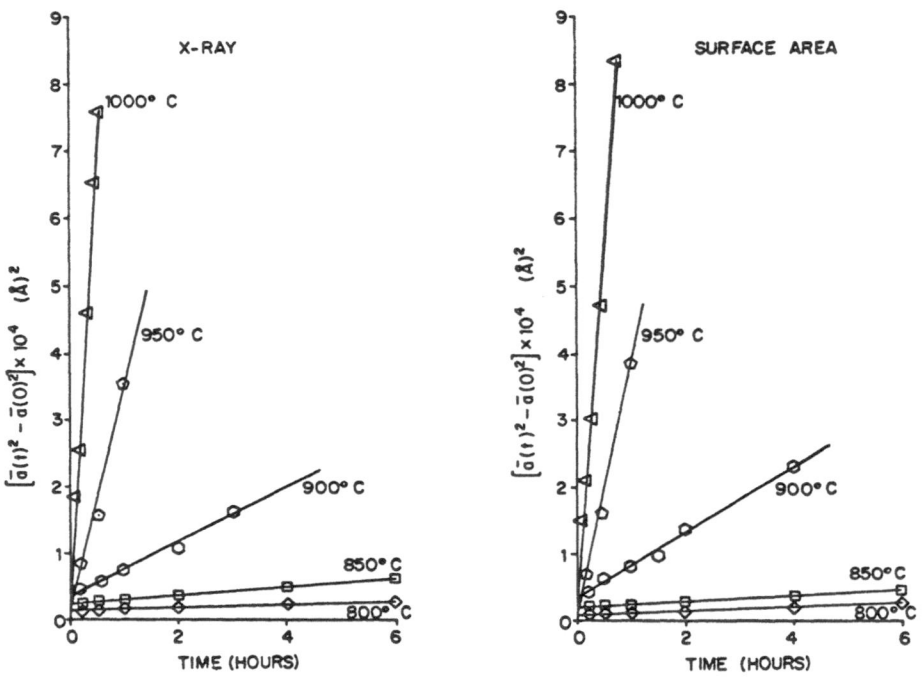

Figure 5. Fit to Phase Boundary Reaction Equation.

the most probable rate controlling process for the growth of RuO_2
particles in the glass by solution-precipitation. This conclusion
was further strengthened by comparing the predicted growth behavior
of the particles by a diffusion controlled solution-precipitation
process with the observed growth behavior. The diffusion coefficients
were calculated from the high temperature viscosity data, using the
Stokes-Einstein equation [11] and substituted in equation (4) to compute the time required for the average particle size to attain a particular value. The calculated values were 10 to 100 times greater
than those observed experimentally, suggesting that the diffusional

process is much faster and hence, cannot be the rate determining process.

The results shown in Figs. (3-5) were obtained with samples containing 30 w/o RuO_2, but similar experiments using samples containing 6 and 18 w/o RuO_2 gave identical results, suggesting that the growth of particles is independent of liquid content. This result also supports the observation that growth of RuO_2 particles occurs with reaction controlled solution-precipitation as the rate controlling process.

The phase boundary reaction constant $\left(K_T\right)$ was determined from the slopes of the plots in Fig. 5 for different temperatures. These values were then plotted as log K_T versus 1/T and from the slope of the straight line thus obtained the activation energy for the phase boundary reaction was computed to be 100 Kcals/mole. As anticipated, this activation energy is higher than that predicted for the diffusional process (63 Kcal/mole). Using the K_T values in equation 3, the times required for the initial stage sintering of RuO_2 to be completed (x/a = 0.3) were calculated at different values of particle size and temperature. At 800°C initial stage sintering should be complete in about one minute for 60Å particles. Since the particle size of RuO_2 powder used for resistor fabrication is expected to lie in this size range it can be concluded that initial stage sintering of the conductive particles does take place during routine firing of RuO_2-lead borosilicate glass resistors in a tunnel kiln. Further heating results in dissolution of smaller particles and decrease in the total number of particles and hence, destruction of the physically continuous network.

Conclusions

Growth of RuO_2 particles in the lead borosilicate glass occurs by a phase boundary reaction controlled solution-precipitation-process. From the predicted kinetics for the initial stage sintering process it can be concluded that during routine firing of the RuO_2-glass resistors in a tunnel kiln, the initial stage sintering of the conductive particles does take place.

References

1. H. C. Angus and P. E. Gainsburg, "Glaze Resistors With Ruthenium Dioxide," Elec. Comp., 84 (1968).

2. V. Ambegoakar, S. Cochran, and J. Kirkijarvi, "Conduction in Random Systems," Phys. Rev. B, 8, 3682 (1973).

3. R. W. Vest, Semi-Annual Technical Report, July 1, 1971 to
    December 31, 1971, for Grant No. DAHC-15-70-G7, ARPA Order
    No. 1642, February 1, 1972.

4. J. Frenkel, "Viscous Flow of Crystalline Bodies Under the
    Action of Surface Tension," J. Phys. U.S.S.R. 915, 385 (1945).

5. W. D. Kingery, "Densification During Sintering in the Presence
    of a Liquid Phase. I. Theory," J. Appl. Phys. 30, 301 (1959).

6. H. Fischmeister and G. Grimvall, "Ostwald Ripening – A Survey,"
    Materials Science Research 6, 119 (1973).

7. A. Prabhu, G. L. Fuller, R. L. Reed and R. W. Vest, "Viscosity
    and Surface Tension of a Molten Lead Borosilicate Glass," J.
    Amer. Ceramic Soc., 58, 144 (1975).

8. A. Prabhu, G. L. Fuller and R. W. Vest, "Solubility of RuO_2 in
    a Pb Borosilicate Glass," ibid, 57, 408 (1974).

9. E. F. Kaeble, "Handbook of X-Rays," McGraw-Hill, Inc., 1967.

10. F. W. Jones, "The Measurement of Particle Size by the X-Ray
    Method," Proc. Ray. Soc. (London), 166A, 16 (1938).

11. G. W. Greenwood, "The Growth of Dispersed Precipitates in
    Solutions," Acta Metallurgica, 4, 243 (1956).

Acknowledgements

The authors express their appreciation for the technical assistance of Robert L. Reed throughout the course of this work. This
work was supported by Delco Electronics Division/GMC, Basil S.
Turner Foundation, and the Advanced Research Projects Agency under
Grants Nos. DAHC15-70-G7, DAHC15-73-G8.

PARTICLE COARSENING IN FUSED SALT MEDIA

R.D. McKellar and C.B. Alcock

Department of Metallurgy and Materials Science

University of Toronto, Toronto, Canada

INTRODUCTION

The growth of particles of a dispersed phase in a solute saturated solid or liquid matrix has been the subject of much interest in recent years. The dispersion of particles coarsens in order to decrease their total surface energy by the transfer of material from the small to larger particles leading to a decrease in the number of particles.

The theory of particle coarsening has been developed by Greenwood[1], Lifshitz and Slyozov[2] and by Wagner[3]. Reviews on the topic have been published by Greenwood[4], Ardell[5] and by Fischmeister and Grimvall[6].

There are two possible rate controlling mechanisms in particle coarsening: diffusion of material between particles or the dissolution and deposition of material at the particle surface. Expressions for the rate of growth of the mean particle size have been derived for the cases of diffusion controlled growth and first order reaction controlled growth[2,3] and recently for second order reaction controlled growth[7]. It has also been shown that after long periods of time, the particle size distribution approaches a stationary shape. Expressions have also been determined for the shape of the stationary particle size distributions for these cases. Therefore, the theory provides two means of determining the rate controlling mechanism in an experimental study.

In most studies to date, there has been only limited agreement between experimental results and theory. Studies on solid

systems invariably lead to a broadening of the particle size distribution while good agreement is usually found for the time dependence of the mean particle radius. There are several factors
which can affect the growth of particles in a solid matrix such as
grain boundaries and elastic interaction which are not considered
in the original theory. By using a liquid matrix, these problems
can be eliminated. Particle coarsening has been investigated in
liquid systems but the particle volume fraction was always quite
high, leading to problems with diffusion zone interactions[8]. For
example: Exner et al[9] studied the growth of particles during liquid phase sintering. The liquid phase in these systems is generally solid and opaque at room temperature. This means that the
only way to measure the particle size distribution is to section
the sample. Consideration must then be made for the greater probability of intersecting a large particle and for that of section-
ing particles off centre. This can lead to large errors in the
final result. By choosing salt as a matrix material, the advantages of a liquid matrix may be obtained and the particle size
measurement can be simplified. This is because the salt may be
dissolved away with water, leaving the water insoluble particles
ready for observation.

     The system chosen for study consisted of a dispersion of lead
sulphide particles in a matrix of sodium chloride - potassium
chloride eutectic. This system was chosen because of the high
melting point of lead sulphide at 1114°C. and the relatively low
melting point of the eutectic at 670°C. The solubility of lead
sulphide in the eutectic has also been accurately determined by
Mohapatra, Alcock and Jacob[10].

EXPERIMENTAL MATERIALS AND PROCEDURE

     The materials used in this study were of high purity. The
sodium chloride and potassium chloride were specified as being
greater than 99.99% pure, the lead greater than 99.999% pure and
the sulphur was a doubly distilled grade. The eutectic salt was
prepared by making a 50-50 mole% mixture. The salt was dried in a
vacuum oven for 24 hours at 100°C. and then 350°C. for five days
under a vacuum of 10^{-3} mm. of mercury. This was necessary to remove any trace of moisture as it has been shown that impurities
can drastically reduce the coarsening rate. The lead sulphide was
prepared by the reaction of the stoichiometric amounts of lead and
sulphur in a quartz capsule at high temperature. The capsule was
held at 250°C. for two days and then raised to 900°C. for five
days. After this preparation, the lead sulphide was ground to
-325 mesh with care taken not to oxidize the particle surfaces.

     The furnace used in this study was a Kanthal wound resistance type which had a 2" even temperature zone that was controlled to ± 2°C. by a Pt/Pt-Rh thermocouple.

A sample was prepared by placing a dispersion of particles in the eutectic salt and sealing this mixture into an evacuated quartz capsule. These samples were then held at a constant temperature for various times. The amount of lead sulphide added was calculated from the knowledge of the solubility at the experimental temperature and the fact that a low volume fraction was desired to minimize diffusion field interaction. While in the furnace, the capsule was contained in a steel jacket in a horizontal position to minimize the temperature fluctuations of the furnace. The capsule itself also contained small quartz rods running the length of the capsule to act as baffles to minimize stirring. If stirring occurs, there is an enhanced probability of aggregation. After the heat treatment, the sample was quenched and the entire contents dissolved in water to liberate the particles. Sodium silicate was added to the water to disperse the lead sulphide particles and then this suspension was filtered through a fine organic filter. This method of sample preparation produced a completely random distribution on the filter and eliminates sampling errors since the entire capsule contents were filtered. The filter containing the particles could then be viewed directly under the microscope. The particles were measured individually using a measuring reticule and were then statistically analyzed.

In this study, the experiments were performed using a 1.0 wgt.% lead sulphide dispersion in the eutectic salt. This level is just above the saturation solubility at 750°C. and gave a particle volume fraction of 5×10^{-4} at this temperature.

EXPERIMENTAL RESULTS AND DISCUSSION

The particles observed in this study were distinctly cubical in shape as can be seen in Figure 1. This regular shape made the particles very easy to measure. An equivalent radius was defined as half the cube edge. In the case where two visible cube edges were not equal, the average was taken following a similar approach by Ardell[8].

For each sample, 800 particles were measured in two groups of 400 each. This was done to determine the consistency of the measuring technique. In each case, the difference between the mean particle radius in each group was less than 5%.

The results of the particle size measurements on these samples are summarized in Table I.

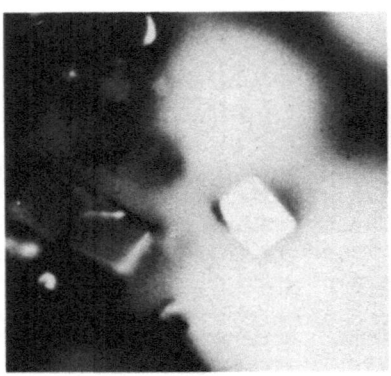 

(a)   (b)

Figure 1: Particles of PbS held for (a) 45 hours and (b) 90 hours
           at 750°C. in (NaCl,KCl) eutectic. ⟵⟶ 10μ.

     The results were plotted as both the cube and the square of
the mean particle radius according to the rate equations for diffusion controlled and interface controlled growth respectively, as
shown in Figure 2. As can be seen, the results can be fit to either of the two relations to produce a linear relationship within
the limits of experimental error.

     The slope of the $\bar{r}^3$ versus time curve, determined by a least
squares analysis gives a value of 1.57×10^{-15} cm^3/sec. This is
very close to the value obtained when values are substituted into
the rate equation for diffusion controlled growth. Values for the
diffusion coefficient and the interfacial energy were estimated at
10^{-5} cm^2/sec. and 150 ergs/cm^2 respectively. The value obtained
for the rate constant using these values is 1.19×10^{-15} cm^3/sec.
which agrees well with the experimental value. This would seem to
indicate excellent agreement with the theory of diffusion controlled growth. However, the linear relationship of the results

TABLE I: Results of particle size measurements on a 1.0 wgt.%
          PbS-(NaCl,KCl) eutectic sample at 750°C.

Sample No.	Time (Hours)	Mean Diameter (A)
1	5	37,380
2	20	101,860
3	45	127,530
4	90	158,200

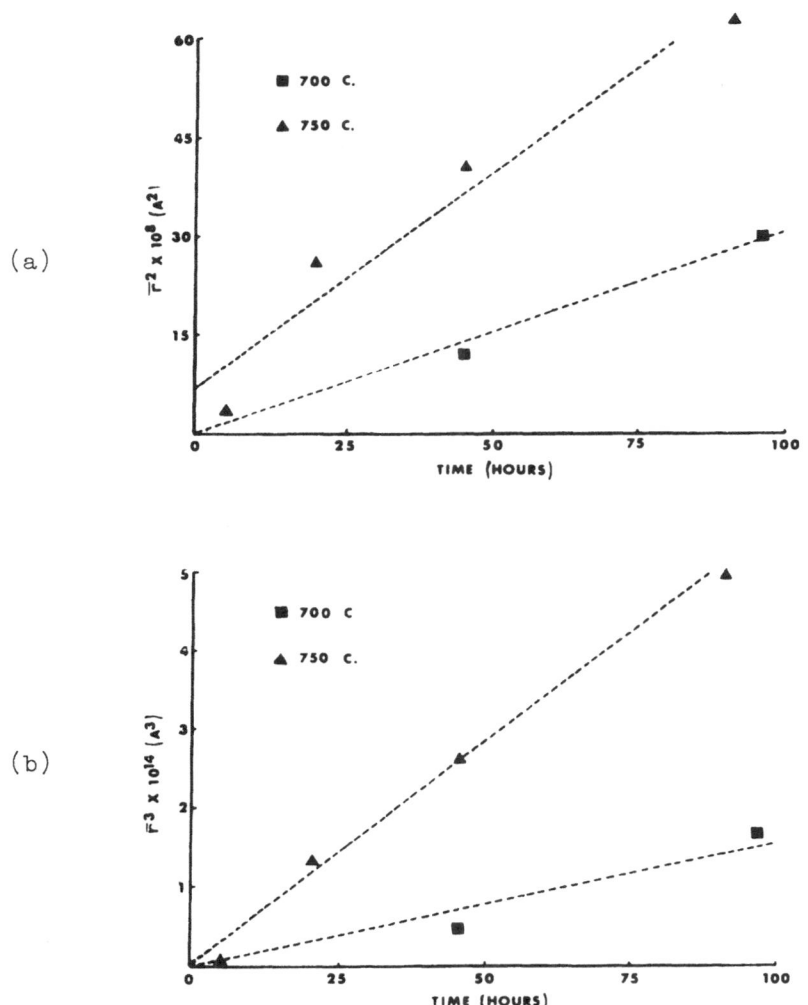

Figure 2:   Mean particle radius (a) squared (b) cubed versus time.

in the $\bar{r}^2$ versus time plot would also indicate good agreement with the theory of first order reaction controlled growth. A theoretical value, however, cannot be obtained due to the unknown value of the transfer coefficient in the rate equation.

A statistical study shows that although the results can be fit to either graph, the results are better fit to the $\bar{r}^2$ versus time curve, even though the growth rate does show excellent agreement to the calculated value for diffusion controlled growth.

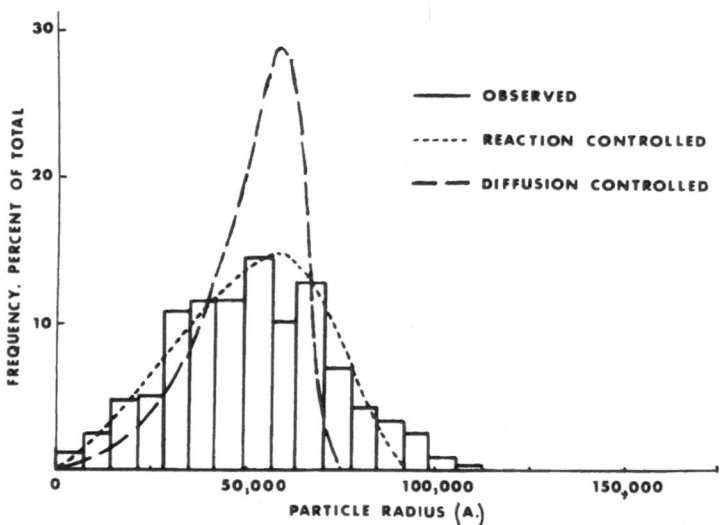

Figure 3:   Observed and predicted particle size distributions for
            PbS-(NaCl,KCl) eutectic sample held 20 hours at 750°C.

        The second method available to determine the reaction mechan-
ism is the particle size distribution. After a short initial
time, the distribution did reach a stationary shape as shown in
Figure 3. Also shown on this figure, are the theoretically predicted distributions for diffusion controlled and first order reaction controlled growth. These distributions were obtained by
substituting the mean particle radius of the experimental results
into the two expressions derived by Wagner for these mechanisms.
It can be readily seen that there is good agreement between the
experimental results and the distribution for first order reaction
controlled growth. A chi-square statistical test was used to determine the goodness of fit between the experimental results and
these two predicted distributions. This method has also been used

TABLE II: Results of the chi-square statistical test for goodness
            of fit between experimental and theoretical particle
            size distributions at 750° C.

Sample No.	Time (Hours)	Reaction Controlled			Diffusion Controlled		
		χ^2	γ	$\alpha,\%$	χ^2	γ	$\alpha,\%$
1	5	95.3	4	0	368	3	0
2	20	7.1	9	65	–	7	0
3	45	12.3	10	28	716	7	0
4	90	8.3	8	40	33	5	0

by Exner et al[9] to distinguish between these two distributions and
to eliminate the effect of statistical errors in particle size
measurement. The calculated values of χ^2, the number of degrees
of freedom, γ, and the subsequent values of α, the percent probability of fit for this test, are given in Table II. The agreement
between the experimental results and the reaction controlled
growth distribution is shown to be quite good as usually a 5% degree of fit is considered significant.

The results were also plotted as the logarithm of the particle
size versus the logarithm of cumulative percent, as shown in Figure
4. Several authors[11,12] have shown that this manner of displaying
results eliminates errors due to statistical variations in measurement. The curves in this figure can be represented by straight
lines in the section between 20 and 80%. If the system is diffusion controlled, the slope of the linear section will be 1/4. If,
on the other hand, the system undergoes reaction controlled growth,
the slope will be 4/9. On this diagram, a line of slope 4/9 was
superimposed on the results and very good agreement was found. It

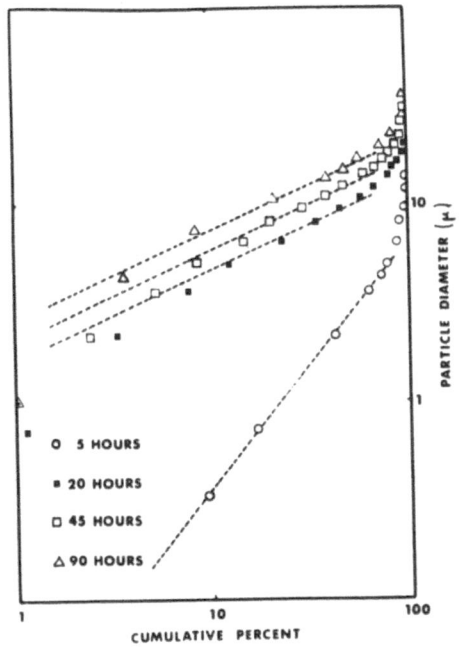

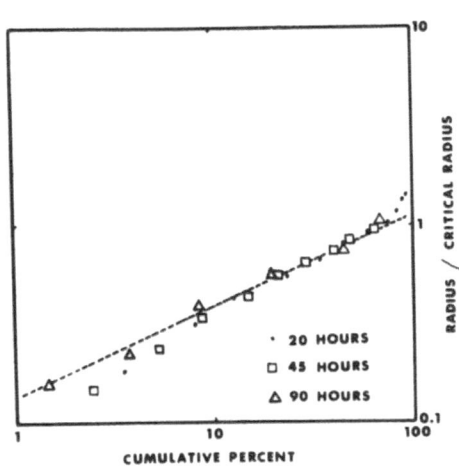

Figure 4: Cumulative particle
size distributions for samples
at 750°C.

Figure 5: Normalized cumulative
particle size distributions for
samples at 750°C.

TABLE III: Results of particle size measurements on a 1.0 wgt.%
             PbS-(NaCl,KCl) eutectic sample at 700°C.

Sample No.	Time (Hours)	Mean Diameter (A)
5	45	71,100
6	95	109,930

can also be seen from this diagram that the distribution does in
fact reach a stationary shape after an initial time period. The
stationary nature of the distributions is readily seen if the axis
on this diagram is normalized as shown in Figure 5. Three sets of
results are all superimposed on the same curve in this diagram.

Two samples were also aged at 700°C for 45 and 95 hours. The
results of the particle size measurements are given in Table III.
Again it was found that the results could be fit to either a cube
or square relation as shown in Figure 2. When the slope of the
cubic relationship is found, it agrees well with that predicted by
the growth equation when the values are substituted in. The same
values of the diffusion coefficient and interfacial energy were
used as before. The experimental value was found to be 4.4×10^{-16}
cm^3/sec. and the calculated value 8.9×10^{-16} cm^3/sec. so that
close agreement was again obtained. However, as found before, a
statistical test verified that the results are better fit by a
square relation indicative of reaction controlled growth.

The particle size distribution again reached a stationary
shape close to that predicted for reaction controlled growth and
this was again verified by a chi-square test. The values obtained
from this test are given in Table IV. It can be seen that the
time required for the particle size distribution to reach the stationary shape is about 45 hours at 700°C while it was about 15
hours at 750°C.

Table IV: Results of the chi-square statistical test for good-
            ness of fit between experimental and theoretical
            particle size distributions at 700°C.

Sample No.	Time (Hours)	Reaction Controlled			Diffusion Controlled		
		χ^2	γ	$\alpha,\%$	χ^2	γ	$\alpha,\%$
5	45	17.3	7	2	–	5	0
6	95	5.1	9	80	103	6	0

TABLE V: Coarsening data required for the determination of the
          activation energy.

	700° C	750° C
$\Delta \bar{r}^3/\Delta t$	4.40×10^{-16} cm^3/sec	1.57×10^{-15} cm^3/sec
$\Delta \bar{r}^2/\Delta t$	4.16×10^{-13} cm^2/sec	9.10×10^{-13} cm^2/sec
C_o	5.31×10^{-5} moles/cc	7.44×10^{-5} moles/cc

Since results are available for two temperatures, the activation energy can be estimated. There are two different methods
used in the literature. The most common method is to calculate
the activation energy from the slope of a plot of the logarithm of
the rate constant versus the inverse absolute temperature[9]. This
method neglects the change in other temperature dependent factors
such as solubility and interfacial energy. The alternate method
is to take the slope of a plot of the logarithm of the rate constant times the temperature and divided by the solubility versus
$1/T$[8]. This method still neglects the change due to the diffusion
coefficient and the interfacial energy so that the calculated
value of the activation energy is actually an apparent value of
the activation energy. The coarsening parameters used in these
calculations are given in Table V. It should be noticed that the
rate constant used in this calculation may be either the slope of
the cube or square relation between mean particle radius and time,
depending on which mechanism is assumed to be rate controlling.
The values obtained from this analysis are given in Table VI. For
all cases, the activation energy is too high for diffusion controlled growth which should be of the order of 2 - 10 Kcal/mole.
These values of the activation energy then appear to support the
case for reaction controlled growth

TABLE VI: Values obtained for the apparent activation energy.

Method	Diffusion Controlled	Reaction Controlled
1	50.7 Kcal/mole	30.9 Kcal/mole
2	30.9 Kcal/mole	19.6 Kcal/mole

SUMMARY AND CONCLUSION

From this analysis of the particle coarsening results from
the lead sulphide - (NaCl,KCl) eutectic system, it appears that

first order reaction controlled growth takes place. However, the results are close to that expected for diffusion controlled growth and therefore, the transfer barrier is not very large. It has been shown that the determination of the reaction mechanism by the exponent of the mean particle radius in the growth equation is of little value. When experimental error is considered, the results can fit either a square or cubic relation and even produce a value of the rate constant equal to the calculated value for diffusion controlled growth. It has been shown that better criteria for the determination of the reaction mechanism are the shape of the particle size distribution after long times and the value of the apparent activation energy.

REFERENCES

1. Greenwood, G.W., Acta Met., 4(1956)243.
2. Lifshitz, I.M. and Slyozov, V.V., J. Phys. Chem. Solids, 19(1961)35.
3. Wagner, C., Z. Electrochemie, 65(1961)581.
4. Greenwood, G.W., Instit. of Metals, London 1969, Monograph 33, p. 103.
5. Ardell, A.J., Instit. of Metals, London 1969, Monograph 33, p. 111.
6. Fischmeister, H. and Grimvall, G., 'Ostwald Ripening - A Survey' in 'Sintering and Related Phenomenon', Material Science Research 6, Proc. Third Inter. Conf. on Sintering and Related Phenomenon, University of Notre Dame, June 5-7, 1972, Plenum Press, New York, 1973.
7. Kahlweit, M., 'Aging of Precipitates' in 'Industrial Crystallization', Proc. of Symposium oy Instit. of Chem. Engineers, London, April 15-16, 1969.
8. Ardell, A.J., Acta Met., 20(1972)61.
9. Exner, H.E., Santa Marta, E. and Petzow, G., 'Grain Growth in Liquid Phase Sintering of Carbides' in 'Modern Developments in Powder Metallurgy', Ed. by H.H. Hausner, vol. 4, Plenum Press, New York, 1971.
10. Mohapatra, O.P., Alcock, C.B. and Jacob, K.T., Met. Trans., 4(1973)1755.
11. Kirchener, H.O.K., Metallography, 4(1971)297.
12. Moreen, H.A., Metallography, 1(1969)349.

THE OXYGEN PROBE AS A CERAMIC SYSTEM

C.B.Alcock
Chairman, Department of Metallurgy
and Materials Science
University of Toronto, Canada

Oxide solid solutions which are based on the elements of Group IV and have the fluorite structure, can be prepared with vacant anion sites when an aliovalent cation oxide such as CaO or Y_2O_3 is mixed with ZrO_2, HfO_2 or ThO_2. In the case of the last-named oxide, which has the fluorite structure in the pure state, it is possible to show that the electrical conductivity of this solid is markedly enhanced and also qualitatively modified when CaO or Y_2O_3 additions are made(1). The pure solid is a semi-conductor, having a p-type region over a wide range of temperatures which is the dominant conduction mechanism from about 10^{-6} to 1 atmos oxygen pressure. Below this pressure range, the oxide shows n-type conductivity which varies as $p^{-1/4}O_2$. In the solid solution range with CaO or Y_2O_3 additions, there is a range of oxygen pressures over which the electrical conductivity is independent of the oxygen pressure, from about 10^{-6} down to 10^{-30} atmos at 1000°C. The n-type component has thus been displaced as the major contribution to the total electrical conductivity in the low pressure region, by the addition of a larger electrolytic conductivity. This is the direct result of the introduction of oxygen ion vacancies. Although the vacancy concentration may readily be extended up to six atomic percent by solid

solution formation, it is found that the electrical
conductivity decreases when the vacancy concentration
passes about 3.5 percent. In order to maximize the
ionic transport number, t_i, which is the fraction of
the electrical conductivity due to ionic migration, it
is necessary to use vacancy concentrations around this
maximum, since the semi-conduction components are
relatively unaffected by aliovalent cation additions to
thoria (Fig.1).

The solid solutions which have been of greatest
commercial significance as solid oxide electrolytes, are
those based on zirconia. These have an electrolytic
conductivity which is an order of magnitude greater than
those of thoria-based electrolytes at a given temperature
and the p-type semiconduction at high oxygen pressures
which is a characteristic of these latter oxides, is
masked by the larger anionic conductivity. The n-type
semiconduction becomes significant at higher oxygen
pressures than in the thoria-based electrolytes, and at
1000°C it becomes important below 10^{-20} atmos oxygen
pressure(2). Pure zirconia does not have the fluorite
structure at temperatures less than about 2400°C, and
only a restricted range of solid solutions containing
anionic vacancy concentrations, the so-called
"stabilized" zirconia, have the electrolytic behaviour.
The phase diagram for the $CaO-ZrO_2$ system due to
Garvie(3) indicates that about 15 mole percent of CaO
is required to produce the fluorite structure at high
temperatures(Fig.2). The corresponding results for the
$MgO-ZrO_2$ system suggest that smaller amounts of MgO are
required to stabilize the electrolytically conducting
structure at high temperatures, but that the fluorite
solid solutions are unstable with respect to decomposition to the component oxides at temperatures below
1400°C (Fig.3)(4-6).

Carter and Roth(7) in a study of the electrical
properties of the $CaO-ZrO_2$ system showed that the
electrolytic conductivity "aged" in a given sample,
with time, and they attributed this to an interaction
between Ca^{2+} ions and the oxygen vacancies. When the

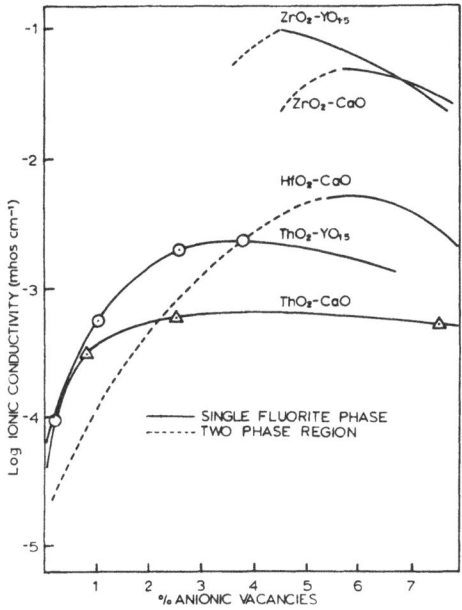

Fig. 1 - Variation of the conductivity of fluorite
structures with vacancy concentration at
1000°C (Ref.1).

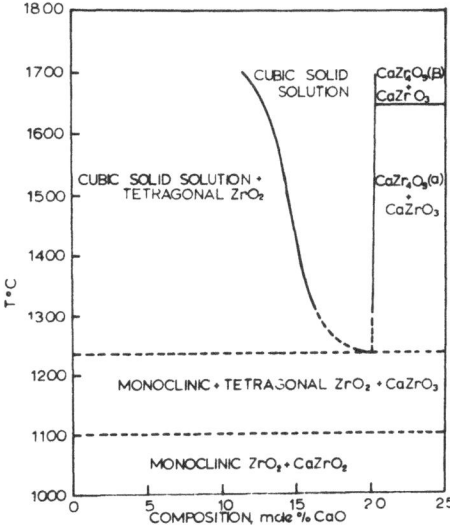

Fig. 2 - Phase relationships in the ZrO₂-CaO system
(Ref.3).

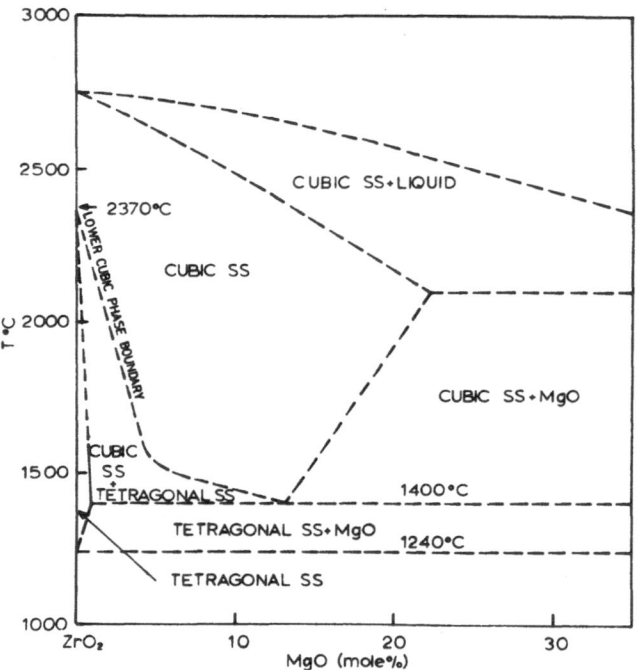

Fig. 3 - Phase relationships in the ZrO₂-MgO system.

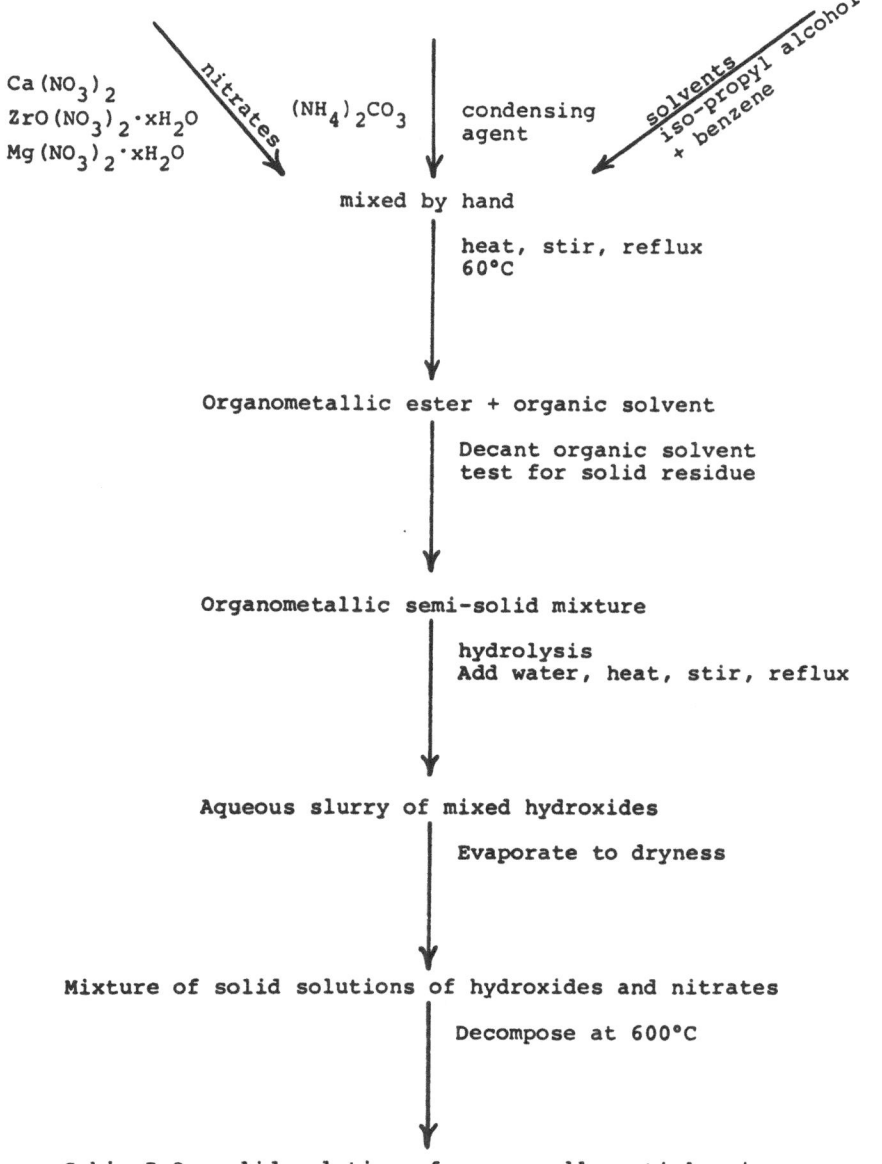

Fig. 4. Schematic diagram of the preparation of cubic zirconia solid solution via an organometallic compound.

solid solution is originally formed, the cations are randomly mixed, but there is a tendency for ordering of the cations with time at high temperatures which leads to a reduction of the conductivity.

Such effects could also be related to the method of preparation of the samples. Because of the very low diffusion coefficients of cations in these solid solutions(8), the preparation of solid solutions by mixing the individual oxides is increasingly avoided in favour of the mixing of solutions containing the appropriate amounts of the two cationic species, followed by the formation and decomposition of organometallic compounds(9). Such preparative routes also provide starting materials which are fine crystals (<1 μm) of a metastable cubic structure at temperatures below about 500°C and lend themselves very well to a variety of sintering procedures(Fig.4).

Commercially manufactured tubes and crucibles of magnesia or lime-stabilized zirconia are now available in relatively impervious form. It has been shown, however, that these materials become increasingly permeable to oxygen at temperatures above about 1200°C(10)(11), and thus transport of oxygen through such materials occurs not only in the ionic form but also as ions accompanied by positive holes. They can therefore only be used in an electrochemical measuring system where a constant oxygen potential can be maintained by replenishment at the electrolyte interface. When this is possible, a simple concentration cell may be constructed which contains a reference electrode of known oxygen pressure, the electrolyte and the electrode of unknown oxygen pressure such as a solution of oxygen in a liquid metal. An inert electronic connection is required from each electrode to a measuring device which records the E.M.F. of the cell. This is given algebraically by the equation

$$E = \frac{RT}{4F} \ln \frac{p_{O_2}}{p'_{O_2}}$$

R the gas constant
T the absolute temperature
F Faraday's constant

and p_{O_2} is the oxygen pressure exerted by the reference electrode at the temperature T whilst p'_{O_2} is that exerted by the electrode of unknown oxygen dissociation pressure.

The construction and utilization of such a cell may be made in a number of ways with quite satisfactory materials of a wide variety providing the temperature is below about 1200°C and the apparatus may be brought to temperature over a period of hours. Fig.5 shows a typical system in which a stabilized zirconia tube is used as the electrolyte, Ni/NiO mixture is the reference electrode and a copper-zinc alloy containing a dilute solution of oxygen is the "unknown" electrode. Electronic contact with the Ni/NiO electrode can be made with platinum wire and a cermet such as $Mo-Al_2O_3$ has been successfully employed as a contact to copper melts at 1300°C.

Such a system for measurement presents minimal problems to the ceramist who must merely provide a relatively pore-free tube of stabilized zirconia which will be treated with great care. The tube is usually fairly coarse grained and shows the presence of a phase segregated during sintering which contains significant amounts of silica (Fig.6). However, the regions of this second phase are relatively well separated by grains of electrolyte, and thus should not contribute to significant electronic short-circuiting in the electrochemical system. The term "system" is here defined in an engineering sense rather than through the normal scientific definition. According to the latter, a system is a part of the universe which has been isolated for the study of a particular property under carefully chosen conditions. In the engineering sense, the author believes that a more appropriate definition of a system is "an assembly of components, which are intended to produce a programmed sequence of events, leading to as near an approximation to a desired result as circumstances will allow".

If the desired result is the measurement of the

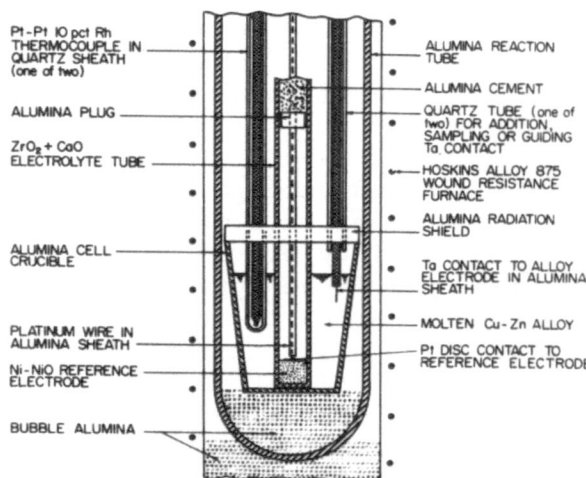

Fig.5 - Diagram of a galvanic cell measuring the oxygen
       solubility of Cu-Zn alloy (Ref. 18).

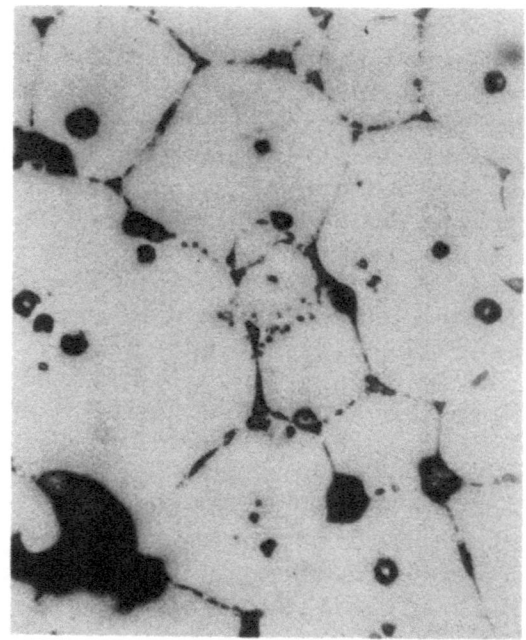

Fig.6 - Micrograph of a section of a commercial ZrO_2-CaO
       tube (Mag. X450).

oxygen content of liquid steel at 1600°C during the steelmaking process on the industrial scale, then the components of the electrochemical system must be entirely different from those shown earlier for laboratory research.

A stabilized zirconia tube cannot be immersed rapidly into liquid steel unless its composition is adjusted to accommodate to the extreme thermal shock, and even then only a very limited geometry can be used. The coefficient of thermal expansion of the pure $CaO-ZrO_2$ solid solution is quite large and increases monotonically with increasing temperature(12). The coefficient of expansion of pure ZrO_2 varies markedly with temperature also, but decreases over range of temperatures immediately above the monoclinic-tetragonal transformation temperature(13). It has therefore been proposed(14) that a mixture of the cubic lime-stabilized phase with pure ZrO_2, so-called "partially stabilized zirconia", would withstand thermal shock better than the pure cubic system because of this mixed behaviour with respect to thermal expansion. Furthermore, it seems possible that a mixture of phases would be less favourable to crack propagation than a single continuous phase, because micro-cracks are formed around the cubic grains which originate in the segregated monoclinic phases.

Although such a proposal would offer hope of overcoming the thermal shock problem, it would introduce an undesirable element from the electrochemical point of view. This is because pure zirconia is a semiconductor at all oxygen pressures and temperatures, and therefore connecting volumes of this phase could act as a short circuit to the electrolyte and bring about a reduction in the E.M.F. and hence an avoidable error in the measurement..

An alternative possibility is to use a material with a greater amount of porosity than was used in the construction of tubes, but this leads to a decrease in mechanical strength.

Both of these possible solutions can be applied with varying degrees of success if the electrolyte is used in the form of a small sintered pellet which is flame-sealed into a Vycor tube (Fig.7). This material softens at steelmaking temperatures and so allows for the expansion of the electrolyte pellet. The tube also provides some mechanical support if a porous material is used as the electrolyte. The major disadvantage of the flame-sealed electrolyte pellet-Vycor tube combination is that the sealing process is difficult to control accurately and, as a result, the wetting of the electrolyte by the flowing Vycor varies around the periphery of the electrolyte. This differential wetting then sets up a peripheral stress pattern after sealing, which is not uniform and probably makes the assembly thermal-shock prone. Also, the interaction between CaO and SiO_2 being much stronger than that between CaO and ZrO_2, there is a tendency for the glass to "leach out" calcium oxide near the seal. This leads to destabilization of the zirconia, and the formation of a peripheral ring of semi-conducting material around the electrolyte. One further disadvantage of the flame-sealed system is that if the probe is to be used at lower temperatures than steelmaking temperatures, the Vycor will not soften and the probe frequently shatters before the measurement is completed.

A number of studies of practical systems for the measurement of oxygen in liquid steel have been carried out by the Electrochemical Sensors Project Group at the University of Toronto. The objective has been to find the most suitable combination of components for an oxygen probe which could be produced on the large scale for industrial application. The study is being made with support from the National Research Council of Canada and in collaboration with Leigh Instruments Ltd. Assembled oxygen probes are made for immersion in 200 lb. melts of liquid iron, and metallographic specimens are made of typical assemblies for optical microscopy and electron microprobe analysis.

The electrolytes which are used are synthesized by

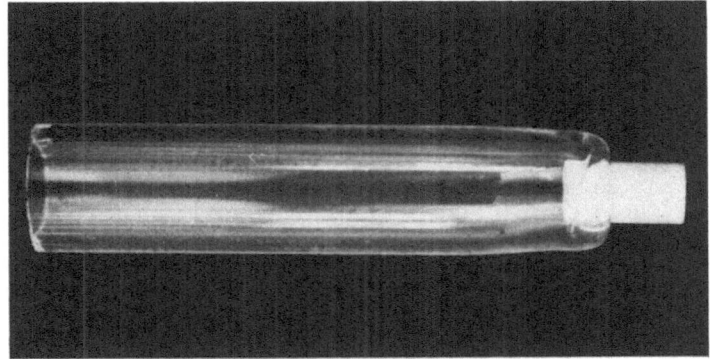

Fig. 7- Oxygen-probe tip; probe manufactured by
        Leigh Instruments Ltd., Canada.

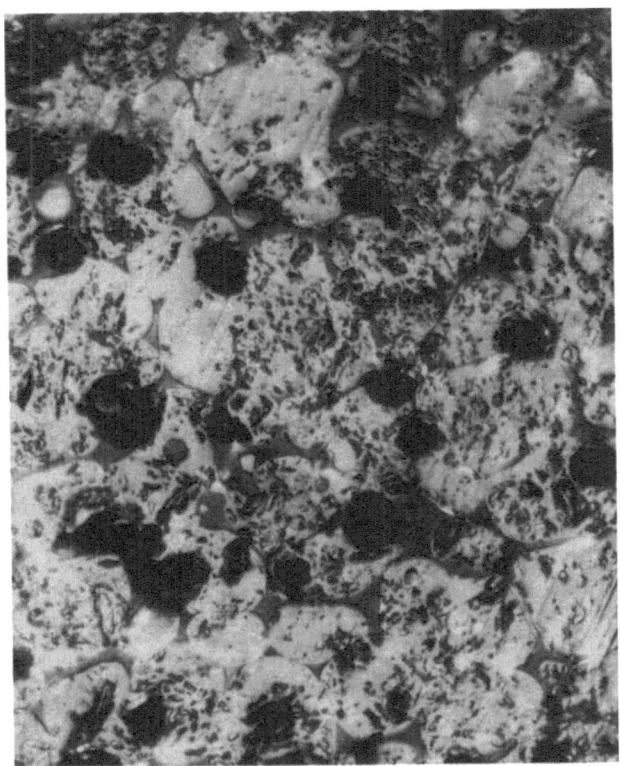

Fig.8 - Micrograph of a section of a pellet made of
ZrO_2-MgO-CaO, after flame-sealing into Vycor (Mag.X600).

the organometallic route outlined above or are made from
commercially available powders. It is found, as might
be expected, that materials of high purity and high
density (95%) are extremely susceptible to thermal
shock and should be avoided. Electrolytes made from
MgO-stabilized material undergo excessive decomposition
during the flame-sealing process (Fig.8) and therefore
yield low EMF's despite their ability to withstand
thermal shock relatively well. This thermal shock
resistance is only useful for small amounts of
segregated phase, probably less than 20%(15).

Mixed electrolytes, containing a controlled
concentration of CaO and MgO, can be readily synthesized,
and by the appropriate heat-treating cycle after sintering, can be produced with a fairly well-defined amount of
zirconia in separated islands, both intra- and inter-
granular. The resistance to thermal shock is better
when the MgO content is increased up to the limits
indicated above, and when an appropriate amount of the
monoclinic phase has been allowed to segregate.
The commercial powder sinters to yield a relatively
porous material (ca 80% theoretical density) and some
CaO is leached from the edges during flame-sealing of
the Vycor. The EMF results are, however, not substantially lower than those obtained with the best laboratory-
made materials, and demonstrate the usefulness of a
porous system vis-a-vis the two phase system. A micrograph of this material in the as-used condition is
shown in Fig.9 and typical EMF traces are shown in
Fig.10.

Since the flame-sealing process appears to prej-
udice the system towards failure by thermal shock, it
appears worthwhile to investigate other methods of
joining the electrolyte pellet to the Vycor tube. This
may be successfully accomplished with a number of
commercial alumina- or silica-based cements, and the
tendency to thermal shock cracking is considerably
reduced, as might be expected. In tests which have
been carried out so far, it has been found that cemented
probes tend to take a longer time to reach thermal

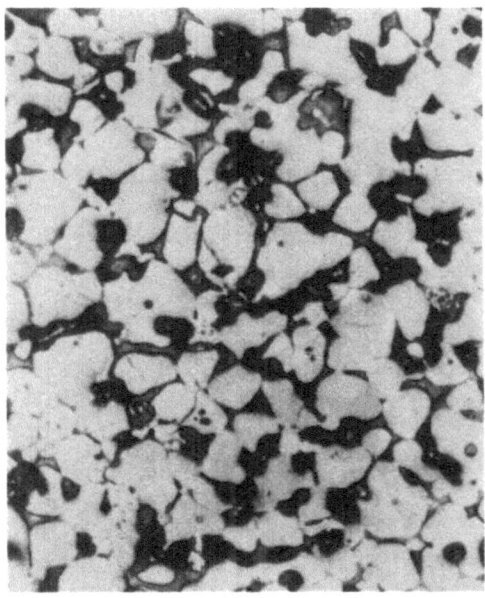

Fig.9 - Micrograph of a section of partially stabilized zirconia, used in the Leigh-probe.

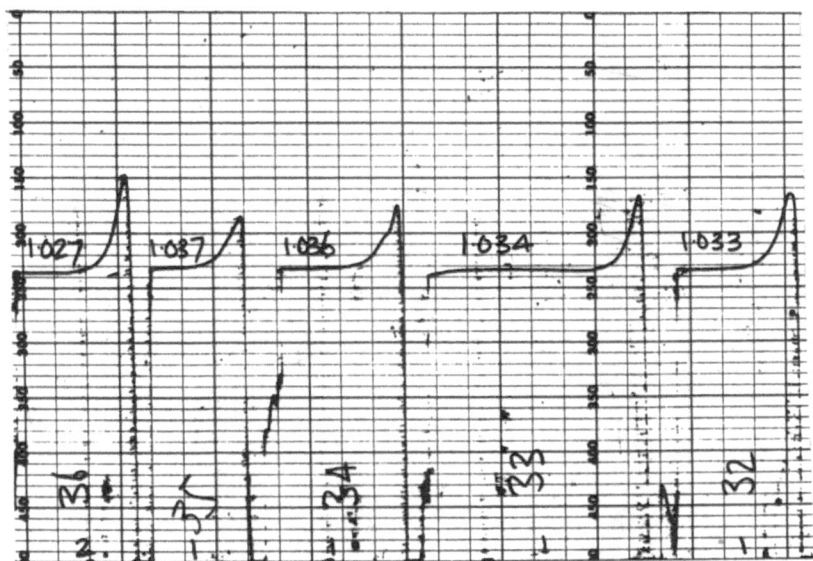

Fig. 10 - Emf traces measured in steel, obtained by the Leigh oxygen probe.

equilibrium, and this is probably due to poorer thermal
conduction to the electrolyte. However, recrystallized
glass systems can also be used which do not suffer from
this lack of thermal transparency.

With respect to reference electrodes, three have
been used successfully in steelmaking oxygen probes.
These are oxygen, which is produced by the high temperature decomposition of an oxide, molybdenum mixed
with MoO_2, and chromium mixed with Cr_2O_3. Although a
corrosion reaction does occur between chromium and
zirconia-based electrolytes during long term exposure
at temperatures above 1400°C(16), in the short time
allowed for oxygen probe measurement, approximately
15 sec., this appears to be insignificant.

One of the major problems which occurs in high
temperature oxygen probe systems is that of finding a
suitable contact material for the completion of the
electrical circuit from the liquid metal to the EMF
measuring device. In the steelmaking oxygen probe, this
has been a steel tube which melts away during the measurement. Longer periods of exposure have been obtained
with lower melting liquid metals through the use of
molybdenum/alumina cermet rods(17), but these have only
a limited time of survival at steelmaking temperatures.

Conclusion

It can be seen from these considerations that the
many factors which contribute to a successful ceramic
system must be considered as a whole in setting the
objectives for research. The major thrust of research
in ceramic systems in connection with sintering seems
to have been towards the realization of theoretical
densities and to the production of controlled grain size
material with a minimum pore content. The work which
has produced a successful steelmaking probe as a ceramic
system indicates that it would be valuable to consider
two-phase systems, since there is another degree of
freedom in the reaction to thermal shock conditions,
and possibly even the introduction of controlled amounts
of pore volume, preferably in the well-distributed

range of sizes.   Once a successful electrolyte material has been developed, it has been the experience in this project that a major consideration in the construction of ceramic electrolyte systems has been the means which are available for joining the components together. Traditional methods which involve fusion can lead to more problems than they can solve, and the need for fundamental studies on ceramic joining methods which lead to impermeable joints would repay further effort.

I wish to acknowledge finally the many contributions of my colleagues at the University of Toronto working in the Electrochemical Sensor Project. These are Doctors Zador, Etsell, Jacob and Chan, who have contributed to the many facets of this research.

Finally, I must express our gratitude as a Group to the National Research Council who have made this study possible through a very imaginative grant system relating fundamental research to its industrial application.

References

1. B.C.H.Steel and C.B.Alcock, Trans.Met.Soc.AIME, 233, 1359 (1965).
2. M.F.Lasker and Robert A.Rapp, Z.Phys.Chem.N.F.49, 198 (1966).
3. R.C.Garvie, J.Amer.Ceram.Soc.51, 553 (1968).
4. D.Viechnicki and V.S.Stubican, J.Amer.Ceram.Soc.48, 292 (1965).
5. T.H.Etsell and S.Zador, publication in preparation.
6. C.F.Grain, J.Amer.Ceram.Soc.50, 288 (1967).
7. R.E.Carter and W.L.Roth, "Emf Measurements in High Temperature Systems" IMM publication, London, 1968, pp.125-144.
8. W.H.Rhodes and R.E.Carter, J.Amer.Ceram.Soc.49 (5), 244 (1966).
9. K.S.Mazdiyasni, C.T.Lynch and J.S.Smith, J.Amer. Ceram.Soc.49 (5), 286 (1966).
10. C.B.Alcock and J.C.Chan, Cndn.Met.Quarterly 11 (4), 559 (1972).

11. A.W.Smith, F.W.Meszaros and C.D.Amata, J.Amer.
    Ceram.Soc.49, 240 (1966).
12. R.C.Garvie, "High Temperature Oxides", Part II,
    Refractory Materials, Vol.5-II, Academic Press
    1970, pp.117-166.
13. R.E.Jaeger and R.E.Nickell, Ceramics in Severe
    Environments, Materials Science Research, Vol.5,
    Plenum Press, New York, 1971, pp.163-184.
14. R.C.Garvie and P.S.Nicholson, J.Amer.Ceram.Soc.55
    (3) 152 (1972).
15. F.J.Kennard, R.C.Brandt and V.S.Stubican, J.Amer.
    Ceramic Soc.57 (10), 428 (1974).
16. R.J.Fruehan, L.J.Martonik and E.T.Turkdogan,
    Trans.Met.AIME 239, 1226 (1967).
17. C.M.Diaz and F.D.Richardson, Emf Measurements in
    High Temperature Systems, IMM publication,
    London, 1968, pp.29-42.
18. R.A.Rapp and D.A.Shores "Physicochemical Measure-
    ments in Metals Research", Vol.IV, Part 2,
    Interscience Publishers, New York, 1970,pp.123-192.

THE SINTERING OF STRONTIUM CONTAINING PEROVSKITES

T.J. Gray, R.J. Routil and M. Rockwell

Atlantic Industrial Research Institute

Nova Scotia Technical College, Halifax, Nova Scotia, Canada

ABSTRACT

Several strontium containing perovskites are of considerable interest as potential electrode materials for MHD, fuel cell and electrochemical applications. Their conductivity is critically dependent on sintering conditions and variations of several orders of magnitude are frequently observed. To a considerable extent these characteristics relate to secondary phase separation at grain boundaries or extreme disorder. Micrographic and electron microbeam probe data are presented in correlation with electrical conductivity and other characteristics for representative samples of particular applicational interest.

INTRODUCTION

With ever increasing research on materials potentially suitable for MHD electrode applications, considerable interest has focussed on conductive perovskite systems. Jonker[1], Meadowcroft[2] and others have reported extensively on lanthanum containing perovskites incorporating transition metal ions from the first or second series and have shown that lanthanum can readily tolerate substitution by strontium.

In 1967/8 one of the world's largest deposits of strontium mineral, celestite, was identified in the Loch Lomond region of Cape Breton, Nova Scotia. The deposit was explored and proven by Kaiser Celestite Mining of Cape Breton who mine and upgrade the mineral, tranship by road to Point Edward, Cape

Breton where it is converted to very high purity (99%+) strontium carbonate and strontium hydroxide, an intermediate product. The National Research Council, Ottawa initiated a major research program on strontium containing compounds in 1971 and one area under critical evaluation was strontium containing perovskites. Preliminary data have already been published (The Proceedings of the International Conference on Strontium Containing Compounds).

When samples of strontium lanthanum chromite and cobaltite were prepared in these laboratories of compositions described in the literature, wide divergences were immediately apparent. Meadowcroft[2] later indicated that very small deficiencies of the transition metal ion increased the electrical resistivity and also catalytic activity abruptly. It was considered essential to study the detailed qualitative sintering characteristics. While sintering per se is an engrossing study, the added complication of sintering a highly complex structure with the possible segregation of other phases restricts the study to one of a qualitative nature.

It is essential to review very briefly the nature of "perovskites"

PEROVSKITES

The term "perovskite" is applied to a broad class of compounds, generally but only approximately defined as ABO_3, typical of which is $CaTiO_3$. Classically they are cubic single cells, but this is rarely the case, with angle and length distortions and with anisotropic strains giving pseudo-cubic, distorted cubic, with or without a superlattice. Multiplication of the unit cell is quite general. In the classic form the larger A cations, comparable in size to the $O^=$ ion, are situated at the corners of a cubic unit cell, with the smaller B cations at the centre and the oxygen at the centre of the faces. Each A ion is in a position of 12-fold co-ordination and each B ion in 6-fold co-ordination. Each oxygen is associated with 4 A cations and 2 B cations, with the larger cation occupying the A site. The valencies of the ions are of only secondary importance and any pairs of ions or combinations having appropriate radii and an aggregate valency of 6 can appear. In summation, the perovskite structure can be regarded as a rigid framework of large oxygen anions interspersed with smaller cations in the various interstices. In such a structure the actual distribution of the cations is primarily statistical and electrical neutrality is preserved by lattice vacancies in the anion and cation lattices. The lattice tolerates two or more chemically different ions occupying crystallographically equivalent sites. While attempts have been made by O'Keefe, Hyde et al[3] to associate these and other compounds with complex multi-layer structures, it is in many respects unrealistic to advocate such an approach in view of the enormous number of variations of position and structure actually encountered.

The present research program, directed towards the development of electrically conducting perovskites suitable for MHD, electrochemical and superconducting electrode applications, has been concerned primarily with strontium modified rare-earth containing perovskites incorporating transition metal ions from the first or second series. Jonker[1], Meadowcroft[2] and others have reported extensively on lanthanum, yttrium and rhenium cobaltites. In these compounds the lanthanum can readily tolerate substitution by strontium, in general with a significant improvement in electrical conduction properties. According to Meadowcroft, changing the rare-earth component has far less effect than changing the transition metal ion. A very large number of strontium containing perovskites has been investigated qualitatively within the present program, many of which are potentially useful in electrode applications. Resistivities of the order of milliohms/cm^2 at room temperature can readily be achieved with satisfactory high temperature stability. It is noteworthy that in general, when a second phase is observed crystallographically in the preparation and fabrication of the electrodes, the resistivity rises, often by many orders of magnitude, with a corresponding decrease in stability. Furthermore, marked improvements can be achieved by leaching at an intermediate stage to remove superficial transition metal oxide which otherwise detracts from stability.

EXPERIMENTAL

Whereas it had initially been intended to study lanthanum strontium chromite, preliminary investigation suggested that more positive results would derive from a study of the copper-modified compound since resistivity changes and new phase formation during sintering would be easier to observe.

The methods used for the preparation of strontium containing perovskites can be roughly divided into two distinct groups on the basis of the means of homogenization: in the first group are the methods employing mechanical procedures which include the classical ceramic techniques of dry mixing and dry and wet ball-milling, while the second group is characterized by chemical processes which yield perfectly homogeneous final products or their calcinable precursors from solutions of suitable salts of the components; among these methods the best known are precipitation, coprecipitation, spray-drying and freeze-drying. Various combinations of both types of method (mechanical and chemical) are often found useful for the preparation of more complex perovskites. The selection of the technique most suitable for a particular compound depends mainly on considerations of chemical composition and the requirements on chemical purity, stoichiometry, homogeneity, degree of structure development, particle size and production costs.

$SrCO_3$, $Sr(NO_3)_2$ and $Sr(COO)_2.H_2O$ are the most common sources of strontium. A novel method developed during the current research program employs $Sr(OH)_2.8H_2O$ which, in mixtures with various metal oxides, reacts readily at relatively low temperatures.

After calcination of the starting mixtures the samples are refired in several steps at gradually higher temperatures and the development of the perovskite structure is investigated by X-ray diffraction analysis after each step. Fully reacted samples are pulverized and pressed into discs or rods, either by end-to-end or isostatic pressing, at pressures ranging from 10,000 to 20,000 psi. If necessary, the shape and size of the pressed samples are adjusted by machining after prefiring to suitable hardness. The final sintering is scheduled according to the results of preliminary tests with samples of the same composition.

The $(La_{0.71}Sr_{0.29})(Cu_{0.71}Cr_{0.71})O_{\sim 3}$ sample was prepared by ball-milling a mixture of La_2O_3, $SrCO_3$, $CuCO_3$ and Cr_2O_3 in ethanol. This mixture was then dried, calcined at $950^\circ C$ for 2 hours and screened through 400 mesh. Discs (25 mm diameter) were pressed at 10,000 psi and after pre-firing at $950^\circ C$ for 2 hours were sintered at various temperatures between $1000^\circ C$ and $1200^\circ C$.

The sintered specimens were photographed before any polishing or testing was undertaken, as shown in Fig. 1.

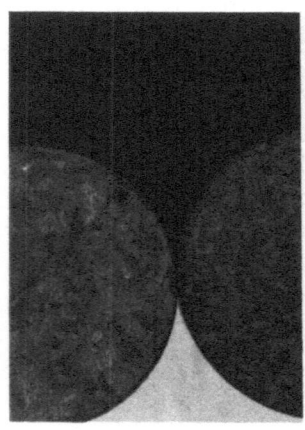  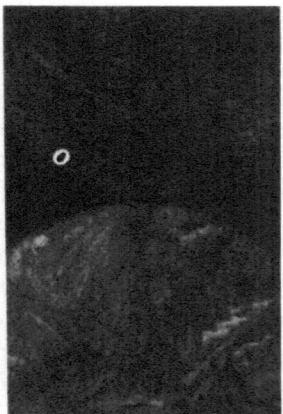

1025° 1050° 1100° 1125° 1200°
        1000° 1075° 1150°

Fig. 1

Samples were next carefully ground plane parallel, carefully polished with diamond grit to 100μ and low-power micrographs taken, which are shown in Fig. 2 and which clearly indicate the dramatic change between 1000° and 1025°C.

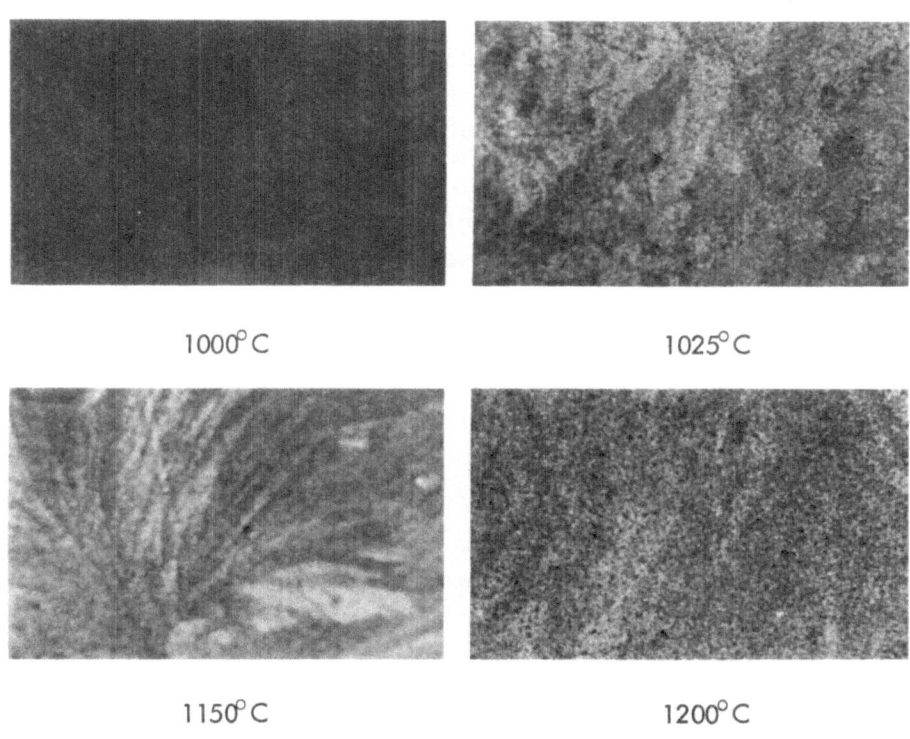

1000°C   1025°C

1150°C   1200°C

Fig. 2

Resistivity measurements were then made between pressure loaded soft platinum electrodes, using a Sullivan squareware Kelvin milliohm bridge in a four probe mode. Data obtained are as follows:-

Temperature of Sintering	1000	1025	1050	1075	1100	1125	1150	1200
Resistivity Ωcm^{-1}	19.7	127.8	20.3	12.2	9.3	7.9	5.9	3.0

Electron microbeam probe studies were then conducted to determine whether there was any extensive segregation. These are illustrated for three temperatures (1050° - 1100° - 1200°) in Fig. 3.

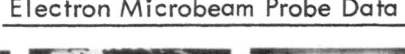

Electron Microbeam Probe Data

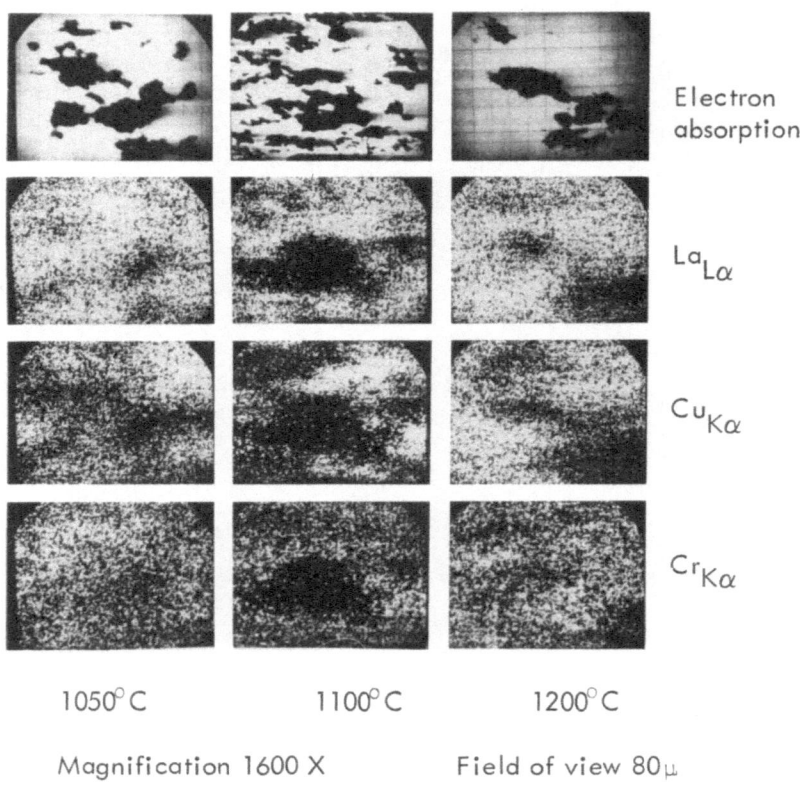

Electron
absorption

La$_{L\alpha}$

Cu$_{K\alpha}$

Cr$_{K\alpha}$

1050°C   1100°C   1200°C

Magnification 1600 X   Field of view 80μ

Fig. 3

DISCUSSION

These preliminary data illustrate the complexity of the sintering process in a perovskite system. The microbeam probe in particular shows radical disturbance in the intermediate temperature range but no significant segregation at maximum temperature. In studies on $La_{0.81}Sr_{0.19}Co_xO_3$ Meadowcroft[2] had observed abrupt increases in resistivity and corresponding increases in catalytic activity which he related to a second phase. In these investigations with similar compositions the "active" phase was positively identified as having the K_2NiF_4 type structure which is related to perovskite[4] while isostructural compounds Sr_2TiO_4, $SrLaAlO_4$ and $CaMnO_4$ have been described by Ruddlesden and Popper[5].

Existence of additional phases in the present system has not yet been fully investigated. In common with the lanthanum strontium chromite series any excess, even very slight, of chromium apparently separates at grain boundaries but can be eliminated by dilute acid electroleach after presintering.

It can be speculated that closer attention to the sintering characteristics of these important materials can aid materially in those improvements required to bring them into widespread use as high temperature electrodes. The authors wish to acknowledge the assistance of Dr. K.M. Castelliz in X-ray characterizations and of the National Research Council, Ottawa for their support and encouragement of this continuing program on strontium compounds.

REFERENCES

1. JONKER G.H. and J.H. van Santen, Physica $\underline{16}$, 337 (1950).

2. MEADOWCROFT D.B., International Conference on Strontium Containing Compounds, 119 (1973).
    Meadowcroft, D.B., Private Communication.

3. HYDE B.G. et al., Ann. Rev. Material Science, 1974.

4. RÜDORFF W., J. Kändler and D. Babel, J. Anorg. Allgem. Chemie $\underline{317}$, 261 (1962).

5. RUDDLESDEN S.N. and P. Popper, Acta Cryst. $\underline{10}$, 538 (1957).

INFLUENCE OF MgO ON THE EVOLUTION OF THE MICROSTRUCTURE OF ALUMINA

J.G.J. Peelen

Philips Research Laboratories

Eindhoven, The Netherlands

INTRODUCTION

Small amounts of additives can have a great influence on the sintering of ceramic powders. The most extensively studied example is, without doubt, the effect of dopants especially MgO, on the sintering behaviour of Al_2O_3. In spite of all these studies many questions are still unsolved since Coble[1] reported that the addition of 0.25 wt % MgO inhibits discontinuous grain growth and allows nearly theoretical densities to be achieved.

Often the role of MgO has been described as that of a grain growth inhibitor, although it does not inhibit normal grain growth but it only inhibits the excessive grain growth. The result is that pores are not trapped inside the grains but can disappear by diffusion of vacancies.

The action of second-phase particles of spinel located preferentially at grain boundaries in pinning these boundaries and decreasing the mobility has generally been realised. Jorgensen and Westbrook[2] have put forward reasons for supposing that MgO dissolves preferentially in the grain boundary region (solute segregation). Haroun and Budworth[3] conclude that MgO is present both as a grain-boundary film and as second-phase particles. However, an additive in solid solution can also suppress discontinuous grain growth by reduction of grain-boundary mobility as has been pointed out by Bruch[4] and by Rossi and Burke[5] and as has been demonstrated for ThO_2 doped Y_2O_3 by Jorgensen and Anderson[6].

443

The mechanism by which MgO can reduce grain-boundary mobility is not well understood. It will be related to the mechanism by which MgO influences the sintering behaviour of alumina. It is an experimental fact that both divalent and quadrivalent cations enhance the initial densification rate of alumina. This has been explained by assuming a Frenkel type defect structure for alumina[7,8]. Incorporation of MgO in the Al_2O_3 lattice would give rise to cation interstitials, while quadrivalent cations for instance, Ti would increase the concentration of cation vacancies. If oxygen diffuses rapidly enough along the grain boundaries[9], an increase in either cation interstitials or cation vacancies can probably increase the sintering rate.

The goal of the present work was to study the evolution of the microstructure of alumina with increasing amounts of MgO, starting far below the solubility limit. Attention has been paid to the density and grain size of the resulting structure. As the porosity, pore size and pore size distribution in sintered alumina determine the in-line transmission of light[10], this quantity has been measured as well. Furthermore, Auger electron spectroscopy has been used to see whether different concentrations of MgO could be found in the boundary and bulk regions. Marcus and Fine[11] found a strong Ca enrichment at the grain boundaries but they could not detect any MgO at all. Analyses of their published data yields a detection limit for Mg which is rather high, so it seemed worth while repeating their experiments. During this work Taylor et. al.[12] reported that they found Mg in enhanced concentration at the grain boundaries using the technique of X-ray photoelectron spectroscopy, which has a somewhat greater sensitivity.

EXPERIMENTAL

All experiments were carried out with a deagglomerated alumina (obtained from Rubis Synthétique des Alpes, A15RZ) with a particle size of $\approx$ 0.3 μm. Impurity content has been determined by spectrochemical and atomic absorption analyses. The results can be summarized as follows:
Fe 5, Ga 4, Mg 0.8, Si 30, Na 10, K 30, Ca $\leqslant$ 10 ppm.

The MgO was added as a solution of reagent-grade Mg-acetate. $4H_2O$ in absolute alcohol to an emulsion of the Al_2O_3 in the same alcohol. After drying and sieving the powder was prepressed into disks 20 mm in diameter and 10 mm thick and isostatically pressed at 100 MN/m^2. The disks were heated in oxygen at 700°C to decompose the acetate to oxide. The sintering took place in an high-purity Al_2O_3 tube heated in a Mo resistance furnace.

Great care was taken to measure the apparent densities of the specimens because of the small difference between the theoretical density of $Al_2O_3^{as}$ determined by the X-ray diffraction method as 3.986 g/cm^3. For the determination of the density the method of Prokic[13] was applied, which takes into account the counterbalancing force of the air. Distilled and boiled out water was used as immersion liquid. Approximately an accuracy in the apparent density of 0.25‰ can be achieved.

After polishing of the specimens the microstructure was revealed by thermal etching at 1450°C for 1.5 h. The mean grain size was calculated using the relation[14]: G = 1.5 l where $\bar{l}$ is the mean grain intercept of random lines on the plane sections. In order to measure the in-line transmission slices with a thickness of 0.5 mm were prepared and polished on both sides. The measurements were carried out as described previously[8].

RESULTS

Influence of MgO on the Density

Sintering experiments have been carried out on Al_2O_3 powders with increasing amounts of MgO dope, starting with the undoped Al_2O_3 powder for comparison. The density of the specimens after sintering at 1630°C for 1.5 h in a humid

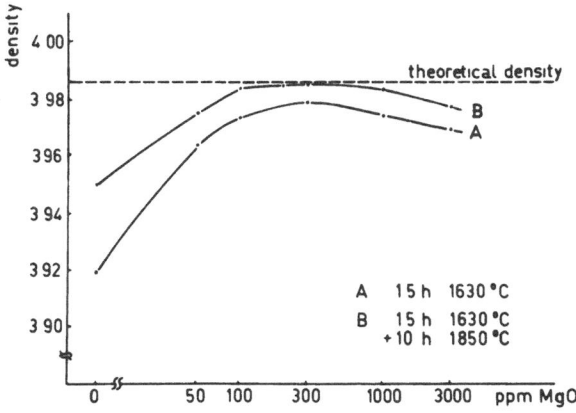

Fig. 1 Influence of MgO on the density of sintered Al_2O_3

H_2-atmosphere (dewpoint 20°C) to prevent volatilization of MgO, is given in Fig. 1, curve A. It appears that doping with 50 ppm MgO is already sufficient to increase the density of Al_2O_3 substantially. The density reaches a maximum at 300 ppm dope level, for larger dope concentrations a decrease in density is found. Second phase particles can be detected in the specimens with 300 ppm MgO and more, in agreement with the data of Roy and Coble[15]. For the solubility of MgO in Al_2O_3 in vacuum they found:

$$\ln X = 8.1 - 30,706/T$$

where X = atomic fraction Mg/Al and T = absolute temperature. This means the solubility of 250 ppm MgO in Al_2O_3 at 1630°C in vacuum and maybe slightly higher in H_2 with a dewpoint of 20°C. The highest density is thus obtained when the amount of MgO corresponds to the solubility limit in Al_2O_3.

All specimens have undergone a second heat treatment at 1850°C for 10 h. This results in an increase in density for all samples, see fig. 1, curve B. There is no distinct peak now because all samples with a MgO content between 100 and 1000 ppm reach nearly theoretical density. Higher MgO contents result in a decrease of the density. According to Roy and Coble[15] the solubility limit at 1850°C is 1350 ppm MgO.

Influence of MgO on Grain Size

The microstructure of the specimens has been evaluated and the mean grain size has been determined for the various

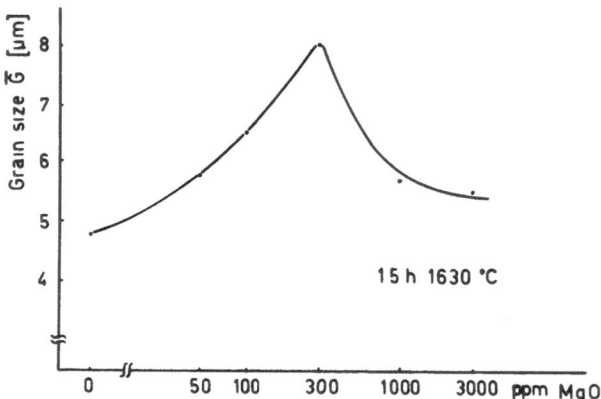

Fig. 2 Influence of MgO on the mean grain size of Al_2O_3 sintered at 1630°C for 1.5 h.

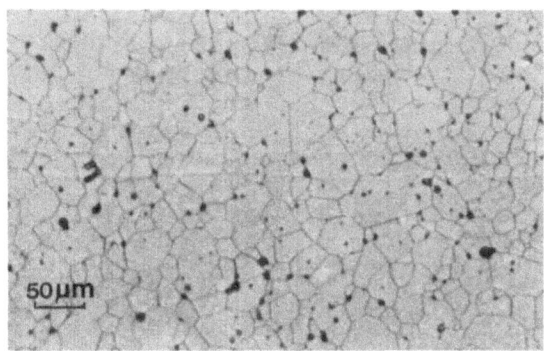

Fig. 3 Undoped Al_2O_3 sintered at $1630^{o}C$ for 1.5 h

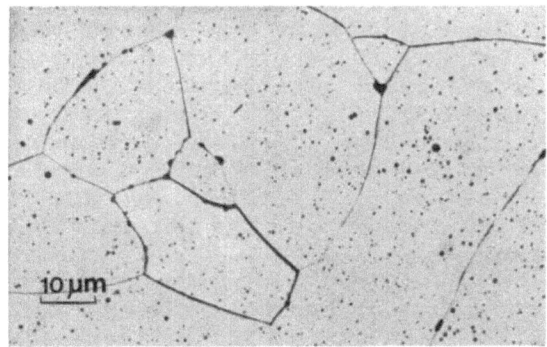

Fig. 4 Undoped Al_2O_3 sintered at $1850^{o}C$ for 10 h

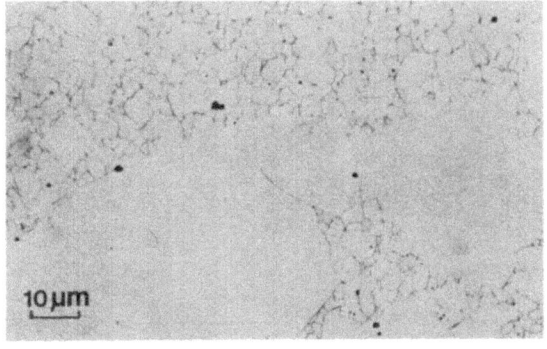

Fig. 5 Al_2O_3 doped with 100 ppm MgO sintered at $1850^{o}C$ for 10 h

amounts of MgO dopant. Fig. 2 gives the result after the
sintering treatment at 1630°C for 1.5 h. No discontinuous
grain growth occurred in the undoped Al_2O_3 as Fig. 3 shows.
It appears that MgO promotes grain growth until the solubility limit is reached. The mean grain size decreases when
more MgO is present because of the dragging effect of the
second phase particles on the grain boundaries.

To explain the increase in grain size in the single
phase region, it must be remembered that at the same time
an increasing density was found. Pores lying on a grain
boundary influence grain growth more or less like a second
phase. The fewer the pores the more freely the grain boundaries can move.

During the second sintering period at 1850°C for 10 h
the well known discontinuous grain growth occurred in the undoped Al_2O_3 sample, see Fig. 4. The specimen doped with 50
ppm MgO contains cracks and shows non-uniform grain growth:
many very large grains are visible within a matrix of much
smaller grains. This grain growth occurs in a stage at which
most pores have disappeared. The large grains now are pore-
free in contrast to the case of undoped Al_2O_3. The same holds,
but to a much less extent, for the sample with 100 ppm MgO.
This sample contains no cracks. The non-uniform grain growth
is illustrated in Fig. 5.

All samples with more MgO show a very regular distribution
of grain sizes. The mean grain size of these samples is given
in Fig. 6. Essentially the same dependence is found as in the

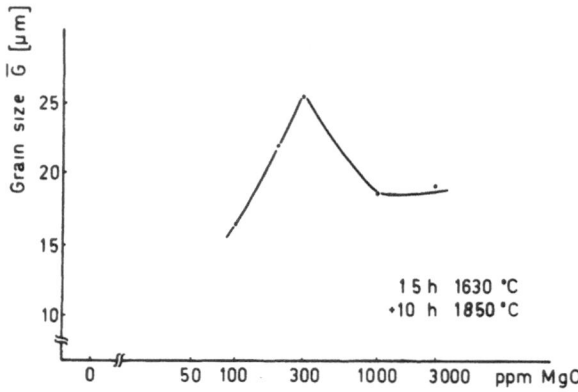

Fig. 6 Influence of MgO on the mean grain size of Al_2O_3 after
       a second sintering period at 1850°C for 10 h

previous case after sintering at 1630°C. The sample with 1000 ppm MgO contains many second-phase particles and the mean grain size is lower than is found in the sample with 300 ppm MgO. Although part of the second phase can originate from precipitation during cooling, the solubility rate of the spinel particles, present at the grain corners, is apparently very low.

Influence of MgO on the In-line Transmission

The in-line transmission of sintered Al_2O_3 is mainly determined by the microstructural parameters porosity, pore size and pore size distribution[10]. Therefore it can be expected that the amount of MgO has a great influence on the in-line transmission. The transmission was measured after the second heat treatment at 1850°C. The sample with 50 ppm MgO containing some cracks was not used for the transmission measurements. Results for some other specimens are given in Table I. The transmission values are for monochromatic light with a wavelength of 0.5 μm. The samples with 300 ppm MgO and more contain an increasing amount of the second phase particles which act as scattering centres. Although the total volume fraction of pores is about the same for the samples with 100 and 200 ppm MgO, the former has the highest transmission. The reason must be the larger mean pore size which is found in this sample with its non-uniform grain growth.

Auger Spectroscopy

To investigate the grain boundary segregation phenomena the powder with already 1000 ppm MgO was doped with 45 ppm CaO via the Ca-acetate. This fitted in another investigation

TABLE I

Measured in-line transmission and MgO content

ppm MgO	transmission
100	29
200	25
300	18
1000	12
3000	4

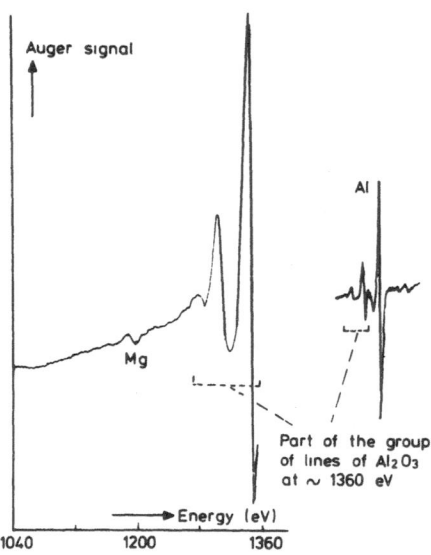

Fig. 7 Part of an Auger electron spectrum of a fracture
       surface of Al$_2$O$_3$ by means of averaging techniques

about the influence of CaO in the sintering behaviour of
Al$_2$O$_3$, on which will be reported later. A sample was
sintered at 1630°C for 1.5 h and 1850°C for 10 h in H$_2$. For
the experiments a commercial Auger-spectrometer was used,
consisting of a Physical Electronics Cylindrical Mirror
Analyzer and a standard Auger ultra high vacuum system
(Ultek-Perkin Elmer). Rods 2 mm in diameter were prepared
from the sample and fractured under ultra high vacuum (10^{-10}
torr). The fresh surface was then analyzed with the Auger
Spectrometer.

The Ca peak at 292 eV was very clear in the Auger
spectrum. The Ca concentration at the surface could be fixed[16]
at about 6 at %. This means an enrichment of the surface by
a factor of 1000.

Since there was no Mg peak in the spectrum signal
averaging techniques were applied to decrease the detection
limit. The resulting spectrum for Mg is shown in Fig. 7.
Quantisation of the Mg peak yields a Mg concentration of
about 1000 ppm (at.). This agrees with the analysis of the
original powder. No grain boundary enrichment could thus be
established. It must be remarked that the accuracy of the
Auger analysis is a factor of 2. A further remark must be
that part of the MgO may have evaporated during the

sintering treatment. A minor enrichment of Mg on the grain
boundaries is therefore still possible.

DISCUSSION

The present work demonstrates that MgO influences both
the densification kinetics and grain growth of Al_2O_3 in the
same way. This indicates that densification and grain growth
are geverned by the same process.

Very small amounts of MgO promote the sintering of
Al_2O_3 and a second phase of spinel is not necessary. The
experiments with Auger spectroscopy do not point to an important grain boundary segregation of MgO, which would decrease grain boundary mobility. Equally, the results shown
in Fig. 2 do not support the theory of grain boundary
segregation. Coble[1] observed that addition of MgO does not
inhibit the rate of normal grain growth and Coble and
Burke[17] list possible reasons for this. Fig. 2 shows that
MgO even promotes normal grain growth.

It seems that the average growth rate is completely
controlled by the volume fraction and size of the pores.
The more MgO is present (below the solubility limit) the
more the sintering rate and thus the rate of pore removal
is enhanced. This means that there are fewer restraints on
the grain boundaries and the grain boundary velocity can
more closely approach the velocity of a free boundary. The
essential action of MgO is then the enhancement of the densification of Al_2O_3 and not the reduction of grain boundary
mobility. Rao and Cutler[8] found that the sintering rate of
Al_2O_3 doped with Fe^{2+} ions increases with the total Fe
content and it seems that the same is true for Al_2O_3 doped
with Mg^{2+} ions.

Above the solubility limit second phase particles are
present and this changes the situation. The second-phase
inclusions lie on grain edges and grain corners and influence
the grain growth process. An increase in the MgO content
corresponds to an increase in the average number of second
phase particles per grain boundary and/or an increase in
their average size. This results in an increasing drag on
the grain boundaries and explains the decreasing mean grain
size found experimentally. Mocellin and Kingery[18] who have
annealed samples of MgO-doped Al_2O_3 also find that second-
phase dragging probably controls the grain growth kinetics.

Another aspect of an increasing amount of second-phase

particles is the decrease in sintering rate which is expressed in lower densities (see Fig. 1). The same phenomenon
has been found in other systems. Reynen[19] has put forward
the hypothesis that grain-boundary sliding is essential for
fast sintering. A dispersed second phase would suppress this
grain-boundary sliding for purely geometrical reasons. If
this hypothesis holds in this case longer sintering times
should give higher densities. This has still to be done.
Another explanation involves an overall decrease of mobility
in the boundary, when the boundary is "blocked" by a dispersed second phase.

Finally a remark on the non-uniform grain growth in the
samples with 100 ppm MgO and less (see Fig. 5). It is
possible that this is caused by a non-uniform distribution
of MgO. More MgO may be present locally and this results in
a higher rate of pore removal and more rapid grain growth.
These large grains can then grow at the expense of others.
This could explain that the large grains are nearly pore free
(in contrast with the large grains in the case of undoped
Al_2O_3) and that large pores are often found between large
and small grains. These large pores are possibly the reason
for the high in-line transmission of this specimen. However,
more detailed studies on mechanisms of grain growth and pore
growth are certainly necessary.

CONCLUSION

The addition of MgO to Al_2O_3 promotes both the densification and grain growth of Al_2O_3 as long as the solubility
limit has not been reached. The most effective dope level
corresponds to the amount that can be brought into solid
solution. The essential action of MgO seems to be enhancement
of the pore removal rate. At higher dope levels, when second
phase particles are present, grain growth is slowed down and
the densification is also negatively influenced. With
Auger spectroscopy no important enrichment of the grain
boundary area with MgO could be found.

ACKNOWLEDGEMENT

Many thanks are due to Dr. J. Verhoeven for performing
the Auger measurements and his interest in the subject.

REFERENCES

1. R.L. Coble, J. Appl. Phys. 32, 787, 793 (1961).
2. P.J. Jorgensen, J.H. Westbrook, J. Am. Ceram. Soc. 47, 332 (1964).
3. N.A. Haroun, D.W. Budworth, Trans. Brit. Ceram. Soc. 69, 73 (1970).
4. C.A. Bruch, Bull. Am. Ceram. Soc. 41, 799 (1962).
5. G. Rossi, J.E. Burke, J. Am. Ceram. Soc. 56, 654 (1973).
6. P.J. Jorgensen, R.C. Anderson, J. Am. Ceram. Soc. 50, 553 (1967).
7. R.J. Brook, J. Yee, F.A. Kroeger, J. Am. Ceram. Soc. 54, 444 (1971).
8. W.R. Rao, I.B. Cutler, J. Am. Ceram. Soc. 56, 588 (1973).
9. A.E. Paladino, R.L. Coble, J. Am. Ceram. Soc. 46, 133 (1963).
10. J.G.J. Peelen, R. Metselaar, J. Appl. Phys. 45, 216 (1974),
11. H.L. Marcus, M.E. Fine, J. Am. Ceram. Soc. 55, 568 (1972).
12. R.I. Taylor, J.P. Coad, R.J. Brook, J. Am. Ceram. Soc. 57, 539 (1974).
13. D. Prokic, J. Phys. D. (Appl. Phys.) 7, 1873 (1974).
14. M.I. Mendelson, J. Am. Ceram. Soc. 52, 443 (1969).
15. S.K. Roy, R.L. Coble, J. Am. Ceram. Soc. 51, 1 (1968).
16. P.W. Palmberg et. al., Handbook of Auger Electron Spectroscopy, ed. 1972.
17. R.L. Coble, J.E. Burke, Progr. Ceram. Sci. 3, 197 (1963).
18. A. Mocellin, W.D. Kingery, J. Am. Ceram. Soc. 56, 309 (1973).
19. P.J.L. Reijnen in: "Problems of Nonstoichiometry", Ed. A. Rabenau, North-Holland Publishing Comp. Amsterdam, 1970.

REACTION SINTERING OF CoO-NiO SYSTEM

S. Bošković and M. Stevanović

Boris Kidrič Institute of Nuclear Sciences

Lab.170,11001 Belgrade, POB 522,Yugoslavia

INTRODUCTION

In the case of parallel development of sintering process and chemical reaction, followed by volume changes of the material, the dimensional changes due to the reaction superimpose on that due to densification. Some examples of reaction sintering in the literature /1,2, 3,4/ in which , during densification, new compound is formed by solid state reaction, show that shrinkage due to sintering and volume change (expansion or shrinkage) due to reaction, take places at the same time. Volume change of the system, due to reaction, affects the sintering process, as well as the total densification.

CoO-NiO system, with molar CoO content higher than 0.2 is a two-phase system: solid solution-spinel /5,6, 7/. In the temperature interval 200-1000°C depending on oxygen partial pressure, chemical composition, initial structure, and the powder characteristics, the reaction of spinel formation followed by oxidation and spinel decomposition followed by reduction, may occur /5/. Either during sintering of pure spinel $NiCo_2O_4$ or during sintering of pure solid solution. $/Co_yNi_{1-y}/_{1-x}O$, at temperatures up to 1000°C, chemical reaction accompanied by densification takes place.

The system under study was CoO-NiO material.

By performing sintering experiments, at two temperatures and in two atmospheres with different oxygen partial pressures (10^3 and 1 atm.), as well as by following the reaction of formation and decomposition of spinel

by X-ray analysis, specific surface area change, pore
volume changes and densification, the attempt was made
to find out the way, in which these two reactions, affect densification during sintering.

EXPERIMENTAL PROCEDURES

Powder of solid solution $/Co_{0.66}Ni_{0.33}/_{1-x}O-$ **(A)**,
was prepared by decomposition of mixture of Co and Ni
carbonates, at $450°C$ in nitrogen, during 3 hours /8/. A
portion of this powder, was exposed to oxygen at $400°C$
for 5 hrs in order to obtain powder having the same cation ratio Co/Ni = 2. Thus powder B was obtained. Properties of initial powders are given in Table 1.

TABLE 1: Characteristics of initial powders

Sample	Chemical formula	Lattice parameter $/Å/$	Specific surface area/m^2/gr/	Theoreti- cal densi- ty/g/cm^3/
Powder A	$Co_{0.66}Ni_{0.33}O$	4.23	14.2	6.564
Powder B	$NiCo_2O_4$	8.10	4.6	5.990

Powders A and B, were pressed at 392.4 & 78.5 MN/m^2
respectively, with the intention to get close values
of green densities (~60%). Sintering was performed at
$500°$ and $800°C$, in argon $/P_{O2} \sim 10^{-5}atm/$ and oxygen, during 2-60 minutes.   During sintering spinel content
(Fig. 1) was followed by X-ray analysis. From these data,
corresponding theoretical densities were evaluated.
Green, and sintered densities were evaluated from weight
and dimension measurements. In Fig. 2 the difference between green and sintered densities in percent of the theoretical density are plotted against the time of sintering. In addition , the pore volume changes during sintering and specific surface area of sintered samples at
$500°C$ are given in Figs. 3 and 4 respectively.

The overall volume change of a sample during re-
action sintering, $\Delta V_{tot} = V - V_o$, is due to changes caused
by chemical reaction, and porosity decrease due to neck
formation between the particles. Volume changes of samples due to development of the reaction, are therefore a
sum of the changes caused by lattice expansion or shrinkage of that part of the mass which undergoes transition
after which the mean theoretical density is changed, as
well as by pore volume increase or decrease, caused by

phase transition.
    Therefore we have:

$$\frac{\Delta V_s}{V_o} = \frac{V_p - V_{po}}{V_o} = \frac{V_{tot} - V_t - V_o + V_{to}}{V_o} =$$

$$= \frac{\Delta V_{tot}}{V_o} - \frac{V_t - V_{to}}{V_o} \qquad\qquad /1/$$

where V_{po}, V_o and V_{to} are the initial pore volume, volume of the sample and theoretical volume respectively V_p, V_{tot} and V_t represent the same quantities after reaction sintering.
    The total porosity change within the samples $\Delta V_s / V_o = \Delta V_p / V_o$ (Fig. 3), was evaluated from the expression

$$\frac{\Delta V_s}{V_o} = \frac{\Delta V_{tot}}{V_o} - \frac{\varsigma_o}{\varsigma_{to}} \left(\frac{\varsigma_{ot}}{\varsigma_t} - 1\right) = \frac{\varsigma_o}{\varsigma} \left(1 - \frac{\varsigma}{\varsigma_t}\right) -$$

$$- \left(1 - \frac{\varsigma_o}{\varsigma_{to}}\right) \qquad\qquad /2/$$

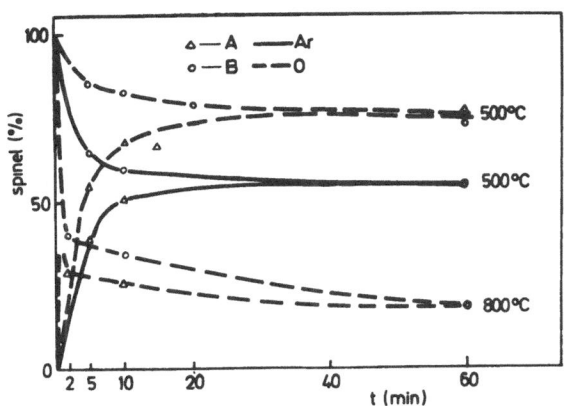

Fig. 1. Phase composition of sintered A and B
        samples

RESULTS AND DISCUSSION

Analysing the phase composition of A and B samples after sintering for 1 hr at chosen temperatures and atmospheres (Fig. 1), one can say that at both $500^{\circ}C$ and at $800^{\circ}C$, almost equilibrium state of spinel formation and spinel decomposition were achieved. Besides, X-ray analysis showed that during the time interval investigated, at $500^{\circ}C$, in the case of A samples, reaction of spinel formation took place. It was more rapid in oxygen. The reaction was accompanied by weight increase. In argon, the rate of reaction was slower, and was accompanied by weight loss. Sintering of B samples at $500^{\circ}C$ was accompanied by spinel decomposition; slower in oxygen than in argon.

At $800^{\circ}C$ in Ar, A samples remained unreacted, while in B samples, very rapid spinel decomposition took place, giving as a reaction product a solid solution of NaCl structure. In B samples heated at $800^{\circ}C$ in O_2 fast decomposition reaction took place. Similarly, in A samples decomposition took place, after very rapid oxidation had occured, during the heating period.

Excluding A samples at $800^{\circ}C$ in Ar, the density difference, as well as the change of relative pore volume after sintering, were caused by two processes: approaching of the particle centres and porosity change as a consequence of the change in phase composition. One should have in mind, that the first process can be influence by the second one.

It can be said from the data represented in Fig. 2 or 3 that the densification rates are always lower in the system in which, during densification, or before it, (B-Ar-$800^{\circ}C$) spinel decomposition took place. Contrary to that the densification level was higher when the sintering of solid solution has been accompanied or preceded by the spinel formation.

In samples A at $800^{\circ}C$ in Ar, the sintering of one component system (of $Me_{1-x}O$ type) took place, promoted by the setting up of the additional driving force, oxygen chemical potential gradient, due to local changes of stoichiometry /8/.

Densification of B samples at $800^{\circ}C$ in Ar, can also be thought of as the sintering of one component system /$Co_{0.66}Ni_{0.33}O_{1-x}$ O , but only after spinel decomposition followed by reduction took place. It is obvious, that decomposition of B samples, brought about the increase of initial porosity (Fig. 3). The overall density increase of B samples at $800^{\circ}C$ in Ar, which is lower than that of A samples under the identical sintering

conditions, is the consequence of the porosity increase
due to lattice volume changes of the system. The data
on $\Delta \varsigma$ and $\Delta v_s/v_o$ of A and B samples at 800°C in Ar
are different and are due to different initial states
of the samples. It could also be added, that the deviation from stoichiometry of the samples A and B during
sintering in Ar, and accompanying the oxygen chemical
potential gradient, which represents the additional driving force for mass transport during sintering, are not
to be the same /8,9/.

Reaction sintering of B samples at the temperature of 500°C both in O_2 and in Ar, represent a case
in which the densification is accompanied by spinel decomposition, a process causing volume decrease of the
material. The development of this process substantially
changes the initial porosity within the samples. The
overall effect,more pronounced in Ar, where spinel decomposition reaction takes place with double rate as
compared to the same in oxygen (Fig. 1), is density decrease and porosity increase. No doubt, this overall
effect is brought about, by porosity increase due to
domination of chemical reaction /10/. The specific sur-

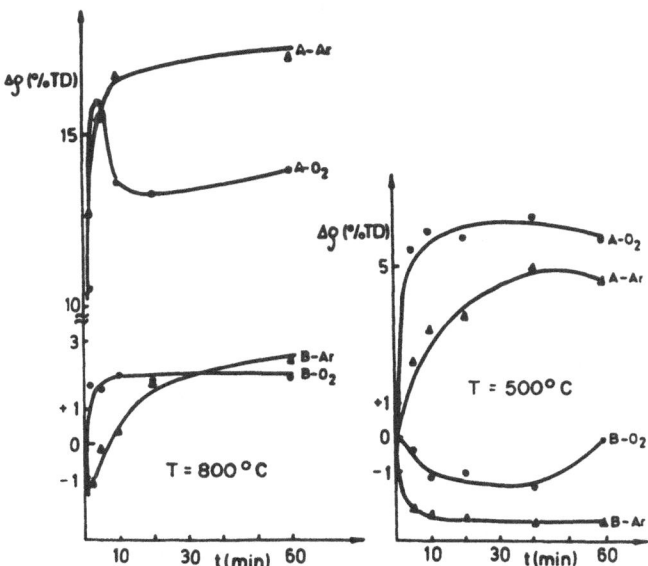

Fig. 2. Green and sintered density difference of
       investigated samples

face area decrease during sintering (Fig. 4) which in
general, is a consequence of rounding of the particles
and pores, and welding together of the particles indicate that, these processes could proceed more rapidly in
Ar than in oxygen. One should bear in mind that in Ar,
new phase is formed under the conditions,of the initial
state deviating more from the equilibrium, than in O_2 ,
and that is why higher surface activity is observed.The
fact, however, remains that reaction sintering of spinel,
during which, spinel decomposition takes places, retards
densification, because of volume changes of the system -
lattice shrinkage and porosity increase within the samples.

Sintering which takes place in A samples, at $500^{\circ}C$
in O_2 and Ar, proceeds together with the parallel development of oxidation reaction of solid solution
$/Co_{0.66}Ni_{0.33}/_{1-x}O$, followed by spinel phase formation.
Density increase (Fig. 2) and porosity decrease (Fig. 3)
after reaction sintering (higher in O_2 than in Ar), are
the consequences of neeck formation between the particles, as well as of the initial porosity change caused by
chemical reaction.

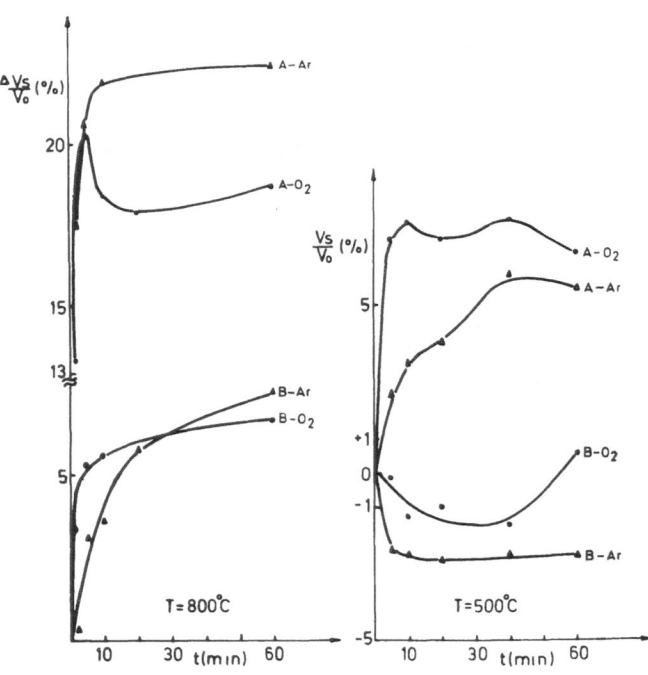

Fig. 3. Pore volume change during sintering of
        A and B samples

        The oxidation of solid solution phase to spinel
is followed by specific surface area decrease.
This is demonstrated by the specific surface area of the
starting solid solution powder - A (14.3 m^2/gr) and the
spinel - B (4.5 m^2/gr) obtained by oxidation of the former. Although in Ar at the temperature of 500°C the spinel is formed in amount of about 50% less, than in oxygen, and the density increase as well as porosity decrease within A samples are lower, than in O_2, specific
surface area change is more pronounced in Ar. On the
basis of this fact, it could be said that spinel formation process, slows down the densification.
        A very important role, during densification of
/Co$_{0.66}$ Ni$_{0.33}$ /$_{1-x}$ O solid solution, accompanied by spinel formation, is played by porosity decrease due to the
appearance of a new phase of larger molar volume than
the initial one. The same effect has been found earlier
/10/.This phenomenon can cause the change of contact
area between particles which are being sintered, as well
as the setting up of the stresses between them. During
sintering at 500°C the porosity decrease due to spinel
formation is the most important phenomenon, and the one,
determining the both overall density and porosity change
within A samples. It seems, however, that the spinel formation itself, especially after a long period of time
when it involves the solid solution particles, already

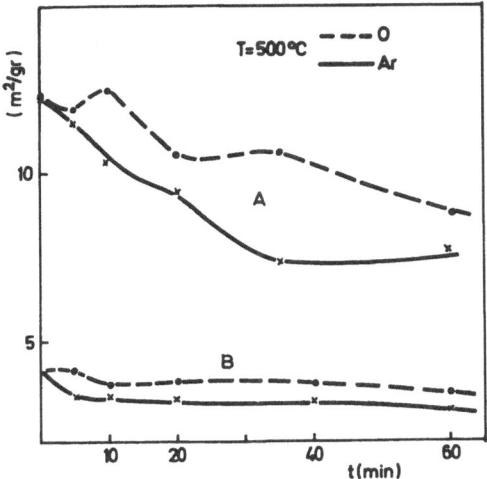

Fig. 4. Specific surface area as a function of tem-
        perature and oxygen partial pressure

welded together, can create the stresses which could
bring about the destruction of the prevously formed
contacts. This can help in explaining the broad maxima
on the density vs. time curves (Fig. 2), porosity
change (Fig. 3) as well as specific surface area changes
(Fig. 4), of A samples, sintered in oxygen for 10 and
40 minutes.

From the Figs. 2 and 3, it can be seen that with
A samples, sintered in oxygen at the temperature $800^{\circ}C$,
for 5 minutes, with parallel reaction of spinel formation (decrease of porosity due to expansion of material)
there is a maximum on the density and porosity curves.
The maximum is followed by minimum, during 20 minutes,
but during that time the decomposition reaction of previously formed spinel compound into solid solution, takes place, bringing about the porosity increase.

Reaction sintering of B samples in oxygen at
$800^{\circ}C$ is also represented in Figs. 2 and 3. The absolute value of shrinkage is higher than pore expansion due
to reaction, contrary to the situation already discussed for the lower temperature.

CONCLUSION

The overall densification, or porosity decrease
of $NiCo_2O_4$ samples, during sintering accompanied or
preceded by spinel decomposition, is the lower, the
more rapid is the spinel decomposition process. During
reaction sintering of solid solution, the spinel formation process intensifies the density increase. In
both cases, porosity and also the density of sintered
samples are determined by porosity change due to the
chemical reactions - increase during spinel decomposition, and decrease during spinel formation reaction.

It seems, from the obtained results, that processes of spinel formation and decomposition, affect
just in oposite way, the welding together of the particles.

REFERENCES

1. F. Thümmler and W. Thomma, Metals and Mat.1(1967)69

2. H.J.S. Kriek, W.F. Ford and J. White, Transactions of the Brit. Ceram. Soc., 58(1959),1

3. D.L. Johnson, and I.B. Cutler, "Phase Diagrams", Ed. Allen M. Alper, Acad. Press, New York, 1970, Vol. II, p. 265

4. E. Kostić, J. Petković and V. Marković, 2nd Int. Symp. on Ceramics, October 1974, Bologna, Italy

5. S. Bošković and M. Stevanović, XIXth Conference ETAN, June 1975, Ohrid, Yugoslavia

6. R.I. Moore and J. White, J.Mat.Sci.,9(1974),1393

7. T. Sakata, K. Sakata, K. Kigoshi, J. Phys. Soc. Japan, 10(1955), 179

8. S. Bošković and M. Stevanović, J. Mat. Sci., 10(1975) 25

9. S. Bošković and B. Živanović, J. Mat. Sci., 9(1974) 117

10. S.J. Kiss, Private communication

OBSERVATIONS ON THE REACTION SINTERING

OF OXIDE SYSTEMS

S.J. Kiss and E. Kostić
Technical College for Ceramics, Arandjelovac
Boris Kidrič Institute of Nuclear Sciences
Beograd

INTRODUCTION

Mass transport processes in reacting multiphase powder systems are due to both, chemical reaction and capillary phenomena /1/. These processes are mutually interdependent and change the course of their kinetics. The mutual distribution of reacting particles is of special importance /2/. In this paper the formation of materials having spinel structure is considered with the emphasis of the mutual positioning of the reactants. For this purpose several model experiments were designed The study was performed on $MO-Fe_2O_3$ and $MO-Al_2O_3$ systems where M stands for Ni, Zn or Mg.

EXPERIMENTAL PROCEDURES

Pellets of 27 mm in diameter, pressed from starting oxide powders, were thereafter sintered up to the density level, that does not bring about considerable change of dimensions during subsequent heat treatment between 1100-1250°C. The spheres, 3-4 mm in diameter were made from these pellets; they were pressed inside the powder of the second reactant /model I/.

Rings were made by pressing all reactants except NiO. Internal diameters of rings were the same as the diameters of above mentioned sintered pellets. So, this model consisted of the sintered pellet of one reactant encircled by pressed ring of the other one /model II/.

Samples of model I and II were treated in the
temperature range 1100-1250°C, during 0-137 hours, together with similar rings without central pellet.
The heating rate was 450°/h. Thus heat treated samples
were polished to reveal the reaction layer and micros-
copically investigated. The densities of rings of model
II samples as well as of the free rings and the samples
of model I were measured.

RESULTS AND DISCUSSION

Density measurements were carried out on the
rings of model II in the systems: Fe_2O_3 (ring)-NiO (pellet), ZnO (ring)-Fe_2O_3 (pellet) and Al_2O_3 (ring)-NiO
(pellet), as well as on the free rings with intention
to compare the sintering conditions. The density measurements of rings with and without sintered pellet in model II showed that in the case of ferric oxide, free
rings were denser. For example, at 1200° - 18 hours the
densities were 4,88 and 5,05 gr/cm³, respectively, but
during the non-izothermal sintering up to 1250° with
heating rate 450°/h, the densities were, 4,96 and 5,14
gr/cm³, respectively. Such differences in density were
not observed in the samples with ZnO and Al_2O_3 rings.
Without considering in detail the mechanism and
kinetics of sintering in each oxide, only two basic conditions for densification of rings: neck formation and
center to center approach, will be pointed out.
In sintering models, usually two spheres are considered, the centres of which approach due to transport
processes without external forces. In the investigated
samples, however, due to unchanged circumference of internal sphere or pellet, this approach is retarded and
may even become impossible. Unchanged diameter of central pellet is only an approximation, because of paral-

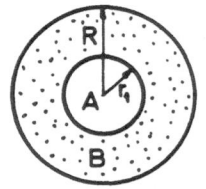 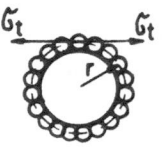

Fig. 1. Shematic representation of the model I

lel development of chemical reaction at the interface
MO/M_2O_3. Let us consider a circular row of fine particles, which surrounds the central pellet, (Fig. 1) The
sintering of a circular row of fine particles without
central pellet would decrease the diameter of the circle, due to approaching of neighbouring particle centres.

Therefore, a tangential tensile stress, σ_t , will
arrise causing macrostresses within the ring, or shell,
independent of the mechanism which controls densification. The above analysis indicates that in the mass transport during sintering stress gradient may play a role.
It can be expected, that the resulting tangential stress
will bring about an additional mass transport independent of the transport conditions towards necks actuated
by capillary forces impeding the translational motion of
the spheres. If the particle centres are fixed, concerning the motion in the tangential direction, as in our
case, their approaching becomes impossible and "densification" should be retarded.

The tangential stress, however, causes hoop stress
which in turn, will cause compressive stress in radial
direction, σ_r . This stress will enhance mass transport
between adjacent particle layers and for this reason
will speed up the densification rate. Therefore, the
conditions under which the necks at particular contacts
are formed, in model samples, are different than those
in the free rings. Overall action of above mentioned
stresses within the pressed pellets of the models I and
II could be different, depending on the system. Of course, the relaxation of these stresses during heat treatment may produce cracks, thus decreasing overall density
which may be difficult to assess. Also the stresses mentioned above may affect the microstructure. The values
of σ_t and σ_r , first of all, depend on material properties - on surface tension, grain size distribution and
packing. It should be pointed out, that σ_t and σ_r values are being changed with development of the sintering
process, due to the change of effective neck surface and
microstructure development.

With ZnO rings a great shrinkage was observed in
both cases /up to 95 % theoretical density at 1200°/,
what pointed to the fact, that these stresses within
ZnO rings did not influence remarkably the sintering
process. With Al_2O_3 rings the volume change in both
cases at 1200° were only about 8 % which hardly can
point out the role of analyzed stresses. With Fe_2O_3
rings, however, the overall effect of the stresses is
retardation of mass transport. At the beginning of

thermal treatment, at the boundary MO/Fe_2O_3 some micro-
cracks were observed in Fe_2O_3 rings, as a result of
stress relaxation, Fig. 2.
    In the model II, when the height of Fe_2O_3 ring
was larger than the sintered NiO pellet, after 137 hours at 1200° remarkable deformation occured both in the
pellet and in the ring, Fig. 3. The observed deformation is probably the consequence of the stress in the
ring caused by sintering, because of the volume decrease
in the reaction zone /3/.
    On the basis of these results, the chemical rea-
ction in the models with Fe_2O_3 shell take place under
the compressive stresses. The reactions in the $MO-Fe_2O_3$
system are investigated mainly on the models which permit oxygen transfer through the gas phase /1,3,4/. If,
in the model I, arround the central MO particle a continuous reaction layer is formed without open chanels,
the transport of oxygen from the boundary region spinel
$/Fe_2O_3$ to the MO/spinel interface is mainly prevented.
In this case the oxidation of the diffused Fe^{2+} ions does
not take place and in the absence of the Fe^{3+} ion diffusion at the MO/spinel interface only the exchange of Fe^{2+}
and M^{2+} is possible. Accordingly, in the spherical models, depending on the conditions of oxygen transfer to
the MO/spinel interface, the mechanism and at the same
time the reaction kinetics may be somewhat different
than in the sandwich samples. Keeping in mind the possible mutual influence of the chemical reaction and
process of mass transfer in the shell /4,5,6/, the
conditions for the reaction can be different depending
on mutual position of the reactants. Different appearance of the reaction zones, Fig. 4, support this assumption. Under the same thermal treatment conditions
the width and character of various zones are different.

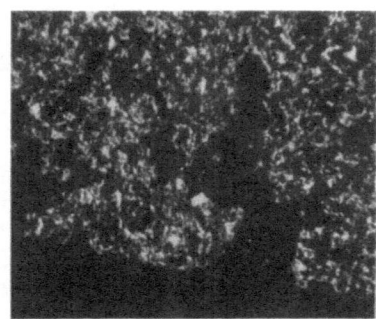

Fig. 2. Microcracks in the Fe_2O_3 ring
        (1100°, 5 min, 360x)

One can see that the pores in the spinel/Fe_2O_3 interface region are mainly parallel to the boundary, probably as a result of the reaction taking place under pressure. At the MO phase surface a great deal of densification occured. This is particularly pronounced in MgO probably due to the formation of solid solution of the (Mg,,Fe)O type. The formation of this type of solid solution was also observed in the ZnO–Fe O system after 135 hours at 1200°C, if ZnO was the sintered central particle in model I. On the surface of the central ZnO particle a marked black 0.5 mm wide layer is formed,(Fig. 5), which has, according to X-ray examination, the ZnO structure. The black color originates probably from FeO being dissolved in ZnO /7/. At the contact interface of the NiO pellet a brown ring was observed. The X-ray examination showed that it consisted of a (Ni,Fe)O solid solution, what is in accordance with the basic mechanism of Ni-ferrite formation /3,4/. The existence of (Fe,M)O type solid solutions during the ferrite synthesis indicates changes in the reaction mechanism in the course of the reaction. If a solubility limit of FeO in MO exists, after reaching this limit in the whole MO particle the exchange of M^{2+} and Fe^{2+} at the MO/spinel interface is prevented.

The formation of solid solution of the·(M,Fe)O type during the ferrite synthesis, makes it difficult to predict the volume changes brought about by the reaction. If on the MO/spinel interface only the exchange of M^{2+} and Fe^{2+} ions takes place, the partial reduction of Fe_2O_3 to Fe_3O_4 at the spinel/Fe_2O_3 interface causes a decreasing of system volume, whereas the evolved oxygen, which is partially retained in the closed pores, increases it. Therefore, the volume change due to the chemical reaction must depend on the reaction conditions as well /3,8,9/.

In the system NiO (pellet) – Fe_2O_3 (shell) – model I – volume increase of 1,3 % was detected after 18

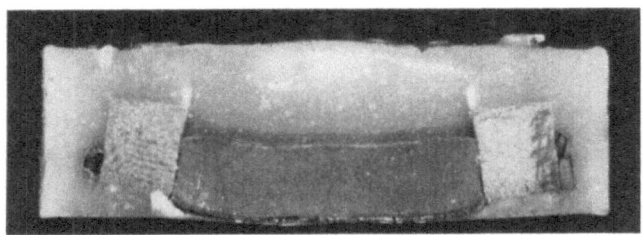

Fig. 3. Deformation in model II: NiO (pellet)-Fe_2O_3 (ring); 1250°; 137 hours

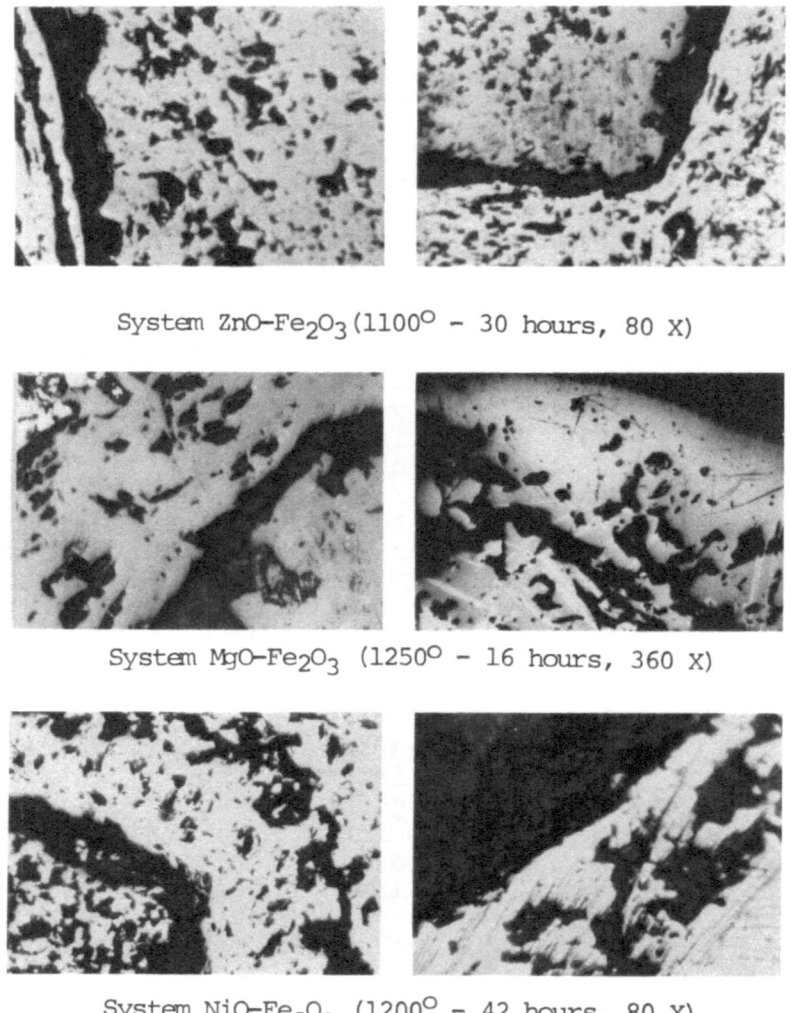

System ZnO–Fe$_2$O$_3$ (1100° – 30 hours, 80 X)

System MgO–Fe$_2$O$_3$ (1250° – 16 hours, 360 X)

System NiO–Fe$_2$O$_3$ (1200° – 42 hours, 80 X)

Fig. 4. Characteristics of reaction zones in the model I

hours isothermal heating at 1200°. It is probably due to the oxygen release in the reaction zone in the presence of nonpermeable sintered Fe$_2$O$_3$ shell.

In general, during cooling a pronounced separation at the MO/spinel interface takes place due to the different coefficients of thermal expansion.

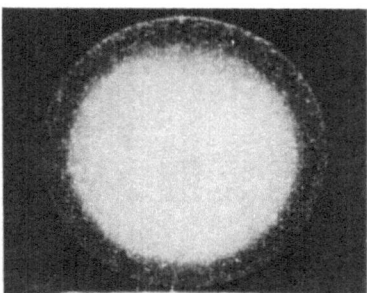

Fig. 5. Cross section of the central ZnO Particle, Model I (1250°,
         137 hours, 3,5 X)

In the NiO-Al$_2$O$_3$ system it was observed that NiAl$_2$O$_3$ layer
forms faster in the samples where NiO is the central particle.
After 72 hours at 1250° C the reaction layer width was 14 and
6 /um respectively, depending on whether NiO was the central
particle or the shell. In the MgO-Al$_2$O$_3$ system there was no pronounced difference in the MgAl$_2$O$_3$ layer width as a function of
the reactant positions. During cooling the MO/spinel interface
was often separated, just like in the MO-Fe$_2$O$_3$ system.

CONCLUSIONS

On the basis of observed changes in the reaction zone region
of the models and the general analysis of the reaction sintering,
it can be concluded:

1. In the reacting multiphase powder systems the chemical reaction
    and the sintering process are mutually dependent.

2. In multicomponent systems MeO-Fe$_2$O$_3$, Fe$_2$O$_3$ being the matrix,
    sintering rate of matrix is lower as compared to the sintering
    rate of equivalent compact without central MO component.

3. During the ferrite synthesis within the ZnO-Fe$_2$O$_3$ system, the
    formation of (Zn,Fe)O solid solution is possible.

4. If the chemical reaction increases the volume of the system,
    the reaction in the spherical particles takes place under the
    condition of radial pressure.

REFERENCES

1. Kuczynski,G.C.,"Sintering and Related Phenomena, Proc.of Int.Conf.Univ.of Notre Dame", Gordon and Breach, 1967

2. Paulus M., in G.C. Kuczynski Ed., Sintering and Related Phenomena, Mat.Sc.Rs.Vol.6, Plenum Press, New York, 1973

3. Paulus M., Eveno P.Y., Reactivity of Solids, Wiley-Interscience, New York, 1968, p. 585

4. Kuczynski G.C.,"Ferrites," Proc.Int.Conf.,University of Tokyo Press, Tokyo, 1971, p. 87

5. Kiss J.S., ITS-47, 1971

6. Kiss J.S., IBK- 171, 1972

7. Toropov N.A., Barzakovskij V.P.,"Dijagrammi sastojanija silikatnih sistemi,"Izd. Nauka, Leningrad, 1969, p. 152

8. Carter R.E., J.Am.Ceram.Soc. 44, p. 116 (1961)

9. Reijnen P.,"Reactivity of Solids," Munich, 1964, Elsevier Publ. Comp. (1965) p. 362

THE DEVELOPMENT OF PRESSURE SINTERING MAPS

D. S. Wilkinson and M. F. Ashby

University Engineering Laboratories
Trumpington Street
Cambridge, CB2 1PZ, England

ABSTRACT

The several mechanisms which contribute to the densification of a powder compact during pressure sintering are described, and equations are listed for the rates of densification that each, acting alone, would yield. The equations are used to construct pressure sintering diagrams. These identify the *dominant mechanism*, for a given temperature, pressure and density. They further display contours indicating the *rate of densification* and the *time to reach a certain density*. Their use in the design and interpretation of experiments is discussed, and their application to practical problems is illustrated by a case study of the pressure sintering of ice.

INTRODUCTION

Several distinct mechanisms contribute to densification when a powder compact is sintered under pressure. The pressure may cause *rearrangement* of the particles[1,2]; it may induce *plasticity* or *creep* in the compact[3-6]; and it may augment the effects of surface tension as a driving force for *diffusion*[7,8]. Most of these mechanisms contribute to the growth of the neck between powder particles, and many of them contribute directly to densification. Which contributes the most (and thus is *dominant*) depends on the conditions: there is no single mechanism which will always account for densification during pressure sintering. In fact the various mechanisms differ in their dependence on the internal parameters (neck size, gas pressure in the pores, compact density) as well as the external variables (pressure and temperature), so their relative contribution changes as sintering progresses.

473

Even the simplest experiments can be properly interpreted only by considering this simultaneous action of several mechanisms. One way of doing this is to construct "pressure sintering maps" using a procedure like that used to synthesise the mechanisms for pressure-less-sintering into "sintering maps"[9]. The basis of the analysis is a set of equations, each describing the densification-rate due to a single mechanism, valid over a certain range of density. These are evaluated (using material properties for a given metal or ceramic) to identify the dominant mechanism; and summed appropriately to give the overall densification rate.

It is the purpose of this paper to outline the present state of development of these maps, and to point out some potential applications for them.

THE STAGES OF SINTERING

It is convenient to think of sintering and pressure sintering as occurring in four stages.

When particles are placed in contact they adhere, though the molecular forces causing them to do this may be weak. If a pressure is now applied, the particles may rearrange by sliding over each other or by local fragmentation, to give a greater packing density. These essentially time-independent processes are poorly understood; though of considerable interest as mechanisms, they are of minor importance in the construction of the maps we describe below. We shall simply refer to them as Stage 0 sintering[9].

This Stage overlaps Stage 1, during which the necks grow rapidly by diffusion (see reviews by Thummler and Thoma[10], Wilson and Shewmon[11], Ramquist[12], and Ashby[9]) or by plasticity[5,6] and the average number of contacts-per-particle increases[13]. At least seven independent mechanisms contribute to neck growth at this stage; they are shown schematically in Fig. 1. Of these, only the first four directly cause densification of the compact; but the remaining three cannot be ignored because, by rounding out the neck they reduce the driving force for densification and so influence its rate.

Carried far enough, Stage 1 sintering leads to connected, roughly cylindrical pores and the compact enters Stage 2. During further sintering these cylinders shrink in radius, any gas in them escaping to the surface. Mechanisms 5, 6 and 7 of Fig. 1 no longer operate, since the driving force for them has disappeared.

Ultimately the cylindrical pores become unstable, and, at a size which is a complicated function of surface and grain boundary

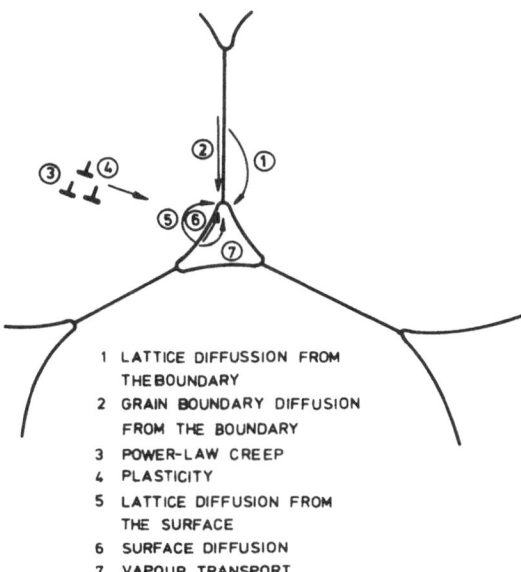

1 LATTICE DIFFUSSION FROM
   THE BOUNDARY
2 GRAIN BOUNDARY DIFFUSION
   FROM THE BOUNDARY
3 POWER-LAW CREEP
4 PLASTICITY
5 LATTICE DIFFUSION FROM
   THE SURFACE
6 SURFACE DIFFUSION
7 VAPOUR TRANSPORT

Fig. 1: Seven distinct mechanisms contribute to pressure sintering
during the initial stage. Only mechanisms 1 to 4 densify directly
but all lead to neck growth, and so influence the rate of densification.

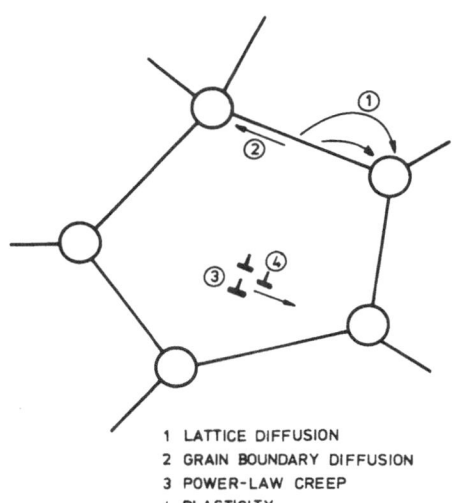

1 LATTICE DIFFUSION
2 GRAIN BOUNDARY DIFFUSION
3 POWER-LAW CREEP
4 PLASTICITY

Fig. 2: Intermediate and final stage densification occurs by
dislocation motion and by diffusion of atoms to the pore surface.

energies[14], they pinch off to form closed, roughly spherical pores.
The sample now enters the final stage of sintering. Pore shrinkage
during this third stage proceeds by mechanisms which closely res-
embles those of Stage 2; they are shown schematically in Fig. 2.
Any gas in the pores is, however trapped; as pores shrink further
the pressure in them rises, with important consequences discussed
below.

We have chosen to develop the technique of constructing pressure sintering maps for Stages 2 and 3. There are fewer mechanisms
involved than in a full treatment, and the unsolved problem of the
kinetics of particle rearrangement is avoided. It is not a serious
restriction: pressure sintering is used industrially where high
density products are required, and it is the final stages of sintering which are of most importance.

THE RATE EQUATIONS

The basis of pressure sintering maps (and indeed of all mechanism maps of this sort) is a set of rate equations. They take the
form

$$\dot{\rho} = \dot{\rho}(\rho, T, p_e, p_o, a, R, \text{physical constants}, \ldots),$$

where

ρ	is the relative density of the compact (i.e. the volume fraction of the compact filled with matter),
$\dot{\rho}$	is the rate of change of ρ with time,
$(1-\rho)$	is the porosity, or volume fraction of space in the compact,
T	is the absolute temperature,
p_e	is the externally applied pressure,
p_o	is the atmospheric pressure surrounding the compact,
a	is the pore radius,
and R	is the average particle space.

In a complete treatment*, a set of rate equations (one describing each mechanism of sintering) is required for each stage of sintering. As mentioned earlier, the final stages are the most interesting if a high density is required. In the maps shown here we
have ignored stages 0 and 1 completely, and have used a single set
of rate equations (developed for Stage 3) to describe the intermediate and final stages of sintering under pressure. With these
simplifications, only four mechanisms need be considered. They describe the densification of a compact containing spherical pores,

---

*We are developing this, and will publish the results elsewhere.

under the combined driving force of surface tension and external
pressure, by
- (a)   Lattice diffusion[8],
- (b)   Grain boundary diffusion[8],
- (c)   Power-law creep (or "dislocation creep")[6],
- (d)   Rate independent plasticity (yielding, or "dislocation
        glide").[3]

It is standard practice, in developing models for Stage 3 sintering, to isolate a unit of the structure by considering a thick
spherical shell of material, centered on a spherical pore of radius
a[15]. The shell has an outer radius b, defined such that the density
of the unit is equal to that of the compact:

$$\rho = \frac{b^3 - a^3}{b^3} = 1 - \left(\frac{a}{b}\right)^3 \tag{1}$$

If the porosity is sufficiently small, then to a good approximation, the pressure on the outer surface of the shell is that
applied externally to the compact: p_e. The pressure on the inner
shell surface is simply the pressure within the pore, p_i. If the
pore shrinks, the unit becomes smaller, the external pressure p_e
does work, and work is done against the internal pressure p_i. In
addition, the surface energy of the unit decreases because the pore
has decreased in size. These changes are correctly taken into account by using a driving force (strictly, a driving pressure) for
sintering which is the same for all four mechanisms of sintering,
and is given by

$$p = p_e - p_i + \frac{2\gamma}{a} \tag{2}$$

where γ is the surface free energy per unit area. (If the pores are
closed, p_i itself depends on pore size. We have included this in
the computation for ice detailed below).

The rate equations for each mechanism are obtained by solving,
for the thick walled sphere, and subject to the boundary conditions
described above, the appropriate boundary value problem: two based
on the diffusive flux equations give densification by lattice and
by boundary diffusion respectively; one based on the constitutive
relation for power-law creep gives densification by creep; and one
based on a yielding criterion gives densification by rate independent plasticity.

Densification by Lattice Diffusion[8]

If vacancies flow, by diffusion, from the surface of the pore
to sinks on nearby grain boundaries, the compact will densify.

Coble[8] assumes that the average distance to a sink is equal to the thickness of the spherical shell defined above, and that the flux has spherical symmetry. With these (reasonable) simplifications, the problem can be solved to give the densification rate due to this mechanism acting alone,

$$\dot{\rho}_1 = 3 \, \frac{D_v \, \Omega}{kT \, b^2} \, \frac{(1 - \rho)^{1/3}}{1 - (1 - \rho)^{1/3}} \, p \tag{3}$$

where D_v is the lattice diffusion coefficient, Ω is the atomic volume, k is Boltzmann's Constant, T is the absolute temperature, and ρ is the relative density of the compact.

(This equation is valid until grain growth occurs. Pores isolated in a large grained matrix shrink more slowly because the source-to-sink distance is now larger than the thickness of the shell wall. Brook[16] has quantified the conditions under which abnormal grain growth is to be expected during sintering. His method and results could be incorporated into maps like those shown here, although we have not yet done this).

Densification by Grain Boundary Diffusion[8]

A model for boundary-diffusion controlled sintering requires a further assumption about the relationship of the boundaries to the pores. We have used Coble's model which supposes three boundaries pass through each pore, and that the average diffusion distance is equal to the shell thickness, as before. The resulting densification rate is*

$$\dot{\rho}_2 = \frac{9}{2} \, \frac{\delta \, D_B \, \Omega}{kT \, b^3} \, \frac{1}{1 - (1 - \rho)^{1/3}} \, p \tag{4a}$$

where δ is the grain boundary thickness when acting as a diffusion path, and D_B is its diffusion coefficient. Using eqs. (1) and (2) we may rewrite this as

$$\dot{\rho}_2 = \frac{3}{2} \, \frac{\delta \, D_B}{a \, D_v} \, \dot{\rho}_1 \tag{4b}$$

---

*In computing the maps shown here, we have assumed that $b = R$, the particle radius. This is consistent with the model which assumes about one pore to every grain in the compact.

As before, this equation is valid until grains grow. There-after, only pores which lie on one or more boundaries can shrink by this mechanism; for the rest, $\dot{\rho}_2$ is zero.

Densification by Power-Law Creep[6]

If the material of the thick spherical shell can creep, an external pressure will cause it to densify slowly. The disloca-tion creep of metals and ceramics can generally be described by a power law relating strain-rate, $\dot{\varepsilon}$, to stress, σ:

$$\dot{\varepsilon} = A \sigma^n \qquad (5)$$

where A and n are material properties, derived from uniaxial creep tests on fully dense material. The boundary value problem out-lined earlier can be solved exactly when this constitutive relation is employed[6]. The result is

$$\dot{\rho}_3 = \frac{3SA}{2} \frac{\rho(1-\rho)}{[1-(1-\rho)^{1/n}]^n} \left(\frac{3}{2n} |p| \right)^n \qquad (6)$$

where S is the sign of the pressure, defined by eq. (2).

The relation (5) describes steady-state creep only. The prob-lem can be solved using a time dependent creep law[6] but since ade-quate data for time dependent creep is not available for most mater-ials, we have not employed it here.

Grain size has little influence on the rate of this mechanism, which will continue to operate even after abnormal growth has occurred.

Densification by Plastic Flow[3]

At low temperatures, or very high strain-rates, creep and diffusion contribute little to the densification of the shell, which behaves more like a perfectly plastic solid than like one which creeps. This behaviour is described by a constitutive law for pla-sticity, which can be written in the form

$$\dot{\varepsilon} = \begin{cases} 0 & \text{if } \sigma < \sigma_y \\ \infty & \text{if } \sigma > \sigma_y \end{cases}$$

where σ_y is the flow strength of the material of which the shell is made. This relationship was used by Torre[3] in developing an expres-

sion for the limiting density reached, by plasticity, in a compact
subjected to a pressure p. His result is

$$\rho_{lim} = 1 - \exp\left(-\frac{3}{2}\frac{p}{\sigma_y}\right)$$

If the density of the compact is less than ρ_{lim}, the application of a pressure (difference) p will cause instantaneous densification to ρ_{lim}; but if (because of the contribution of the other
mechanisms), the density already exceeds ρ_{lim}, this mechanism
contributes nothing:

$$\dot{\rho}_4 = \begin{cases} 0 & \text{if } \rho > \rho_{lim} \\ \infty & \text{if } \rho < \rho_{lim} \end{cases} \qquad (7)$$

In reality, densification will proceed at a rate governed by
practical considerations such as the velocity or response of the
press itself. In constructing the maps we have used a large, though finite value of $\dot{\rho}$ when $\rho < \rho_{lim}$.

THE CONSTRUCTION OF PRESSURE-SINTERING MAPS

We now ask: if the four mechanisms listed above act independently, over what range of pressure, temperature and density is each
dominant? For pressureless sintering the question can be answered
by constructing two-dimensional maps with density (or size of neck
between sintering particles) as one axis and with temperature as
the other (see Ashby[9]). The map is divided into fields, each showing the range in which a single mechanism is dominant, meaning that
it proceeds faster than any other mechanism.

The application of pressure introduces a new external variable.
If it is shown as a third axis, orthogonal to those of density and
temperature, the fields become three-dimensional. But since sintering is commonly carried out at either constant pressure or constant temperature, the appropriate section through this three-
dimensional map conveys the necessary information.

Consider first a map at constant pressure, shown in Fig. 3.
Its axes are relative density ρ, and homologous temperature T/T_M
(though an absolute temperature scale is shown across the top
of the diagram). It was constructed using data for silver (see
Table 1), for a constant external pressure of 4.7 MN/m^2, ($p_e/\mu_0 =$
1.8 x 10^{-4}) and a particle size of 50 μm. It shows three
fields, corresponding to the ranges of ρ and T/T_M for which sintering is dominated by boundary diffusion, power-law creep, and lattice
diffusion. Formally, the field boundaries are constructed by equating the rate equations (shown on the Figure) in pairs and solv-

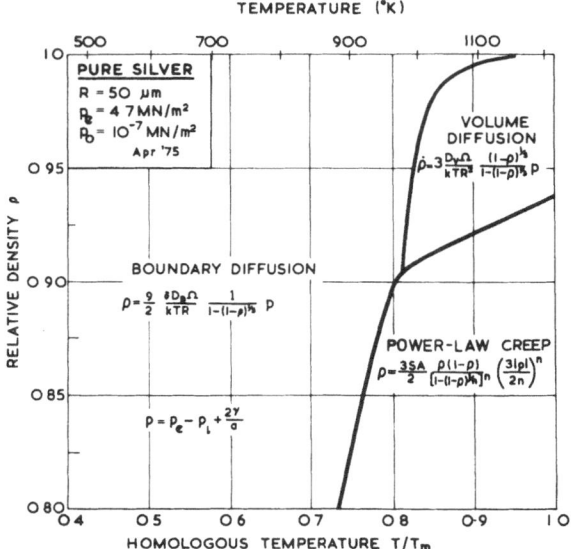

Fig. 3: A pressure sintering map for 50 μm silver particles at constant applied pressure of 4.7 MN/m² illustrating the dominant mechanisms and rate equations describing each mechanism.

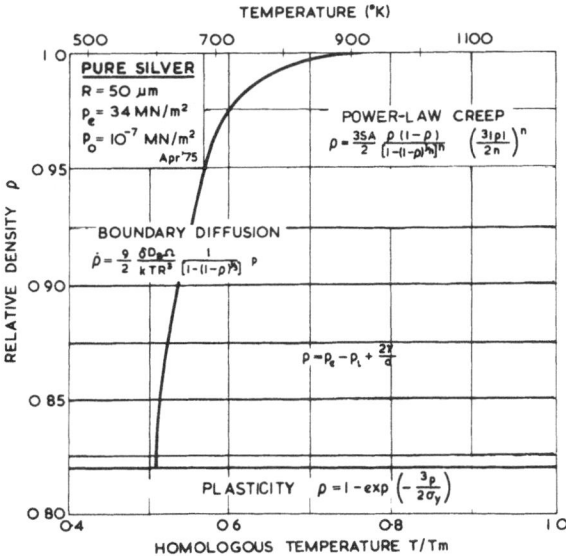

Fig. 4: A pressure sintering map at a higher applied pressure (34 MN/m²) than Fig. 3.

ing for ρ as a function of T. This means that a field boundary is a line along which two mechanisms contribute equally to densification; at the point where three boundaries meet, three mechanisms contribute equally.

If the same material were sintered under a larger pressure, the fields would change in size. This is illustrated by Fig. 4, which is constructed for a larger (constant) pressure than Fig. 3: 34 MN/m^2 instead of 4.7 ($p_e/\mu_o = 1.3 \times 10^{-3}$). The compact now denifies rapidly by plasticity to a density of 0.82 and thereafter by power-law creep or boundary diffusion. The increased pressure has enlarged the power-law creep field, and eliminated that for lattice diffusion.

The other section through the three-dimensional map is shown in Fig. 5: its axes are relative density ρ and normalised pressure p_e/μ, where μ is the shear modulus of the material (and absolute pressure is shown along the top of the diagram). It is constructed using the same equations and data as Figs. 3 and 4, but it is for a constant temperature of 1210 ^oK (T/T$_M$ = 0.98). It matches these two figures at the pressures of 4.7 MN/m^2 and 34 MN/m^2 respectively, and at the temperature of 1210 ^oK. It should be clear from these three figures that the field boundaries are surfaces in density/temperature/pressure space, and that Figs. 3, 4 and 5 are sections of this space.

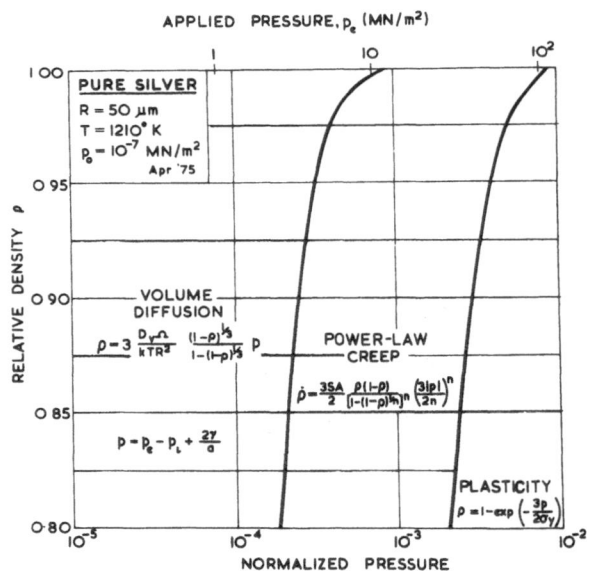

Fig. 5: A pressure sintering map for 50 μm silver particles at a constant temperature of 1210 ^oK.

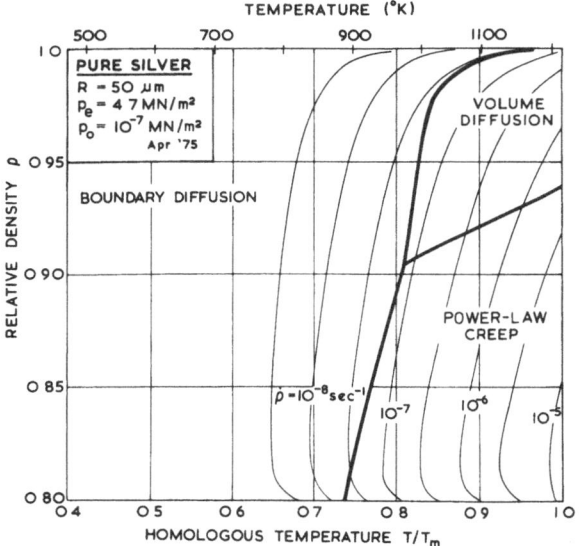

a

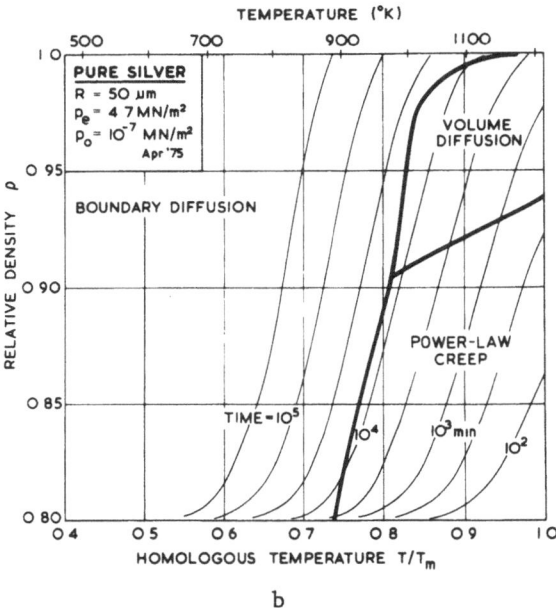

b

Fig. 6: The same map as Fig. 3 with the addition of (a) constant densification rate contours, and (b) constant densification time contours.

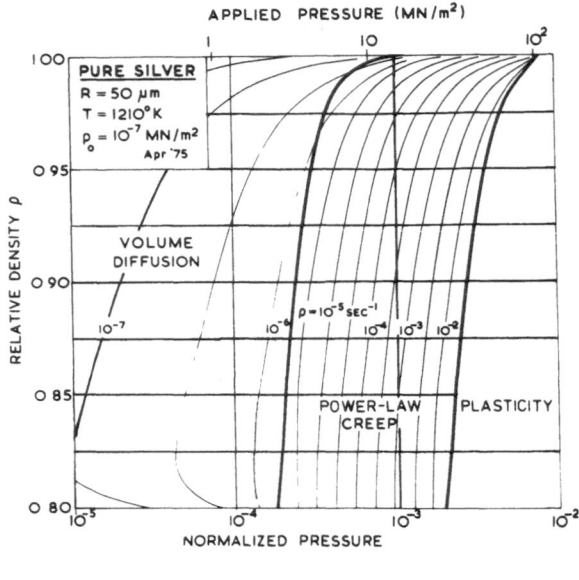

a

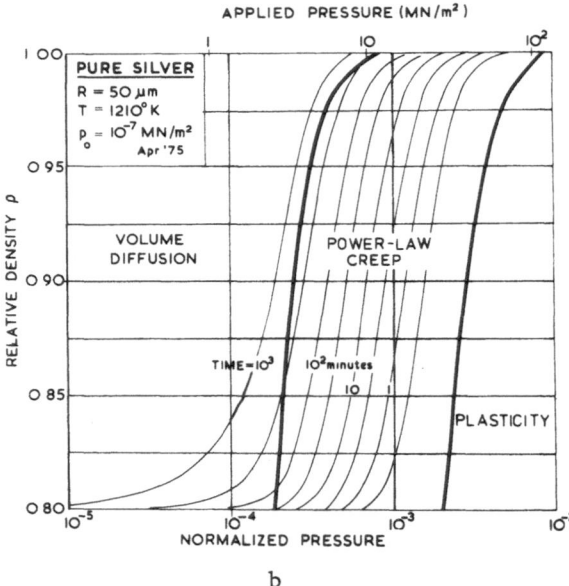

b

Fig. 7: The same map as in Fig. 5 with the addition of (a) constant densification rate contours, and (b) constant densification time contours.

Having established the field boundaries, the maps can be carried one stage further. The net rate of densification is the sum
of the contributions of the four mechanisms, assuming them to be
independent:

$$\dot{\rho} = \sum_{1}^{4} \dot{\rho}_i \tag{8}$$

where the four $\dot{\rho}_i$ are given by eqs. 3, 4, 6 and 7. This can be
evaluated and plotted onto the maps as *contours of constant dens-*
ification rate. Figures 6a and 7a show Figs. 3 and 5, modified in
this way: they allow both the dominant mechanisms of sintering,
and the rate, to be read off.

But it is often more useful to know the density reached in a
certain time. If eq. (8) is integrated to give density as a function of time, it permits contours of time to be plotted onto the
original field-maps: these are shown in Figs. 6b and 7b.

It is obviously possible to construct other sorts of maps,
using time, particle size, or pore pressure as axes, for example.
But we feel there are overwhelming advantages in restricting the
horizontal axes to the two independent macroscopic variables of
pressure and temperature, so that all maps are sections through a
single three-dimensional space.

APPLICATIONS AND A CASE STUDY: PRESSURE-SINTERING OF POLAR ICE

The Interpretation and Design of Experiments

The maps help in the *interpretation* of experiments. Not infreq-
uently, conclusions drawn from an experiment covering a small range
of pressure and temperature are assumed to apply to all pressures
and temperatures. The figures of this paper illustrate that this is
not so. There is no single mechanism of pressure sintering: the
mechanism which is dominant for one combination of particle size,
temperature, pressure, and density may make a negligible contribution for another. The maps show the limits of extrapolation for
experimental data obtained in a limited region, and frequently
suggest a rational interpretation of confusing measurements of sintering kinetics (see the examples quoted in an earlier paper[9]).
They can also indicate the relation of the various areas of powder
processing: sintering, hot pressing, and powder forging.

The map helps also in *designing* experiments. If densification
by power-law creep is to be studied, the appropriate map indicates

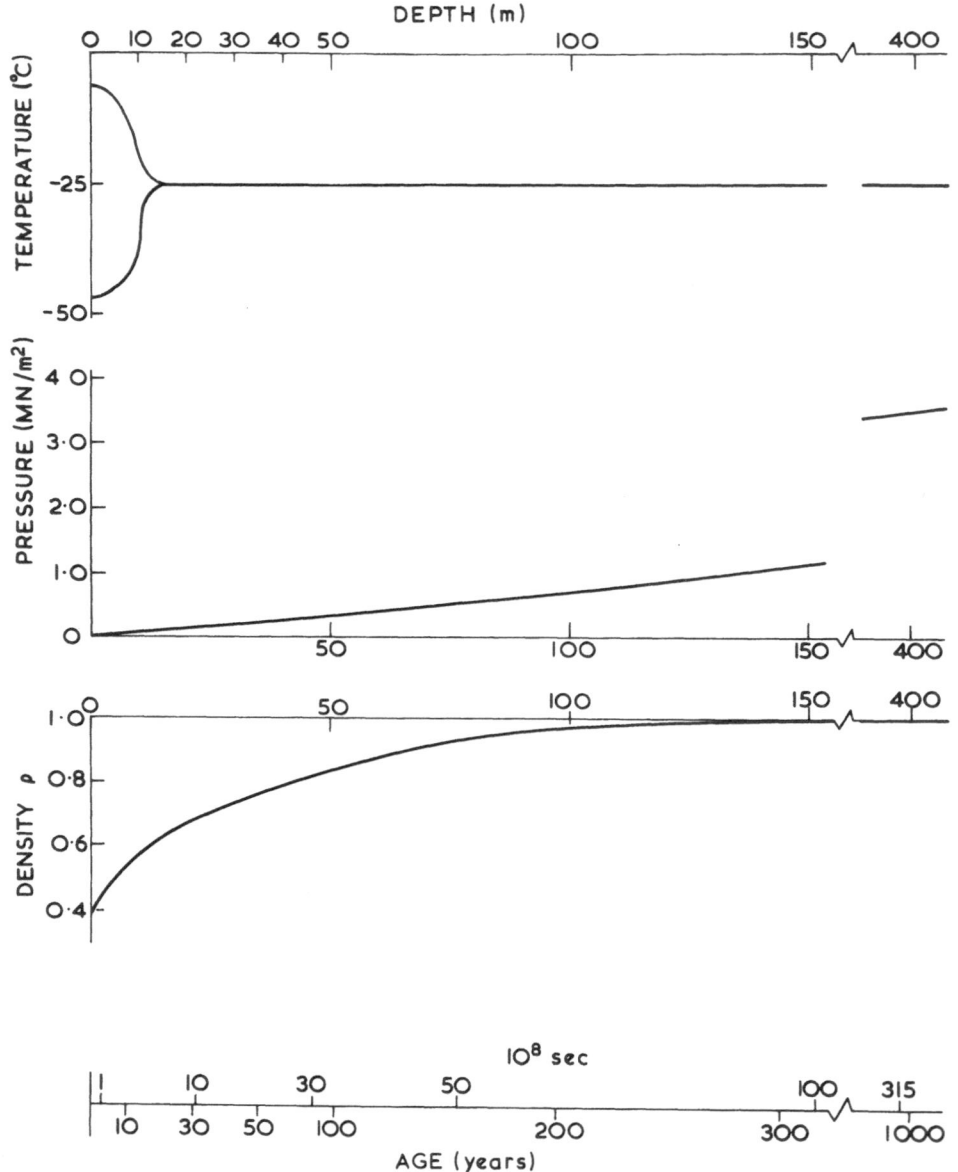

Fig. 8: Data for the variation of temperature, pressure, density, and age with depth for polar ice deposited at Site 2, Greenland[19].

the range of temperature and pressure which must be applied to the material in order to make this the dominant mechanism, and the rates of densification (or the sintering times) which will be involved. Our reason for using normalised variables (relative density, homologous temperature, and normalised pressure) is to extend this application of the maps. Normalised in this way, maps for a pure face-centred cubic metal such as silver are broadly typical of all face-centred cubic metals; maps for a body-centred cubic metal are broadly typical of other such metals, and so on. This means that a map for silver gives general guidance in designing experiments for lead; and that for one oxide should help in designing an experiment for another with a similar structure (see Ashby and Frost[17] for further discussion).

A Case Study: The Pressure Sintering of Polar Ice

Ice, in essentially pure form, condenses from the atmosphere, and precipitates in the polar regions as fine spherical particles. These sinter under the combined driving force of surface tension and the pressure due to the weight of particles deposited on top of them. We have used data derived from two widely separated sites: Byrd Station in Antarctica[18], and Site 2 in Greenland[19]. Both these sites lie in the dry snow zone, where no melting occurs in the summer months, and both are on parts of an ice field which undergoes very little shear. At both sites the way in which the temperature, density, and age of the ice depends on depth has been measured (see Fig. 8).

The top few meters of the compacted ice are very porous, and within this zone (which is undergoing Stage 0 and Stage 1 sintering) the temperature fluctuates with the seasons as shown in Fig. 8. But below about 20 m, where the relative density of the ice is greater than 0.7, the individual ice particles are no longer evident, the pores are cylindrical and the temperature stabilises to a constant, season-independent, level. From 20 m to 400 m, the pressure rises almost linearly with depth, and the ice (which is now undergoing Stage 2 and 3 pressure-sintering) increases in relative density to a limiting value of .997. Since the age of the ice at each depth is known, the approximate densification rate can be calculated (by dividing the density difference between two adjacent points in a column of ice by their difference in age).

This data is plotted in Fig. 9, using the same axes as those we have used for the maps. The two curves show the density as a function of pressure (from Fig. 8) for each site, and the numbers show the $\log_{10}$ of the observed densification rate.

In Fig. 10, a pressure sintering map for ice has been super-

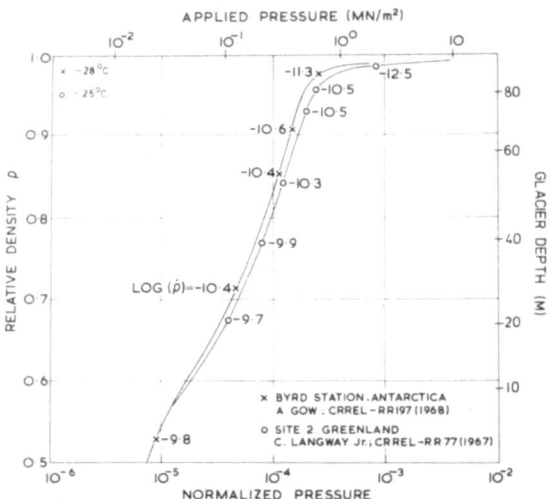

Fig. 9: The densification behaviour of ice particles in two polar
ice sheets. The numbers are $\log_{10}$ of the densification rate. They
show that the densification rate is constant at about 10^{-10}/sec
over a large range of density, but decreases rapidly above a density
of about 0.97.

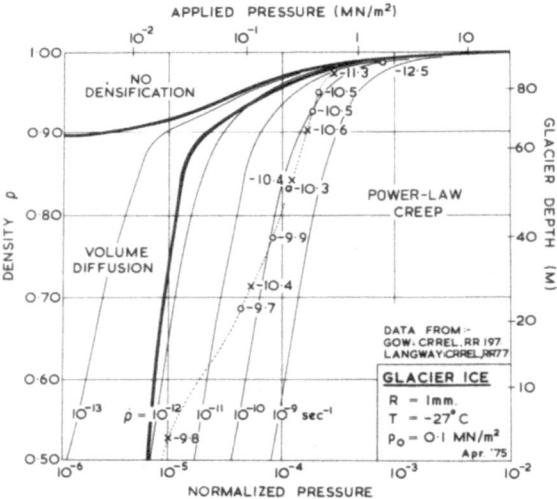

Fig. 10: A pressure sintering map for 1 mm ice particles at a
constant temperature. The data of Fig. 9 is superimposed on the
map to show good agreement between experimental and theoretical
densification rates below about 20 m depth. The map shows that
power-law creep is the dominant mechanism of densification.

TABLE 1

Data used to generate the maps in this paper

	Silver	Ice
Atomic volume, Ω (m³)	1.71×10^{-29}	3.27×10^{-29}
Burger's vector, b (m)	2.89×10^{-10}	4.52×10^{-10}
Melting point, T_M (°K)	1234	273.1
Shear modulus at 300 °K, μ_o (MN/m²)	2.64×10^4 (a)	2.98×10^3 (f)
Temp. coeff. of μ (°K⁻¹)	4.75×10^{-4} (a)	7×10^{-4} (f)
Theoretical density (kg/m³)	1.05×10^4	917 (g)
Surface energy, γ (J/m²)	1.12 (b)	0.1 (h)
D_o for lattice diffusion (m²/sec)	4.4×10^{-5}	1.0×10^{-3} (i)
Activation energy for lattice diffusion, Q_v (kcal/mole)	44.3 (c)	14.1 (i)
Boundary diffusion width, δ (m)	3.0×10^{-10} (d)	9×10^{-10} (j)
D_o for boundary diffusion (m²/sec)	1.2×10^{-5}	1.0×10^{-3} (k)
Activation energy for boundary diffusion, Q_B (kcal/mole)	21.5 (d)	9.4 (k)
Power-law creep constant*, A' (m²/sec)	1.39×10^{-2} (e)	1.25×10^8 (l)
Power-law creep exponent, n	4.3 (e)	3 (l)
Activation energy for power-law creep, Q_{cr} (kcal/mole)	44.3	29.0 (l)

*The constant in eqn. (5) is temperature dependent and is given by

$$A = A' \frac{\mu b}{kT} \frac{1}{\mu^n} \exp\left(-\frac{Q_{cr}}{kT} \right)$$

References
a.   Y. A. Chang and L. Himmel, J. Appl. Phys., 37, 3567 (1966).
b.   H. Jones, Met. Sci. J., 5, 15 (1971).
c.   N. L. Peterson, Sol. State Phys., 22, 409 (1968).
d.   R. E. Hoffmann and D. Turnbull, J. Appl. Phys., 22, 634 (1957).
e.   A. J. Mukherjee, J. E. Bird, and J. E. Dorn, Trans. A.S.M., 62 (1969), 155.
f.   G. Dantl, Phys. Kondens. Materie., 7, 390 (1968); using $\mu = (\frac{1}{2} c_{44} (c_{11} - c_{12}))^{\frac{1}{2}}$.
g.   at -27°C.
h.   N. H. Fletcher, "The Chemical Physics of Ice", Cambridge Univ. Press, (1970), p. 123.
i.   R. O. Ramseier, J. Appl. Phys., 38 (1967), 2553.
j.   estimated as twice the Burgers vector.
k.   estimated as $D_{o,B} = D_{o,v}$ and $Q_B = Q_v$.
l.   from P. Barnes, D. Tabor, and J. C. F. Walker, Proc. Roy. Soc., A324 (1971), 127; valid below -8°C.

imposed on the experimental data. The map itself is based on data (see Table 1) obtained in laboratory tests on pure ice[17], and is constructed for a particle size of 1 mm and a temperature of − 27 ^{o}C − conditions which are close to those observed in the polar ice. The initial pore pressure we have used, 0.1 MN/m^2, corresponds to atmospheric pressure. But as the relative density ρ, passes the permeability limit at about 0.9, the pores become isolated, and the pressure in them rises, until it becomes so high that sintering stops completely; this is the origin of the field labelled "NO DENSIFICATION" on the figure. The other two fields − those for the lattice diffusion and the power-law creep controlled mechanisms − we have encountered before.

The map may now be used to interpret the data. First, the densification rates predicted by the maps agree well with those observed, below about 20 m* (p $\gtrsim$ 0.1 MN/m^2) indicating that the sintering models, plus laboratory data, can adequately account for field observations. Second, the map shows that the dominant mechansim of densification of polar ice is power-law creep, not diffusion. And finally, the end point density at a depth of 400 m is predicted to be .997, identical with that observed in the field.

Grain growth is of little consequence here since it serves only to decrease the rate of sintering due to diffusion, without affecting power-law creep. The map shown in Fig. 10 uses a particle size of 1 mm, the smallest one is ever likely to find in polar ice.

The techniques outlined in this paper have some predictive uses. If, for instance, the data for ice is trustworthy (and the case study quoted above suggests that it is reasonable) then maps can be constructed for other temperatures which would allow the density profile in other, unstudied, ice bodies to be predicted.

SUMMARY

(1) We have presented a method by which sintering and pressure sintering models can be synthesized into diagrams. The diagrams illustrate the ranges of pressure, temperature and density over which a specific mechanism dominates the densification. They also show the total densification-rate and the time taken to reach a given density.

(2) The use of the diagrams is illustrated by considering the densification of polar ice. The maps indicate that power-law

*Above 20 m, temperature fluctuations, and phenomena associated with Stage 0 and Stage 1 sintering lead to a more rapid rate of densification than our maps predict.

creep is the dominant mechanism, and allow a quantitative understanding of the variation of density with depth.

ACKNOWLEDGEMENT

We are grateful to the Science Research Council for their
support of this research, under contract number B/RG/4021.

REFERENCES

1. P. J. James, Powder Met. Intl. 4, 1 (1972).

2. H. E. Exner, G. Petzow, and P. Wellner, "Sintering and Related
    Phenomena", Plenum Press, 351 (1973).

3. C. Torre, Berg-Huttenmann, Montash, Montan. Hochschule Leoben,
    93, 62 (1948).

4. R. L. Hewitt, W. Wallace, and M. C. de Malherbe, Powder Met.,
    16, 88 (1973).

5. A. K. Kakar and A. C. D. Chaklader, Trans. A.I.M.E., 242, 1117
    (1968).

6. D. S. Wilkinson and M. F. Ashby, to be published in Acta. Met.

7. M. S. Koval'chenko and G. V. Samsanov, Poroshkhov. Met., 1, 3
    (1961).

8. R. L. Coble, J. Appl. Phys., 41, 4798 (1970).

9. M. F. Ashby, Acta. Met., 22, 275 (1974).

10. F. Thummler and W. Thoma, Met. Rev., 115, 69 (1969).

11. T. L. Wilson and P. G. Shewmon, Trans. A.I.M.E., 236, 48 (1966).

12. L. Ramquist, Powder Met., 9, 1 (1966).

13. R. L. Eadie, W. A. Miller and G. C. Weatherly, Scripta Met., 8,
    755 (1974).

14. W. Beere, Acta. Met., 23, 139 (1975).

15. J. K. Mackenzie and R. Shuttleworth, Proc. Roy. Soc. B, 62,
    833 (1949).

16.   R. J. Brook, J. Am. Ceram. Soc., $\underline{52}$, 56 (1969).

17.   M. F. Ashby and H. J. Frost, "Constitutive Relations in
      Plasticity", Ed. A. Argon, M.I.T. Press, 1975.

18.   A. J. Gow, U.S. Army Cold Regions Research and Engineering
      Laboratory (CRREL) Report, RR 197 (1968).

19.   C. C. Langway, Jr., U.S. Army Cold Regions Research and Eng-
      ineering Laboratory (CRREL) Report, RR 77 (1967).

INTERPRETATION OF HOT PRESSING KINETICS BY DENSIFICATION

MAPPING TECHNIQUES

M. R. Notis, R. H. Smoak and V. Krishnamachari

Materials Research Center

Lehigh University, Bethlehem, Pa. 18015

ABSTRACT

Interpretation of the kinetics of final stage hot pressing has been accomplished through the use of deformation (or "densification") mapping methods. Creep relations, modified to account for the presence of porosity and for constrained dies, were used to model the pressure-sintering densification rate behavior. These relations are displayed graphically such that boundaries delineate the regions in which any densification mechanism is dominant, and where contours of constant densification rate or constant stress exponent may be compared to experimental values. Examples are shown for the densification of cobalt monoxide. This method, and subsequent manipulation of the computer generated data, enables the quantitative analysis of hot pressing kinetics during final stage densification under conditions where either linear-viscous or multimechanism plastic flow is observed.

I. INTRODUCTION

The hot pressing process is of critical importance to a number of material applications. Hot pressing is currently used to fabricate a large number of fine grained, high density ceramic materials. In many cases, these new ceramics possess optimum combinations of properties such as high strength, excellent electrical characteristics and good optical transparency. This same process is now found to play a major role in the densification and microstructural rearrangement that occurs in ceramic nuclear fuel materials during in-reactor use. There is thus an increasing need for a better understanding of the mechanisms responsible for densification during pressure sintering.

The most fruitful approach to the development of a model for
the hot pressing process has been to relate the kinetics of densification during pressure sintering to the kinetics observed during
normal creep-deformation of a solid. The major complications that
arise are that proper account must be made of the porosity-effec-
tive stress relation and the constrained nature of the hot pressing
die. A primary aim of the present work is to examine both the creep
and hot-pressing behavior of a typical ceramic oxide and to try to
develop approaches whereby either creep information can be used to
predict hot-pressing behavior, or vice versa.

Because it is the final stage of densification that is most
directly related to the production of high densities, because the
isolated pore stage is the configuration most commonly observed in
nuclear fuel rods, and because it is the most straightforward microstructural arrangement for the formulation of quantitative modeling,
(i.e., the morphology is well defined), the description of hot pressing in this paper shall be limited to this stage of the process.

II. DENSIFICATION MECHANISMS

A large number of mechanisms have been proposed to be responsible for densification during the later stages of the sintering
and hot-pressing processes. The most significant of these are:

1. Nabarro-Herring diffusional creep [1,2] where deformation
    is the result of vacancy motion through the bulk lattice
    along a stress-induced concentration gradient from grain
    boundaries in tension to boundaries in compression. The
    strain rate may be expressed as:

$$\dot{\varepsilon}_{NH} = \frac{13.3 \; D_L \; \Omega \; \sigma}{k \; T \; d^2} \tag{1}$$

    where d is the grain size, D_L is the lattice diffusion
    coefficient of the slowest moving species, Ω is the
    vacancy volume, k is the Boltzmann constant and T is
    the temperature. The temperature dependence of the
    lattice diffusion may be expressed as:

$$D_L = D_L^o \; e^{-Q_L/kT} \tag{1a}$$

    where D_L^o and Q_L are the pre-exponential and the
    activation energy for the lattice diffusion process.

2. Coble boundary diffusional creep [3], also a stress-directed diffusion flow, except that the diffusional path is predominantly along the grain boundaries. The strain rate may be expressed as:

$$\dot{\varepsilon}_C = \frac{47.5 \; \delta \; D_B \; \Omega \; \sigma}{k \; T \; d^3} \tag{2}$$

where δ is the grain boundary width and D_B is the grain boundary diffusion coefficient. The temperature dependence for boundary diffusion may be expressed as:

$$D_B = D_B^o \; e^{-Q_B/kT} \tag{2a}$$

where D_B^o and Q_B are the pre-exponential and the activation energy for the boundary diffusion process.

3. Dislocation creep where both stress and temperature are high enough so that deformation is accomplished by both climb and glide of dislocations [4]. A generally accepted semi-empirical form for the creep rate in this case is given by Bird, Mukherjee and Dorn [5] as:

$$\dot{\varepsilon}_C = \frac{K \; D_C \; \mu b}{kT} \; (\frac{\sigma}{\mu})^n \tag{3}$$

where K is a material constant called the Dorn parameter, n is the stress exponent, μ is the shear modulus, b is the Burger's vector, and D_C is the diffusion coefficient for dislocation creep. The temperature dependence for dislocation creep may be expressed as

$$D_C = D_C^o \; e^{-Q_C/kT} \tag{3a}$$

where D_C^o and Q_C are the pre-exponential and the activation energy for dislocation creep.

In the final stage of hot pressing, processes such as particle rearrangement by fragmentation and sliding, pore surface diffusion, and vapor transport are not expected to contribute to densification [6]. Finally, because of the high yield strength of most ceramic materials, deformation by pure dislocation glide may be ignored.

III.   DENSIFICATION MODELS

The variety of modeling approaches that have been used in the
literature range from basic phenomenological descriptions to atomistic interpretations of the mechanisms involved. For example,
Murray, Livey and Williams [7] have attempted to characterize the
hot pressing behavior of a number of materials in terms of phenomenologic linear viscous plastic flow, while Vasilos and Spriggs [8]
have analyzed the densification of MgO in terms of a Nabarro-Herring
diffusion creep model. A major criticism of the phenomenologic
models is that they provide no insight into the mechanisms involved,
and therefore cannot be extrapolated outside the limits of experimental observations. Rossi and Fulrath [6] have combined the atomistic approach of Vasilos and Spriggs with that of Murray, et al.
to describe the final stage pressure sintering behavior of alumina.

The hot-pressing rate equation of Murray et al. [7] was derived from the continuum-sintering theory of MacKenzie and Shuttleworth [9], and treats densification behavior as a Newtonian-viscous
deformation process. If the densification rate due to the externally
applied pressure is much greater than the densification rate due to
sintering (that is, the rate due to the influence of the pore surface
energy effects) the Murray equation has the form

$$\frac{d\rho}{dt} = \frac{9\,A}{4}\,\sigma_a(1-\rho) \qquad\qquad (4)$$

where $A = 1/3\ \eta$, η is the viscosity, σ_a is the applied stress, and
ρ is the relative density.

Since the fractional porosity, P, is given as $P = 1 - \rho$, and
$dP = -d\rho$, equation (4) may also be expressed as:

$$\frac{dP}{dt} = -K'P. \qquad\qquad (5)$$

An expression similar to Equation 5 has been used by Rummler and
Palmour [10] in their study of the hot-pressing behavior of spinel.

The integrated form of Equation 5 is

$$\ell n(P/Po) = -K''t. \qquad\qquad (6)$$

This relationship has been used by Bratton and coworkers [11] to
analyze the hot-pressing deformation behavior of $MgAl_2O_4$ and also
by Terwilliger and Lange [12] in a hot-pressing study of silicon
nitride.

The relationship of Murray et al. predicts a linear dependence
of densification rate upon applied pressure as indicated by Equation

4. This type of linear relationship is not always observed [11,13] indicating a contribution from a non-linear viscous plastic flow mechanism, such as dislocation creep, to the densification process.

An approach to the modeling of hot-pressing behavior more general in scope than that of Murray et al. [7], has been derived by Wolfe [14]. Wolfe utilized the same pore geometry model as that of MacKenzie and Shuttleworth [9] and that of Murray et al. [7], i.e., a homogeneous body containing a random array of pores of uniform size. Each pore was considered to be surrounded by an incompressible spherical shell whose size was determined by the relation

$$r_p = r_s \, P^{1/3} \tag{7}$$

where r_p is the pore radius and r_s is the outer radius of the shell. The strain rate was assumed to follow the relationship

$$\dot{\varepsilon} = A \, \sigma^n \tag{8}$$

where the stress exponent, n, could be greater than 1, and A is a temperature dependent material constant.

Following the approach used by McClelland [15], Wolfe incorporated a stress correction factor to account for the effect of the porosity on the stress acting on the surface of the sphere. The effective stress was given as

$$\sigma_e = g \, \sigma_a \tag{9}$$

where g is the stress correction factor and is a function of porosity. A number of functional relations have been given in the literature [6,8,15,16] for the stress correction factor, g; the dependence of each of these upon porosity level is shown in Figure 1. For final stage densification the majority of these functions approach a value of 1, and in any case the scatter of most experimental data in the final stage makes it difficult to test the validity of one function over any of the others. For this reason, the discussion related to the current work presented here assumes a value of g = 1 unless noted.

Wolfe correctly pointed out that for a more general description that would account for both the effect of surface tension (γ) at the pore, and of the internal pressure (**p**) of trapped gases, the effective stress should be given as:

$$\sigma_e = g \, \sigma_a + \frac{2\gamma}{r_p} - p \tag{10}$$

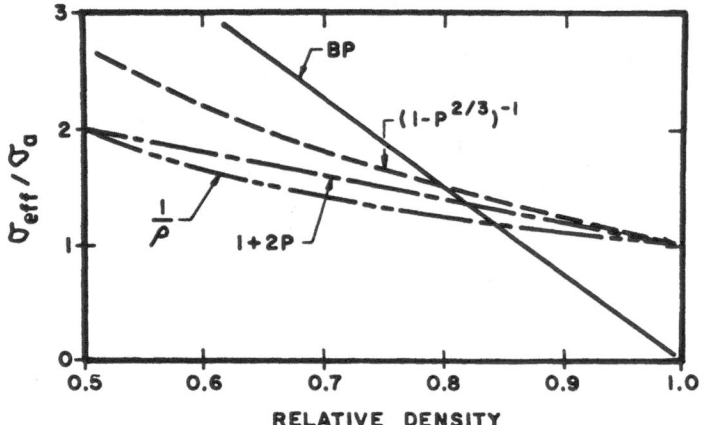

Fig. 1 Stress Correction Factor (g) as a Function
        of Relative Density.

The rate of pore closure was obtained by equating the rate of work
done by the effective stress acting on the shell during pore closure
to the rate of energy dissipated in deforming the solid. The rate
of change of pore radius was found to be

$$\frac{dr_p}{dt} = - \frac{A \; 3^n \; \sigma_e^n \; r_p}{n^n 2^{n+1} (1-P^{1/n})^n} \tag{11}$$

The pore radius was expressed in terms of volume fraction porosity
as

$$r_p = (\tfrac{4}{3} \pi N)^{-1/3} (\tfrac{P}{1-P})^{1/3} \tag{12}$$

where N is the number of pores per unit volume of solid material.
The resultant expression was differentiated with respect to time,
with the result:

$$\frac{dr_p}{dt} = \frac{r_p}{3P(1-P)} \; \frac{dP}{dt} \tag{13}$$

An equation for the densification rate was then obtained by combining equations (11) and (13) and is given as:

$$\frac{dP}{dt} = -A\left(\frac{3}{2}\right)^{n+1} \left(\frac{\sigma_e}{n}\right)^n \frac{P(1-P)}{(1-P^{1/n})^n} \qquad (14)$$

The rate expression may be stated in terms of relative density rather than fractional porosity:

$$\dot{\rho} = A\left(\frac{3}{2}\right)^{n+1} \left(\frac{\sigma_e}{n}\right)^n \left\{\frac{\rho(1-\rho)}{[1-(1-\rho)^{1/n}]^n}\right\} \qquad (15)$$

In its most general form this expression should be combined with equation (10) and the resultant expression thus incorporates surface tension and internal pore pressure effects into the densification model. For the case where there is no porosity correction to the applied stress (i.e., g = 1) equation (15) may be seen to be identical to that given by Wilkinson and Ashby [17].

For the special case where n = 1, and the effective stress, σ_e, is equal to the applied stress, σ_a, Wolfe's equation (15) reduces to

$$\frac{dP}{dt} = \frac{9}{4} A \sigma_a (1-\rho) \qquad (16)$$

which is the equation of Murray, et al. [7] for densification following a linear viscous flow law.

Finally, if equations (8) and (9) are incorporated into equation (15) we obtain:

$$\dot{\rho} = \dot{\varepsilon} \left(\frac{3}{2}\right)^{n+1} \left(\frac{g}{n}\right)^n \frac{\rho(1-\rho)}{[1-(1-\rho)^{1/n}]^n} \qquad (17)$$

Inspection of this equation indicates that the relation has the desired form such that the densification rate is related to the macroscopic creep rate modified by a density function. Further, the densification rate will approach zero as the body under compaction approaches theoretical density.

Hart [18] has recently reported that Equation 15 has been tested over a limited range of data by Wolfe and Kaufman [19] for UO_2. These investigators have obtained good agreement between creep data and hot pressing data for UO_2 at 1850°C using n = 4.5 and g = $[1-P^{2/3}]^{-1}$.

For a constant stress level, and for a given stress exponent, n, a plot of $\dot{\rho}$ versus the bracketed function on right-hand side of Equation 15 should produce a straight line relation. This method has recently been used by Smoak and Notis [20] to calculate the effective stress exponent demonstrated by $MgAl_2O_4$ during hot pressing.

IV. MULTIMECHANISM MODELING

Even a cursory examination of the hot-pressing or sintering
literature indicates that densification rarely, if ever, occurs by
one mechanism alone. Therefore, in order to develop a more realistic densification model, it is necessary to take into account the
various contributions of each separately identifiable densification
mechanism. The approach taken here is to assume that these densification mechanisms can act simultaneously and in an independent
manner. We therefore consider Equation 17 to represent the general
form for a series of densification rate equations, each one related
to a particular creep mechanism; these creep equations, in turn, are
given by rate equations such as Equations 1, 2 and 3. The total
densification rate observed under a given set of experimental conditions may thus be expressed as:

$$\dot{\rho}_T = \dot{\rho}_1 + \dot{\rho}_2 + \dot{\rho}_3 + \dots \tag{18}$$

If predictions of hot pressing kinetics are to be useful over wide
ranges of stress, temperature and grain size, etc., a large number
of possible mechanisms must be taken into account. To a large degree, a specific model is only as good as the details within it, but
as the detail increases, its utility and the physical understanding
gained by its use may decrease. Our emphasis has therefore been the
formulation of procedures to quantitatively handle and to visualize
complex multimechanism processes rather than pursue the detailed
analysis of each individual mechanism.

Recent work on deformation mechanism mapping developed by Ashby
and coworkers [21,22] has shown one way of expressing quantitative
relations for creep while at the same time gaining a conceptual
understanding of the processes involved. These investigations have
shown that the various deformation mechanisms may be combined graphically and their relative contributions can be predicted under a
variety of conditions. We have adopted this approach and have applied it to the final stage of densification during hot pressing[13].

At any given stress and temperature, one mechanism for densification is dominant. This means that on a graphical plot with
stress as one axis and temperature as the other, fields may be determined which show the range of stress and temperature over which
a particular mechanism predominates. In addition, contours of constant densification rate may be shown. This means that if any pair
of the three variables, stress, temperature and densification rate
is known, the third may then be predicted.

The boundaries of the fields are obtained by equating consecu-
tive densification rate equations, i.e.,

$$\dot{\rho}_1 = \dot{\rho}_2$$

$$\dot{\rho}_1 = \dot{\rho}_3 \qquad\qquad (19)$$

$$\dot{\rho}_2 = \dot{\rho}_3$$

(each mechanism thus contributes 50% of the densification rate at a boundary) and solving for stress as a function of temperature. Contours of constant densification rate can be found by summing all of the strain rate contributions (Equation 18) and solving for the stress that results in some constant densification rate value as temperature is varied. These calculations can be easily made and the densification maps may be plotted by a computer if the necessary material constants are known. Maps may be made, for example, for a wide range of stress, temperature, and structural parameters such as grain size, and the results can be examined visually. Conversely, overlays of experimental data onto the densification map quickly enable the calculation of the relative contribution of each densification mechanism to the total densification rate. Similarly, experimental conditions necessary for the observation of a particular densification mechanism may be easily predicted.

V. DENSIFICATION DURING HOT-PRESSING OF CoO

Recently, we have reported studies of the final stage densification during hot pressing [23] and the creep behavior [24] of CoO. These studies were performed over a stress range of 1000-10,000 psi and a temperature range of 950-1100°C. Densification rates were carefully monitored during hot pressing, the samples were hot pressed to near theoretical density, and compression creep testing was performed on creep specimens cut from the larger hot pressed sample. At low stresses we found a linear relation between creep strain rate and stress; the activation energy was found to be very close to the published values for cobalt cation self-diffusion [25]. Interpretation of the data in terms of a Nabarro-Herring creep mechanism (Equation 1) gave good agreement between the calculated effective diffusion coefficients and the magnitude of published cation tracer diffusion rates. It was concluded that the rate controlling mechanism for creep was cation lattice diffusion and that oxygen anion diffusion was enhanced at the grain boundaries. At high stresses, the stress exponent was found to be $n \cong 4.5$, and the activation energy for creep was found to be considerably higher than the cation tracer value but somewhat lower than the published oxygen anion activation energy of 95 k cal/mole [26]. The stress exponent, activation energy and the magnitude of the effective diffusion coefficient in this high stress range are in excellent agreement with results recently reported for creep of single crystal CoO [27]. It was therefore concluded that the deformation was controlled by a dislocation climb-glide process (Equation 3) that involved either oxygen anion bulk diffusion or some type of sub-boundary or dislocation pipe diffusion mechanism. The appropriate material constants derived from the compression creep tests and used to evaluate the hot pressing kinetics are given in Table I.

TABLE I

Input Data for CoO Densification Maps

Atomic Volume, $\Omega \times 10^{23}$(cm)	1.95
Burgers Vector, $b \times 10^8$(cm)	4.26
Melting Temperature, T_M(°K)	2208.
Shear Modulus, $\mu \times 10^{-11}$(dynes/cm^2)	7.08[a]
Activation Energy for Lattice Diffusion, Q_L(k cal/mole)	38.4[b]
Preexponent for Lattice Diffusion, D_L^o(cm^2/sec)	1.5 x 10^{-4}[c]
Activation Energy for Dislocation Creep, Q_C(k cal/mole)	72.7
Preexponent for Dislocation Creep, D_C^o(cm^2/sec)	27.0
Activation Energy for Boundary Diffusion, Q_B(k cal/mole)	32.4[d]
Preexponent for Boundary Diffusion, D_B^o(cm^2/sec)	1.5 x 10^{-3}[e]
Grain Boundary Width, $\delta \times 10^8$(cm)	50[f]
Stress Exponent, n	4.5
Dorn Parameter, K	10^{+4}[g]

Notes

[a]calculated from $\mu = E/2(1+\nu)$ using $E = 17\times10^{11}$dynes/cm^2 [Phys. Rev. 87, 1143 (1952)] and $\nu = 0.2$ [Phys. Met. Metal. USSR, 12, 139 (1961)].

[b][Phys. Rev. 186, 887 (1969)].

[c]from data of [J. Metals 6, 1244 (1954)] assuming $(p_{O_2})^{1/4}$ dependence and extrapolating to 1.38×10^{-5} atm.

[d]approximated using $Q_B = Q_L - \Delta H_F/2$ and data from [J. Phys. Chem. Solids 29, 1597 (1968)].

[e]assuming $D_B^o \cong 10 (D_L^o)$

[f]assumed value

[g]from correlation of [Scr. Met. 7, 115 (1973)] and value of n = 4.5

A densification map for CoO with a relative density of $\rho = 0.975$ and a grain size of 5 microns is shown in Figure 2. The boundaries

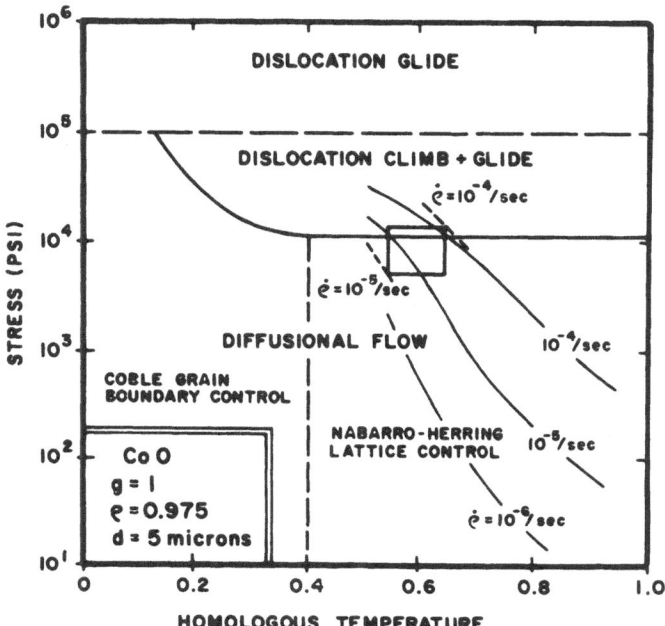

Fig. 2   Densification Map for CoO, ρ = 0.975, Showing
Contours of Constant Densification Rate.

between dominant mechanisms and the contours of constant densification rate shown in the figure were calculated by a computer plotting routine based on the mapping procedure outlined in the preceding
section of this paper. The boxed-in area near the center of the
figure delineates the experimental stress-temperature conditions used
for the hot-pressing experiments. This area is seen to overlap the
regions where densification by Nabarro-Herring-type diffusion and
dislocation plastic flow are expected to contribute to the densification process. The dashed curves show the experimentally determined
densification rate contours corresponding to the extremes of hot-
pressing conditions; the agreement between experimental and predicted
rates, especially at the higher rates, is seen to be quite good.

Figure 3 is a composite densification map showing how both the
mechanism boundaries and the constant densification rate contour of
$\dot{\rho} = 10^{-4}$/sec move as a function of density level. The boundary between Nabarro-Herring flow and dislocation plastic flow is observed
to move upward as theoretical density is approached. Thus the map
demonstrates that the relative contribution of diffusion flow to
densification increases rapidly as full density is approached; also,
at constant stress and temperature, the densification rate decreases
as densification proceeds.

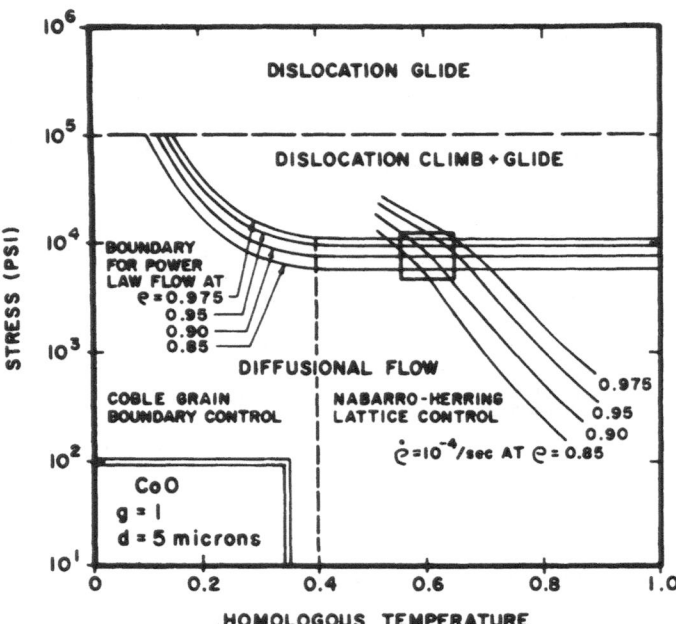

Figure 3   Map Demonstrating Motion of Mechanism
Boundaries and a Contour of Constant
Densification Rate as a Function of Density

Although boundaries between controlling mechanisms are shown
quite easily, a major problem with the densification (or deformation)
maps developed so far, is that the relative contribution of each
mechanism is difficult to visualize as the boundary is approached.
In order to incorporate this type of information into the maps we
have developed plotting routines that superimpose contours of constant effective stress exponent on our densification maps. For any
set of stress, temperature, grain size, and porosity (density) conditions a total densification rate can be computed, and this total
densification rate can be expressed in terms of stress and an effective (experimentally measurable) stress exponent, i.e.,

$$\dot{\rho}_T = \dot{\rho}_1 + \dot{\rho}_2 + \dot{\rho}_3 = A_{eff} \, \sigma^{n_{eff}} \qquad (20)$$

The derivative of the logarith of the densification rate with
respect to the logarithm of stress thus yields:

$$\frac{d \log \dot{\rho}_T}{d \log \sigma} = n_{eff} \qquad (21)$$

An auxiliary sub-rout;   :o the densification mapping computer pro-

gram therefore is setup to calculate total densification rates as a
matrix across the σ - T plot and then to plot contours of constant
effective stress exponent as defined by Equation (21). A densification map for CoO having a density of ρ = 0.9 is shown in Figure 4
with contours of constant stress exponent superimposed. Examination
of the figure shows, for example, how the stress exponent tends to
1 well within the diffusional flow regions, and tends to approach
4.5 within the plastic flow region. The range of conditions where
neither extreme is observed appears to be quite large. Since the
experimental σ - T conditions (boxed-off area in Figure 4) for our
studies of CoO fall within this transition range, change of stress
experiments [28] were carried out in order to verify the predictions
of the densification modeling procedure used here. The applied
stress during hot pressing was changed from 6000 psi to 7000 psi
when the specimen density was ρ = 0.9, the quasi-steady densification
rate was measured before and after the stress change and the stress
exponent was calculated by approximating Equation 21 and was found
to be n = 1.82. This is in excellent agreement with the predictions
of the densification map and gives good confirmation of the relative
contributions of each mechanism to the total densification rate.

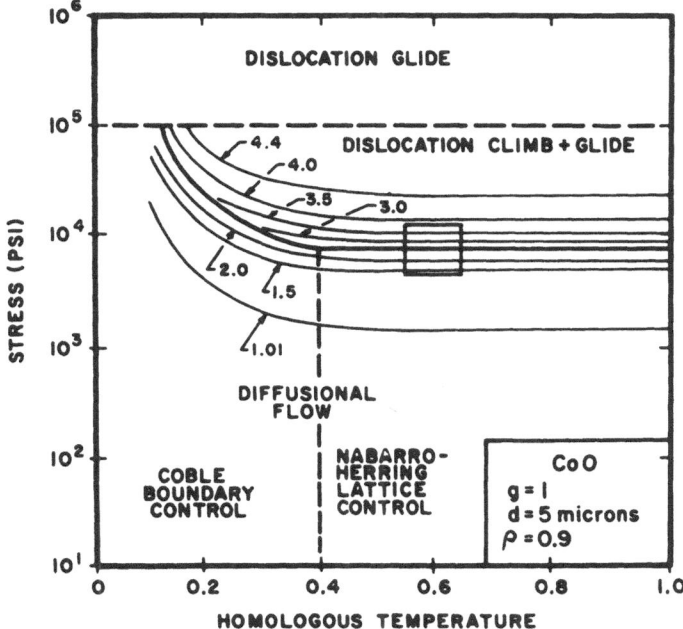

Figure 4   Densification Map for CoO, ρ = 0.9, Showing
           Contours of Constant Effective Stress Exponent (n).

CONCLUSIONS

1. The non-linear viscous flow model derived by Wolfe [14] was found to be the basis for a series of densification rate equations, each one of which could be related to a specific creep deformation-rate equation.

2. The densification rate equations could be combined and displayed on a "densification map" delineating regions of density, applied stress, and temperature over which a particular densification mechanism was dominant. Contours of either total densification-rate or of constant stress-exponent could be overlaid on the densification map; these contours could be used to test the experimental validity of the model assumed as the basis for the map.

3. Creep results for CoO were used as the necessary input data for the generation of a densification map; the predictions of the densification map were then used to compare densification and creep-deformation kinetics in a quantitative manner. The experimental conditions extant during hotpressing lead to densification by a combination of Nabarro-Herring diffusional flow and dislocation creep.

ACKNOWLEDGEMENT

The authors are grateful to ERDA for its support of this project under contract E(11-1) 2408.

REFERENCES

1. F. R. N. Nabarro, "Report of a Conference on the Strength of Solids," Physical Society, London, 75-90 (1948).
2. C. Herring, J. Appl. Phys. 21, 437 (1950).
3. R. L. Coble, J. Appl. Phys. 34, 1679 (1963).
4. J. Weertman, Trans. ASM 61, 681 (1968).
5. J. E. Bird, A. K. Mukherjee and J. F. Dorn, p. 255 in Quantitative Relations Between Properties and Microstructure, D. G. Brandon and A. Rosen, eds. Israel Univ. Press (1969).
6. R. C. Rossi and R. M. Fulrath, J. Am. Ceram. Soc. 48, 558(1965).
7. P. Murray, D. T. Livey and J. Williams, p. 147 in Ceramic Fabrication Processes, W. D. Kingery, ed., Wiley (1958).
8. T. Vasilos and R. M. Spriggs, J. Am. Ceram. Soc. 46, 493(1963).
9. J. K. Mackenzie and R. Shuttleworth, Proc. Phys. Soc. (London) 62, 833 (1949).
10. D. R. Rummler and H. Palmour, III, J. Am. Ceram. Soc. 51, 320 (1968).

11. R. J. Bratton, G. R. Terwilliger and S. M. Ho, J. Mat. Sci. $\underline{7}$, 1363 (1972).
12. G. R. Terwilliger and F. Lange, J. Am. Ceram. Soc. $\underline{57}$, 25(1974).
13. M. R. Notis, p. 1 of Deformation of Ceramic Materials, R. C. Bradt and R. E. Tressler, eds., Plenum (1975).
14. R. A. Wolfe, Bull. Am. Ceram. Soc. $\underline{46}$, 469 (1967).
15. J. D. McClelland, p. 157 in Powder Metallurgy, W. Leszynski, ed., Interscience (1961).
16. R. L. Coble, J. Appl. Phys. $\underline{41}$, 4798 (1970).
17. D. S. Wilkinson and M. F. Ashby, presented at this conference.
18. P. E. Hart, J. Nucl. Matls. $\underline{51}$, 199 (1974).
19. R. Wolfe and S. Kaufman, USAEC Rept. WAPD-TM-587 (1967) NTIS order No. N6817241(DA753707).
20. R. H. Smoak and M. Notis, Bull. Am. Ceram. Soc. $\underline{53}$, 319 (1974).
21. M. F. Ashby, Acta. Met. $\underline{20}$, 887 (1972).
22. H. J. Frost and M. F. Ashby, "A Second Report on Deformation Mechanism Maps," ONR Final Report, Harvard University, August 1973.
23. P. Urick and M. Notis, J. Am. Ceram. Soc. $\underline{56}$, 570 (1973).
24. V. Krishnamachari and M. Notis, to be published.
25. W. K. Chen, N. L. Peterson and W. T. Reeves, Phys. Rev. $\underline{186}$, 887 (1969).
26. W. K. Chen and R. A. Jackson, J. Phys. Chem. Sol. $\underline{30}$, 1309 (1969).
27. V. Krishnamachari, J. Am. Ceram. Soc. $\underline{57}$, 506 (1974).
28. R. A. Penty, D. P. H. Hasselman and R. M. Spriggs, Bull. Am. Ceram. Soc. $\underline{52}$, 692 (1973).

Subject Index